2025 32nd International Conference on Mixed Design of Integrated Circuits and System (MIXDES 2025)

Szczecin, Poland
26-27 June 2025

IEEE Catalog Number: CFP25MIX-POD
ISBN: 979-8-3503-9291-3

Copyright © 2025, Lodz University of Technology
All Rights Reserved

*** *This is a print representation of what appears in the IEEE Digital
Library. Some format issues inherent in the e-media version may also
appear in this print version.*

IEEE Catalog Number: CFP25MIX-POD
ISBN (Print-On-Demand):   979-8-3503-9291-3
ISBN (Online):   978-83-63578-28-2

Additional Copies of This Publication Are Available From:

Curran Associates, Inc
57 Morehouse Lane
Red Hook, NY 12571 USA
Phone: (845) 758-0400
Fax: (845) 758-2633
E-mail: curran@proceedings.com
Web: www.proceedings.com

Mixed Design of Integrated Circuits and Systems

MIXDES 2025

LODZ UNIVERSITY OF TECHNOLOGY
WARSAW UNIVERSITY OF TECHNOLOGY

MIXED DESIGN OF INTEGRATED CIRCUITS AND SYSTEMS

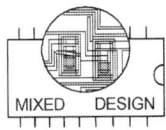

MIXDES 2025

Department of Microelectronics and Computer Science,
Lodz University of Technology, Poland

Institute of Microelectronics and Optoelectronics,
Warsaw University of Technology, Poland

in co-operation with:
Poland Section IEEE - ED & CAS Chapters

Section of Microelectronics & Electron Technology
and Section of Signals, Electronic Circuits & Systems
of the Committee of Electronics and Telecommunication
of the Polish Academy of Sciences

Commission of Electronics and Photonics
of Polish National Committee
of International Union of Radio Science – URSI

Edited by
Wojciech Tylman

Lodz University of Technology
Department of Microelectronics and Computer Science
ul. Wólczańska 221
93-005 Łódź, Poland

Phone: +48 42 631 27 27
Fax: +48 42 638 03 27
E-mail: mixdes@dmcs.p.lodz.pl
Web: http://www.mixdes.org

Preface

This monograph presents recent works on micro- and nanoelectronics design methods, modelling, simulation, testing and manufacturing technology in diverse areas including embedded systems, MEMS, sensors, actuators, power devices and biomedical applications. It contains the selected submissions related to the topics discussed during the 32nd International Conference "Mixed Design of Integrated Circuits and Systems" – MIXDES 2025, held on June 26–27, 2025, in Szczecin, Poland. As this was the first edition of the conference following the passing of its founder, Professor Andrzej Napieralski, the monograph includes a commemorative chapter honouring his life and achievements.

The conference program consisted of two days of sessions including five invited talks:

- *AI for Processors, Processors for AI: Going New Ways for Processor Architectures*
  M. Hübner (Brandenburg Univ. of Techn. Cottbus - Senftenberg, Germany)

- *Electronic Control Systems for Ion Trap Quantum Computers*
  G. Kasprowicz (Warsaw Univ. of Techn., Poland)

- *Mixed Mode: More than Analog and Digital*
  R.S. Murphy - EDS Distinguished Lecturer, R. Torres (INAOE, Mexico)

- *Modern Challenges in Hardware Design*
  M. Zmuda (Intel Technology, Poland)

- *Video-assisted Dentistry with Deep Neural Networks*
  D. Węsierski (Gdansk Univ. of Techn., Poland)

The program of MIXDES 2025 also included two special sessions:

- *Advancing FOSS Compact Modelling: From OTF Transistors to Mott Memristors*
  organized by A. Kloes, M. Schwarz (Technische Hochschule Mittelhessen, Germany),
  W. Grabiński (GMC, Switzerland) and D. Tomaszewski (Lukasiewicz - IMiF, Warsaw, Poland)

- *Artificial Intelligence in Electronic Systems*
  organized by T. Stefański (Gdańsk Univ. of Techn., Poland) and R. Długosz (Bydgoszcz Univ. of Science and Techn., Poland)

All the regular papers were reviewed and selected from submissions from 12 countries. The conference organisers would like to thank all the distinguished scientists who have supported the conference by taking part in the International Programme Committee.

I hope that we will meet together next year in Poznań (June 25 – 27, 2026), one of the oldest and largest cities in western Poland.

Łódź, July 2025

Wojciech TYLMAN
Department of Microelectronics and Computer Science
Lodz University of Technology, Poland

A Tribute to Professor Andrzej Napieralski:
His Life, Work, and Legacy

Professor Andrzej Napieralski passed away on September 29, 2024. As the founder and driving force behind the MIXDES conference for three decades, his contributions were instrumental in shaping its legacy. This year's conference opened with a special presentation dedicated to commemorating his life and achievements. The presentation offered a retrospective on the conference's history, highlighting key moments and milestones, and served both as a tribute for those who knew him and an introduction for those who did not. This chapter aims to summarize the key points of that commemorative presentation.

Andrzej Napieralski was born on December 29, 1950 in Łódź. He attended I Liceum Ogólnokształcące im. Mikołaja Kopernika, the oldest and most prestigious high school in Łódź. He passed his matura (A-level) exam in 1967.

In 1967, he enrolled at the Lodz University of Technology (TUL) in the Faculty of Electrical Engineering (currently the Faculty of Electrical, Electronic, Computer and Control Engineering).

In 1973, he defended his Master's thesis and began working at TUL. He defended his PhD thesis titled "Modelling of Four-Layer Semiconductor Structures for the Purposes of Numerical Analysis" (supervised by Professor Zbigniew Korzec) with distinction in 1977.

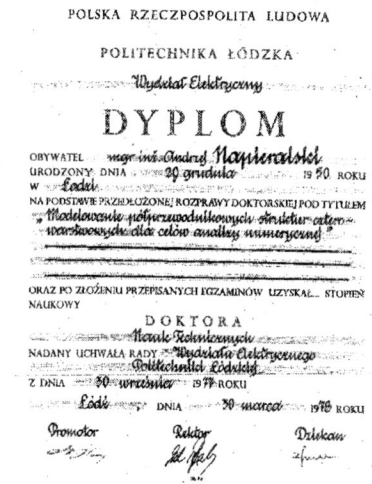

He quickly recognized the importance of international collaboration. His industrial placement took place at Brown Boveri Company in Switzerland, and in 1975 he completed a seven-week visit to GEC Machines Limited in the United Kingdom.

The next significant stage in his career began in 1983 with a six-month internship at Laboratoire d'Automatique et d'Analyse du CNRS in Toulouse, France. Following the internship, he was offered a position as Professeur Associé at the same laboratory. From 1984 to 1995, he lectured at l'Institut des Télécommunications d'Oran in Algeria.

In 1985, he was simultaneously offered positions at LAAS du CNRS and l'Institut National des Sciences Appliquées de Toulouse – he accepted both. During his six-year stay in France, he worked on thermal and electro-thermal simulations of power devices and circuits. He developed numerous computer programs for these simulations, with a significant portion of the research conducted for industrial partners such as Alcatel, Société Anonyme de Télécommunications, SGS-Thomson, and Motorola. His work in France was instrumental in obtaining his DSc degree: first in Poland in 1988, and the following year in France.

In 1991, Professor Napieralski returned to Poland. In 1992, he became Deputy Director of the Institute of Electronics. His research primarily focused on computer-aided design of VLSI circuits, and he also continued his work on the modeling of power devices. He successfully secured funding through European programs such as TEMPUS and ESPRIT, which not only accelerated his research but also enabled young researchers to undertake internships abroad. In 1992, he graduated his first doctoral student, Marek Turowski, followed by Jean Louis Noullet in 1993.

In 1994, the MIXDES conference was established under the name "Mixed Design of VLSI Circuits". The inaugural event took place in Dębe from April 5 to 9. Although the full name has evolved, the acronym and core

concept have remained consistent for over thirty years. From the outset, the conference had no fixed location; instead, it was hosted in various cities

across Poland to enhance its appeal, particularly for international attendees. In 1996, it was held for the first time in Łódź, and photos from that event are included here.

In recognition of his achievements, in 1995, Professor Napieralski was awarded the title of Professor by the President of the Republic of Poland, Lech Wałęsa.

Due to the extensive scope of his research, Professor Napieralski required a strong support team. In 1995, he became Head of the Microelectronics and Computer Science unit within the Institute of Electronics. In 1996, this unit was transformed into an independent department. The Department of Microelectronics and Computer Science (DMCS) began operations in January 1997, with Professor Napieralski as its head – a position he held until 2021. To this day, DMCS remains one of the most innovative and

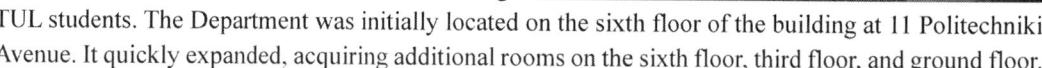

successful departments at TUL. From its inception, it was known by its English name, even among TUL students. The Department was initially located on the sixth floor of the building at 11 Politechniki Avenue. It quickly expanded, acquiring additional rooms on the sixth floor, third floor, and ground floor.

In 2002, he was elected, and in 2005 re-elected, as Vice-Rector for Promotion and International Cooperation at TUL.

In 2003, a new chapter in Professor Napieralski's scientific career began with his first collaboration with Deutsches Elektronen-Synchrotron (DESY), one of the world's leading accelerator centers. His department provided cutting-edge hardware and software solutions, many based on reprogrammable circuits (FPGAs).

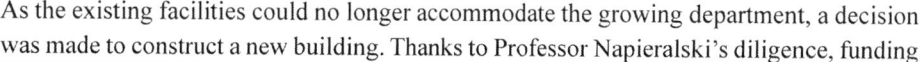

As the existing facilities could no longer accommodate the growing department, a decision was made to construct a new building. Thanks to Professor Napieralski's diligence, funding

was secured from both European and Polish ministries, and an old industrial building was converted into a modern research center. The department moved into its new premises in 2008. Today, it is internationally recognized as a key provider of technologies for large-scale physical experiments, collaborating with institutions such as DESY, ITER, ESS, W7-X, and Fermilab.

The new department building was not the only infrastructure project credited to Professor Napieralski. Between 2009 and 2015, he led the construction of the Information Technology Center. This project encompassed the design, construction, and equipping of a facility for teaching computer science at the university, with a total budget of 50 million PLN.

In 2014, Professor Napieralski graduated his 50th doctoral student, Adam Skurski.

He also served as the supervisor for three honorary doctorate recipients at Lodz University of Technology:

- Prof. Augustin Martinez (1993)
- Prof. Anatolij Gawrikow (2009)
- Prof. Andrzej Jakubowski (2014)

In 2017, by order of the President of the Republic of Poland, he was awarded the Commander's Cross of the Order of Polonia Restituta – the second-highest state order in Poland.

In 2020, Professor Napieralski received an honorary doctorate from Gdynia Maritime University.

Throughout his career, Professor Napieralski:

- published 1149 papers
- participated in 112 scientific projects, many as project leader
- contributed to 36 educational projects
- graduated 57 doctoral students
- mentored 672 engineers and master's degree holders
- served on 29 scientific committees and fellowships
- participated in 55 conference committees
- received five Polish and two international state honours
- was the recipient of 100 awards for scientific achievements

Professor Napieralski should be remembered not only as an outstanding scientist, but also as a mentor, an inspiring

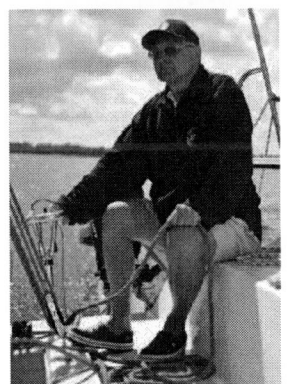

educator, a person of boundless energy, and a true friend. To those of us who had the privilege of working with him, he was not merely a leader – he was like a father to the

Department. One example of his unique approach to building community is the department's football team, which he founded and in which he actively participated.

Beyond his professional life, Professor Napieralski found great joy in sailing, skiing, and sharing good food with family and friends. This is how I wish him to be remembered – with warmth, vitality, and a deep sense of humanity.

Wojciech Tylman

Table of Contents

Preface 3

A Tribute to Professor Andrzej Napieralski: His Life, Work, and Legacy 5

Table of Contents 9

General Invited Papers

AI for Processors, Processors for AI: Going New Ways for Processor Architectures 15
M. Hübner (Brandenburg Univ. of Techn. Cottbus - Senftenberg, Germany)

Electronic Control Systems for Ion Trap Quantum Computers 18
G. Kasprowicz (Warsaw Univ. of Techn., Poland)

Mixed Mode: More than Analog and Digital 21
R.S. Murphy, R. Torres (INAOE, Mexico)

Modern Challenges in Hardware Design 26
M. Zmuda (Intel Technology, Poland)

Advancing FOSS Compact Modelling: From OTF Transistors to Mott Memristors

A Generic Approach for Compact Modeling of Variability and Low-Frequency Noise in Organic Thin-Film Transistors 33
A. Kloes, G. Darbandy (THM Univ. of Applied Sciences, Germany), B. Iñiguez (Univ. Rovira i Virgili, Spain), A. Nikolaou (THM Univ. of Applied Sciences, Germany and Univ. Rovira i Virgili, Spain)

Extraction of Open-Access-PDK Active Inductance Parameters with FOSS Tools 36
M. Brinson (London Metropolitan Univ., UK)

Artificial Intelligence in Electronic Systems

Anomaly Detection on the Edge: Comparison of Reconstruction and Classification Based Approaches 45
Ł. Grzymkowski, T. Cejrowski (Arrow Electronics, Poland), T. Stefański (Gdansk University of Technology, Poland)

Application of Dual-Q TQWT for Atrial Fibrillation Detection with Three-Layered Neural Network 51
T. Pander (Silesian Univ. of Techn., Poland)

Application of Modified Particle Swarm Optimization Algorithm in FIR Filter Design 57
K. Pipka (Gdańsk Univ. of Techn., Poland), T. Talaśka (Gdańsk Univ. of Techn. and Bydgoszcz Univ. of Science and Techn., Poland),
R. Długosz (Bydgoszcz Univ. of Science and Techn. and Aptiv Services, Poland), W. Pedrycz (Univ. of Alberta, Poland)

Design Flow for AI-driven Medical Systems Demonstrated through an Example in Dental Imaging Analysis 62
M. Fechner, K. Śniatała, P. Śniatała (Poznan Univ. of Techn., Poland), S. Ren (San Diego State Univ., USA), R. Śniatała, T. Pawlaczyk (Poznan Univ. of Medical Sciences, Poland)

Edge Computing of Human Poselet 68
T. Byrwa, J. Kłopotek Główczewski, M. Czubenko (Gdansk Univ. of Techn., Poland)

Evaluating Device Variability in RRAM-Based Single- and Multi-Layer Perceptrons 74
A. Blumenstein (THM Univ. of Applied Sciences, Germany and Univ. Rovira i Virgili, Spain), E. Pérez, C. Wenger (IHP Frankfurt (Oder) and Brandenburg Univ. of Techn. Cottbus - Senftenberg, Germany), N. Dersch (THM Univ. of Applied Sciences, Germany and Univ. Rovira i Virgili, Spain), A. Kloes (THM Univ. of Applied Sciences, Germany), B. Iñiguez (Univ. Rovira i Virgili, Spain), M. Schwarz (THM Univ. of Applied Sciences, Germany)

1   Design of Integrated Circuits and Microsystems

CMOS OTA for Detector Readout Electronics Integrator in the ALICE FIT Project 81
J. Miszczyński, P. Otfinowski, A. Laczewski, M. Grzegorzek, I. Brzozowski, C. Worek, P. Wiącek, P. Russek, J. Kitowski (AGH Univ. of Krakow, Poland), J. Otwinowski (Inst. of Nuclear Physics, Poland)

Design Considerations for Integrated SiGe BiCMOS Phase-Locked Loops in the Millimeter-Wave Band 87
F. Herzel, A. Ergintav, C. Carta, G. Fischer (IHP Frankfurt (Oder), Germany)

Design and Optimization of OTA-C Filters with Shared CMFB and Output Stages: Performance, Power, and Area Analysis 93
H. Aleksiuk, O. Bogucki, P. Halman, B. Pankiewicz (Gdansk Univ. of Techn., Poland)

Design of the Charge-Sampling Multiplying PLL in CMOS 40 nm 98
J. Zając, P. Kmon (AGH Univ. of Krakow, Poland)

Enhancing Test-Driven Development for Reconfigurable Hardware through High-Level Synthesis and Early-Stage Validation 103
R. Diachok, H. Klym (Lviv Polytechnic National Univ., Ukraine)

FSMLock: Sequential Logic Locking Case Study 107
J. LaPietra, M. Kurdziel (L3Harris Technologies, USA), M. Łukowiak (Rochester Inst. of Techn., USA)

Implementation of a PLL Loop Circuit for Frequency Synthesis in 65 nm CMOS Technology 112
M. Tymińska (Warsaw Univ. of Techn., Poland), M. Kucharski (OmniChip Sp. z o.o., Poland), W. Pleskacz (Warsaw Univ. of Techn., Poland)

Optimum Design of a Mostly-Digital Fleischer-Laker Switched-Capacitor Bilinear Bandpass Filter in Standard CMOS Technology 117
H. Serra, J.P. Oliveira, J. Goes (UNINOVA-CTS and NOVA FCT, Portugal)

Practical Implementation of Voltage-to-Current and Current-to-Voltage Converter in High Voltage SOI Technology 121
M. Jankowski (Lodz Univ. of Techn., Poland)

Recording Channel Parameters Influence Analysis on Time-Related X-ray Based Measurements in CMOS 40 nm 126
F. Księżyc, P. Kmon (AGH Univ. of Krakow, Poland)

SHA-256 Hash Generator in Verilog HDL 131
B. Rulka, P. Pieńczuk (Łukasiewicz - Inst. of Microeletronics and Photonics and Warsaw Univ. of Techn., Poland), W. Pleskacz (Warsaw Univ. of Techn., Poland)

SYNAPSE - A New Approach to Semi-automated Design of Ultra-low-power Application-specific Embedded Processors 135
X. Ji, T. Kazmierski, B. Halak (Univ. of Southampton, UK)

2   Analysis and Modelling of ICs and Microsystems

Fractional Spurious Tones Analysis of the Space-Time Averaging PLL 143
R. Wiliński, P. Gryboś (AGH Univ. of Krakow, Poland)

High-Level Modeling of RF Power Amplifiers and Antenna Arrays for Efficient Over-the-Air Power Combination in RF Transceivers 150
M. Diacu (Univ. Nova de Lisboa, Portugal), J. Guerreiro (Univ. Nova de Lisboa and Inst. de Telecomunicações, Portugal), J.P. Oliveira (Univ. Nova de Lisboa and UNINOVA-CTS, Portugal), P. Montezuma (Univ. Nova de Lisboa, Inst. de Telecomunicações and Koala Tech, Portugal), P. Viegas (Koala Tech, Portugal)

Reliability Analyses of Ultra-Low Voltage Analog Spiking Neurons 156
G. Brandsteert, L. Van Brandt, D. Flandre (Univ. Catholique de Louvain, Belgium)

3   Power Electronics

A Thermal Behavior of Lateral (VESTIC) BJTs on SOI Substrate 163
P. Mierzwiński (Warsaw Univ. of Techn., Poland)

Considerations on the Importance of Proper Modeling of Heat Transfer Coefficient Values 168
M. Janicki (Lodz Univ. of Techn., Poland)

Influence of the Cooling System on Characteristics of Power LEDs in COB Packages 172
K. Górecki, P. Ptak, D. Płokarz (Gdynia Maritime Univ., Poland)

4   Signal Processing

Azure Kubernetes Service Design Principles in Machine Learning Systems 179
Y. Bershchanskyi, H. Klym (Lviv Polytechnic National Univ., Ukraine)

High-Accuracy ECG Signal Acquisition Using a Power-Efficient 6-bit Level-Crossing ADC 184
A. Amini (Univ. of Pavia, Italy), H. Norouzi Kalehsar (Urmia Univ., Iran)

Low Voltage, High Power Electronic Load Design for FPGA Current Draw Reproducing 188
S. Przybył, P. Sarna, Z. Kulesza, M. Zubert (Lodz Univ. of Techn., Poland)

Recurrent LSTM Neural Networks for Language Modelling and Speech Recognition 193
P. Kłosowski (Silesian Univ. of Techn., Poland)

5   Embedded Systems

A Survey and Practical Application of Ethernet-APL, PROFINET Network and HMI 201
A. Lugli, A. Aragão, E.R. Neto, G.A. Vizotto, J.A. Barbosa, J.P. Paiva (INATEL, Brazil), T. Pimenta (Univ. Federal de Itajuba, Brazil)

Analysis of Selected Cryptographic Algorithms for Data Transmission in Airborne Networks 206
S. Baliński, P. Śniatała, M. Sobieraj, A. Grocholewska-Czuryło (Poznan Univ. of Techn., Poland), J. Xie, S. Ren (San Diego State Univ., USA)

Comparative Survey Between Industrial Communication Protocols Applied in Hazardous Areas 211
A. Lugli, A. Teixeira, J.P. Henriques, J.P. Paiva, J. Azevedo (INATEL, Brazil), T. Pimenta (Univ. Federal de Itajuba, Brazil)

Matlab Simulations in Performance Analysis of Storage Area Networks 216
J. Nazdrowicz (Lodz Univ. of Techn., Poland), M. Tuszyńska (Cracow Univ. of Techn., Poland)

Index of Authors 223

General Invited Papers

14

Mixed Design of Integrated Circuits and Systems – MIXDES 2025

AI for Processors, Processors for AI:
Going New Ways for Processor Architectures

Michael Hübner

Brandenburg University of Technology Cottbus – Senftenberg
Cottbus, Germany
michael.huebner@b-tu.de

Abstract—The rapid evolution of artificial intelligence (AI) is prompting a paradigm shift in processor architecture design. Traditional processors in the embedded computing domain were originally optimized for special-purpose computing, but they are becoming increasingly suboptimal for meeting the performance, energy efficiency, and scalability demands of applications with restricted resources. Concurrently, AI techniques are being employed to enhance and even automate processor design, establishing a reciprocal innovation cycle between AI and hardware. Furthermore, AI technology can be deeply implemented into processor hardware architecture to enable adaptation based on AI inference results.

Keywords—Embedded Systems, Processor Architecture, Run-Time Adaptivity

I. INTRODUCTION

Current processor units are not flexible enough to adapt their hardware architecture dynamically to the runtime state. The proposed approaches will increase this flexibility, enabling us to achieve an increase in computational performance. However, to exploit this ability, we will need sophisticated monitoring, information fusion, and evaluation of the state within a processor architecture. Machine learning algorithms with trained networks will predict the future state of the processor as many clock cycles as possible. This information will then be used to adapt the processor architecture. Advances in these areas will strengthen the foundation for dynamic online adaptations and optimize the use of additional flexibility. Currently, there is a lack of new methods to monitor the state of a processor or system of processors using novel data sampling methods. These methods must utilize new high-level data sources, such as monitoring pipeline stalls, cache hits/misses, etc., to detect complex relationships at the chip level. This differentiates the approach from other research achievements. Furthermore, this approach requires and enables the development of new methods for processing and evaluating recorded data. Monitored data must be collected and evaluated, and finally, decisions must be made regarding processor adaptation at the chip level. This will be accomplished using a combination of data analysis and machine learning methods.

One particularly promising approach to realizing highly adaptive processor architectures is run-time adaptive hardware. New approaches, such as FPGA-based GPUs ([2],[16]) and highly flexible iCores [4], demonstrate how inference results from AI models can guide the adaptive reconfiguration of processor hardware during operation [1]. Rather than being static, these architectures monitor workload characteristics and dynamically adjust their configuration, such as parallelism, precision, memory hierarchy, and execution units, to operate at the optimal point on the power-performance curve at any given time (see figure1).

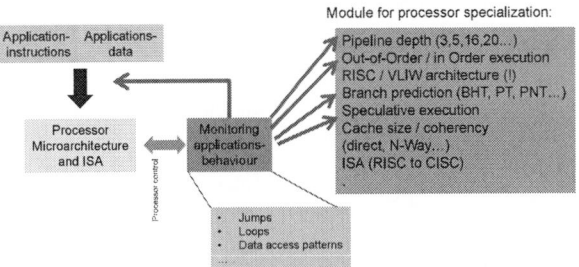

Figure 1. Monitoring and selection of processor mode

This runtime adaptability, informed by AI-driven inference and control, enables systems to achieve higher performance while reducing power and energy consumption simultaneously [3]. This represents a shift from the traditional worst-case provisioning approach to intelligent, context-aware optimization that evolves with the workload in real time, reaching the most beneficial point of the power-performance curve at any given moment.

II. STATE OF THE ART AND NOVEL TRENDS

The amount of effort required to manage processor units in order to operate them more efficiently is increasing in respect of all its subtasks, starting with the monitoring of the system state, followed by its evaluation, and the actions to be taken based on the evaluation. This increase is caused by different aspects of the systems themselves and their application. One key aspect is the mobility and ubiquity of today's computing systems. The processor unit inside such systems has to consume as little power as possible, while still providing the necessary processing performance. Moreover, the required processing performance is increasing over time, even in single application scenarios, as different applications have different phases [6]. Another aspect is that the increasing performance of processor units enables the consolidation of several applications onto a single processor unit. This leads to new scenarios in which a variety of applications with different criticalities are sharing the same processor unit. Moreover, other applications may be loaded and unloaded during runtime, e.g., by requests of other entities such as a user. Thus, dynamic mixed-critical systems are a direct consequence [7].

Besides these application-related aspects, there are other reasons for developing better and more complex processor management, such as, for instance, the heterogeneity of today's computing systems. Processor units can contain dedicated accelerators in order to increase their efficiency and performance. There are several examples of this trend in the academic world as well as in industry. Research projects such as *ERA*, *FlexTiles* or *EURETILE* have been developing new heterogeneous architectures. But also commercial products, as for instance ARM's *bigLITTLE* technology [8] or AMD's FPGA chips [9] are heterogeneous in order to enable their increased efficiency and performance.

The heterogeneity and reconfigurability in software and hardware enable adaptive processing, which allows modifications to be made in the field. These modifications can be executed by the unit itself or by another entity, and thus, it can be optimized for a specific state or mode of operation. The first key step will be to develop a management system that is able to adapt the processor unit and that has the capability to both monitor the processor unit and to evaluate the monitored state. For the efficient monitoring of a processor unit it is necessary to have detailed knowledge about both its software and hardware architecture and their reconfiguration capabilities. This is also the case for the evaluation of the monitoring results. The different approaches for reconfiguration are discussed in the next subsection.

Adaptive computer systems exploit the heterogeneity to dynamically execute tasks more efficiently than in a homogeneous system [10] with respect to the runtime state. Moreover, the reconfigurability allows the adaptation of the system dynamically on the hardware level. Such adaptations may include dynamic voltage and frequency scaling (DVFS) or a modification of the instruction set (IS) to better fit the executed application. A fine grained adaptation of the processor architecture, e.g. the number of pipeline stages, parallel execution units, adaptive branch prediction and an adaptive cache memory is not currently supported in the state of the art processors in industry or in academic research.

The authors of [11] have proposed a reconfigurable soft-core processor architecture, which utilizes the internal configuration access port of AMD FPGAs to configure the logic during runtime and to additionally write and read processing data to and from the FPGA internal memory. Thus, the architecture of the presented general-purpose processor can be optimized for the execution of specific application. The authors have and [12] have proposed approaches similar to the previous one: the invasive Core (i-Core), respectively the extensible microprocessor without interlocked pipeline stages (eMIPS). The i-Core implements a core IS and an extension for the IS. The functionality of the IS extension is implemented in a reconfigurable fabric and prepared during compile time. The IS extension can also be executed as a sequence of the IS core. Therefore, the decision as to whether to use the IS extension or not can be made during runtime. By optimizing the application execution with this method a speedup factor of 22 has been achieved for a single task video encoder for the H.264 codec in comparison to the state of the art implementation. A similar approach has been followed within the eMIPS project. The application's instructions are tracked during execution and the system tries to identify the most often used basic blocks in order to create a new IS extension which is optimized for its execution. The eMIPS approach also utilizes reconfigurable fabric in order to adapt dynamically at runtime. In a recent work of Al Kadi et. al., a scalable and run-time adaptive Graphics Processor Unit (GPU) was presented [13]. In this work an average speedup of 3.9, and a maximum speedup of even 13 with the run-time adaptive GPU in comparison to the hard-wired ARM9 CPU including the NEON arithmetic accelerator has been achieved. The approaches presented in these research works are a first step in the direction of achieving highly dynamic processor architecture, but the deployment with specific processor architectures such as Superscalar and Very long instruction word (VLIW) is missing. Furthermore, the essential topic of a minimal or better non-invasive monitoring of processor status and its realization is not addressed at all. Also the usage of machine learning algorithms, respectively trained neuronal network to predict a processor state and to derive out of this information a decision to adapt the processor is not present, but is a very important research topic.

Monitoring of processors and processor systems is a well understood and investigated topic. Halsall and Hui described the rationale and the design of a performance monitoring system for embedded computing systems [14]. System-related and application-specific events can be recorded through software functions. Today, advanced hardware-support for monitoring is available; enabling more sophisticated monitoring with less overhead and new parameters as for instance power consumption. There are several other projects looking into the information extraction and evaluation in processor units and systems, as well as how to process these types of information. For instance, the *System-Physician-on-a-Chip (SPOC)* project, funded by DFG is investigating the monitoring of integrated circuits on transistor and physical level and their online adaptation [15].

III. CONCLUSION AND OUTLOOK

The rapid advancements in artificial intelligence are fundamentally transforming the landscape of processor architecture design, particularly in the embedded computing domain. As traditional processors encounter challenges in meeting the demands of modern AI-driven applications, particularly in terms of performance, energy efficiency, and scalability, it becomes apparent that a new design paradigm is necessary. AI is a driving force behind the need for architectural innovation and offers the tools to realize it, creating a synergistic loop where AI benefits from and contributes to hardware evolution.

Moving forward, the integration of AI into processor design is expected to deepen. Emerging architectures will likely feature AI-driven adaptability at the hardware level, enabling systems to dynamically optimize their operation based on inference results and contextual requirements. Furthermore, the integration of AI into design processes has the potential to expedite development cycles and generate highly specialized, efficient processors tailored to specific application domains. As this reciprocal relationship between AI and hardware continues to evolve, it holds the promise of unlocking new levels of computational efficiency and intelligence in resource-constrained environments.

REFERENCES

[1] S. Mahmood, M. Huebner, M. Reichenbach: „A Design-Space Exploration Framework for Application-Specific Machine Learning Targeting Reconfigurable Computing", International Symposium on Applied Reconfigurable Computing 2023, 371-37

[2] Hernandez, Fricke, Al Kadi, Reichenbach, Hübner: „Edge GPU based on an FPGA Overlay Architecture using PYNQ", 2022 35th SBC/SBMicro/IEEE/ACM Symposium on Integrated Circuits and Systems Design (SBCCI)

[3] M. M. Goncalves, F. Benevenuti, H. Munoz, M. Brandalero, M. Hubner, F. Kastensmidt, J. R. Azambuja, Investigating Floating-Point Implementations in a Softcore GPU under Radiation-Induced Faults, In: IEEE International Conference on Electronics, Circuits and Systems (ICECS), 2020

[4] Michael Hübner, Diana Göhringer, C. Tradowsky, J. Henkel : „Adaptive Processor Architecture", International Conference on Embedded

[5] Computer Systems: Architectures, Modeling, and Simulation (SAMOS XII), 2012, Samos, Greece

[6] Srinivasan, S.; Kumar, R.; Kundu, S.: Program phase duration prediction and its application to fine-grain power management: IEEE Computer Society Annual Symposium on VLSI, 2013.

[7] Heiser, G.: Virtualizing embedded systems - why bother?: Design Automation Conference, 2011

[8] Chung, H.; Kang, M.; Hyun-Duk, C.: Heterogeneous Multi-Processing Solution of Exynos 5 Octa with ARM big.LITTLE Technology. http://www.arm.com/ja/files/pdf/Heterogeneous_Multi_Processing_Solution_of_Exynos_5_Octa_with_ARM_bigLITTLE_Technology.pdf, accessed August 2015

[9] Al Kadi, M. et al.: Dynamic and partial reconfiguration of Zynq 7000 under Linux: International Conference on Reconfigurable Computing and FPGAs, 2013

[10] Padoin, L. E.; et al.: Performance/energy trade-off in scientific computing: the case of ARM big.LITTLE and Intel Sandy Bridge, 2015, 9

[11] Huebner, M.; et al. Eds.: Adaptive Processor Architectures, 2012, ISBN: 978-1-4673-2295-9

[12] Pittman, Richard Neil et al. "eMIPS, A Dynamically Extensible Processor." (2006) https://api.semanticscholar.org/CorpusID:13567995

[13] Muhammed Al Kadi, Benedikt Janssen, and Michael Huebner. 2016. FGPU: An SIMT-Architecture for FPGAs. In Proceedings of the 2016 ACM/SIGDA International Symposium on Field-Programmable Gate Arrays (Monterey, California, USA) (FPGA '16). Association for Computing Machinery, New York, NY, USA, 254–263

[14] F. Halsall and S.C. Hui: Performance monitoring and evaluation of large embedded systems, Software Engineering Journal Volume 2, Issue 5 https://doi.org/10.1049/sej.1987.0024

[15] DFG: System-Physician-on-a-Chip (SPOC). Chip Health-Monitoring Infrastructure IP and Run-Time Adaptation. http://gepris.dfg.de/gepris/projekt/269724693, accessed 1 Dec 2015

[16] Langhammer and Constantinides: A Statically and Dynamically Scalable Soft GPGPU, FPGA '24: Proceedings of the 2024 ACM/SIGDA International Symposium on Field Programmable Gate Arrays Pages 165 – 175, https://doi.org/10.1145/3626202.36375

Electronic Control Systems for Ion Trap Quantum Computers

Grzegorz Kasprowicz
Warsaw University of Technology
Warsaw, Poland
Grzegorz.Kasprowicz@pw.edu.pl

Abstract—Ion trap quantum computers are complex optoelectronic systems that require several electronic subsystems orchestrated to sub-1ns precision, both low-noise and high speed. WUT developed a modular control system in EEM/DIOT form factor called SINARA (Standardised Instrumentation Architecture for Research Applications) [1]. SINARA became the de facto standard in atomic and Molecular laboratories worldwide; a few leading quantum computer manufacturers also use it. Its success, which can be attributed to the open-source, open-hardware licensing model, has significantly advanced the field of quantum computing by providing a reliable and adaptable control system. SINARA was primarily developed for lab applications but recently has been transformed into an industrial form factor (DIOT [5]) because many lab projects have turned into spin-offs and startups.

The scaling of quantum computers creates several technological issues. Most topologies are enclosed in a vacuum and use cryogenic temperatures. Future quantum computer architectures need to implement electronic control systems capable of steering hundreds of thousands of electrodes with high precision and relatively high voltage of a few dozen V. Within the framework of the Quantera SIQCI [3] project, a collaborative effort aimed at advancing the field of quantum computing, we are building a demonstrator of such a scalable control system. An essential step of the design process is characterising a standard 180nm high voltage process at temperatures below 10 K. The next step is updating the process PDK and designing the test structures using updated models. This paper presents the requirements for such a control system, the first results of the measurements and the design methodology.

Furthermore, this talk will unveil the first coherent operations results performed on ions trapped in the WUT labs using the SINARA system, marking a significant milestone in our system's practical applications.

Keywords—FPGA, ASIC, Cryogenics, AMO, Trapped Ions, ARTIQ

I. INTRODUCTION

Scalable, fault-tolerant quantum computers (QCs) will solve certain classes of problems significantly faster than their best classical counterparts. While the ultimate realisation of a fault-tolerant device is considered a long-term goal, noisy intermediate-scale quantum (NISQ) devices are already available today and are expected to be used, for example, in optimisation tasks, novel material design, or critical processes in logistics, healthcare, and finance. However, current quantum devices cannot be scaled to 1000 qubits or efficiently programmed due to the lack of suitable control systems and software. As a result, the roadmap for building large-scale, universal computing devices faces a key technological "bottleneck." **The WUT team develops dedicated control systems for ion trap quantum computers. Since 2022, our quantum computer infrastructure has been built based on developed technology [5].** Several key technologies are necessary to create a multi-thousand-qubit ion trap quantum computer:

- **Advanced cryogenic ion trap** with integrated optics (photonic integrated circuit), ion transport electrodes (QCCD) and readout sensors, capable of trapping and manipulating of hundreds of thousands of ions

- **Scalable, low-noise electronic circuits** tailored for ion transport operations such as moving, splitting, and merging sub-registers within segmented traps at room temperature and under cryogenic conditions;

- **Efficient compilation of quantum algorithms** based on low-level instruction sets for segmented QPU units, optimizing algorithms with more than 100 qubits.

II. SINARA CONTROL SYSTEM

A. Introduction

SINARA (Figure 1) is a modular, open-source measurement and control hardware ecosystem designed for quantum information processing applications that require deterministic, high-precision timing. It is based on industrial standards and consists of over 50 digital and analogue input and output modules. The hardware is controlled and managed by ARTIQ [2], an open-source software system for experimental control that provides nanosecond timing resolution and sub-microsecond latency via a high-level programming language. The measurement and control systems

Several grants and contracts funded this research: NESTER (MAZOWSZE/0153/1900), DOB-SZAFIR/01/A/023/01/2020, IDUB-POB-FWEiTE, QUANTERAII/1/80/SIQCI/2022, POIR.01.01.01000553/20

   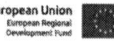

used in quantum physics experiments suffer several problems. In general, an improvised solution is built in-house without enough consideration for good design, reproducibility, testing, and documentation. It makes those systems unreliable, fragile, difficult to use, maintain, and reproduce in other labs, and hard to repair. It also duplicates work in different laboratories. Also, the performance and features of the existing systems (e.g., regarding pulse shaping and synchronisation abilities) are becoming insufficient for some experiments.

Figure 1. Sinara chassis - example configuration.

Sinara and ARTIQ projects address the above issues by providing a collaborative hardware and software environment that is both open-source and commercially available from multiple vendors. The community involved in the Sinara project has successfully developed over 100 boards and modules over the last eight years. Some are already commercially available, with most of the rest to follow. Nearly all modules were developed at the WUT, most of which were designed by the Author. The design of SINARA is indebted to much prior work on control hardware and software by the ion-trapping community. Currently, SINARA is the leading control system in the AOM community, and several quantum computing vendors use it as a foundation for their systems.

B. SINARA ecosystem

The Sinara ecosystem offers a nearly complete set of modules performing the following functions:

- System controllers
- Analogue to digital and time to digital converters
- Digital to analogue converters
- Servos and regulators
- Fast digital Inputs and Outputs
- Camera and optical sensor interfaces
- Arbitrary Waveform Generators (AWG)
- Radio Frequency and DDS synthesisers
- RF, low frequency, and HV amplifiers
- Low-voltage and high-voltage power supplies
- Various adapters and converters

III. ION TRAP QUANTUM COMPUTER SCALING

Practical, real-life problem-solving algorithms require hundreds of thousands of qubits. While the standard, modular SINARA system addresses most quantum computer needs, it fails to deliver signals for future ion trap processing units. Existing and near-term quantum computers up to 1000 qubits are built on the existing SINARA ecosystem, including 32-channel DAC cards, but such a solution is not scalable due to several reasons:

- Difficulty in delivering millions of signals to cryogenic QPU
- Physical space and power needed for hundreds of electronic racks
- Noise limitation of room-temperature electronics, essential for high-fidelity qubits
- Signal integrity and power integrity issues associated with transfer of control signals to cryostat

Due to those constraints, WUT develops dedicated ASICs (Figure 2, Figure 3) within the Quantera SIQCI project.

Those ASICs need to address the following issues:

- Operation at 4K in ultra-high vacuum
- Offer scalability to millions of DAC channels
- Provide high voltage operation of at least +/-20V
- Consume very low power, the entire cooling budget at 4K stage is a few Watts
- Provide ultra-low noise, at least 16-bit precision and very low drift (1ppm or better)
- Integrate with the SINARA/ARTIQ ecosystem

The WUT team developed and succeeded in tape-out and cryogenic measurements of the first test ASIC, which includes DAC and test structures. The second, multi-channel DAC is being finalised and will be sent for tape-out in June.

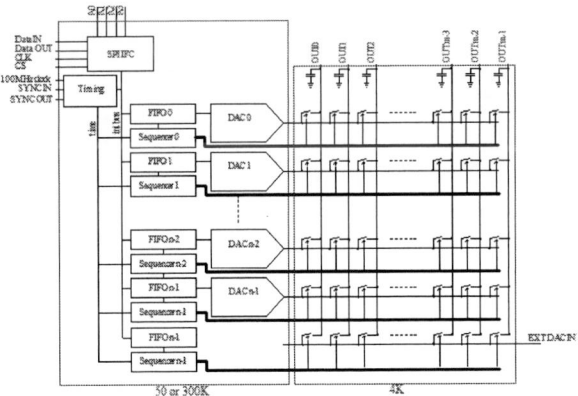

Figure 2. CRYO-ASIC concept.

Figure 3. First batch of CRYO-ASIC-I

Figure 5. CRYO-ASIC-I installed at the 4K stage of cryostat

Several tests were performed in both 300K and 9K temperatures. For low voltage transistors increase of drain current was observed - 40% for 1.8V NMOS and 150% for 3.3V and HV NMOS structures. This effect is likely caused by reduction in phonon scattering, leading to an improved carrier mobility. However, in case of HV PMOS and NEDIA NMOS transistors characteristics at 9K differ substantially from those at room temperature (Figure 4). Such characteristics render them useless as switches and muxes in high voltage ASICs. For this reason, a new architecture needs to be proposed. We plan to use embedded 14-bit DACs followed by S&H high voltage amplifier.

We also performed tests of integrated 10-bit DAC. No important differences were measured between 300K and 9K operation (Figure 6).

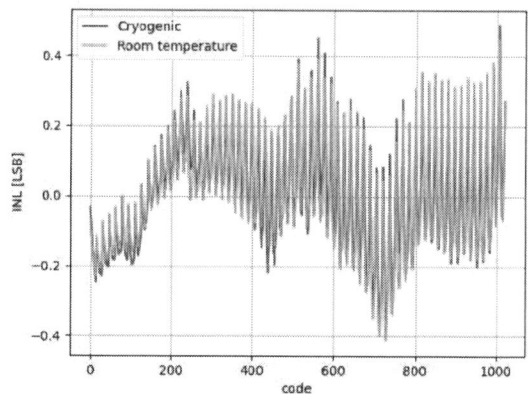

Figure 6. DAC INL at 300K and 9K

ACKNOWLEDGMENT

I want to thank the entire Sinara and the ARTIQ community, without whom this project wouldn't exist. Special appreciation to: David Allcock, Chris Ballance, Tim Ballance, Sebastien Bourdeauducq, Joe Britton, Ken Brown, Stanisław Hanasz, Tom Harty, Robert Jordens, Anna Kaminska, Marcin Kiepiela, Paweł Kozakiewicz, Paweł Kulik, Jakub Matyas, Jonathan Mixrahi, David Nadlinger, Christian Ospelkaus, Krzysztof Pozniak, Maciej Przybysz, Drew Risinger, Daniel Slichter, Ana Sotirova, Mikołaj Sowinski, Filip Switakowski, Zbigniew Wawrzyniak, Weida Zhang, Konrad Norowski, Krzysztof Siwiec, Dominik Kasprowicz, Tomasz przywózki, Adam Borkowski, and others involved.

REFERENCES

[1] "Sinara Open Hardware Project", 2025. [Online]. https://sinara-hw.github.io/.

[2] S. Bourdeauducq et al., "Zenodo", 2021. https://doi.org/10.5281/zenodo.1492176.

[3] "SIQCI project", 2022. https://quantera.eu/siqci/.

[4] "MIKOK project", https://www.mikok.pl/.

[5] "DI/OT project", 2025. https://gitlab.com/ohwr/project/diot/-/wikis/home.

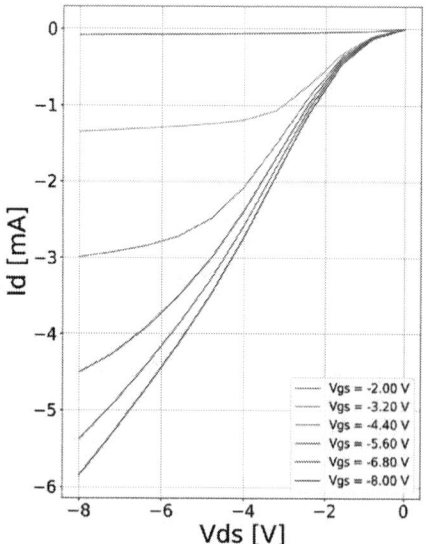

Figure 4. HV PMOS characteristics

Mixed Mode: More than Analog and Digital

Roberto S. Murphy, Reydezel Torres

Instituto Nacional de Astrofísica, Óptica y Electrónica
Tonantzintla, Puebla, México
rmurphy@ieee.org, reydezel@inaoep.mx

Abstract—**In this paper we present some research results of the endeavors of the High Frequency Laboratory at the INAOE, in Puebla, Mexico. For over 35 years, the group has been dedicated to the modeling and characterization of active and passive devices for RF applications, both on-chip and on PCB. Passive devices encompass inductors, interconnects, and antennas. Modern mixed-mode circuits are made of more than analog and digital components; for RF applications, they must include several passive devices as well. We also present some important considerations on high-frequency de-embedding of structures.**

Keywords—**High frequency, modeling, transistors, passive devices, de-embedding.**

I. INTRODUCTION

In its origin, the term "Mixed Mode" referred to integrated circuits (ICs) made from digital and analog components. Mixing these types of circuits opened the field of IC design to include a vast scope of functionalities, paving the way for versatile circuits including system on chip (SoC), and lab on chip (LoC), amongst many more.

Nowadays, due to the evolution of technologies, manufacturing processes including transistors that reach cut-off frequencies in the hundreds of GHz range are readily available [1], providing sophisticated bases for the design and development of a slew of wireless circuits impacting telecommunications, the Internet of Things (IoT), industrial and medical applications, and many more.

All these wireless circuits require, besides transistors, resistors and capacitors, inductors, transmission lines on chip, through silicon vias (TSVs) for 3D integration, and antennas to transmit and receive all sorts of data. Hence, mixed mode circuitry includes many passive devices working together with active ones to be able to meet the stringent requirements imposed on these types of circuits.

In this paper we present the work done in the modeling, measurement and characterization of active and passive devices at the High Frequency Laboratory of the INAOE, Puebla, in the last 35 years. The text is organized as follows. Section II addresses specific challenges related to active devices, particularly transistors, which require the consideration of additional effects when increasing the operating frequency or modifying their structure. Section III presents the modeling of integrated inductors, followed by a summary of our work with coplanar waveguides (CPWs) and through silicon vias (TSVs) in Section IV. Section V presents some results for integrated antennas. In Section VI we discuss the importance of de-embedding the measurements, especially for frequencies approaching 100 GHz. The paper closes with some general conclusions and future perspectives.

II. ACTIVE DEVICES

Many research groups worldwide have dedicated their time and efforts to model active devices for RF applications throughout the years. The references are too many; it suffices to mention that a literature search will provide thousands of returns. This is a very dynamic field due to the constant evolution of technology, which rapidly renders models obsolete as new designs and manufacturing processes become available. Notwithstanding, very accurate and reliable compact models are available for IC designers nowadays; however, there are some effects that these models so not take into consideration. An important one is the dependence of drain current on the number of fingers of an interdigitated gate regardless of the area. Fig. 1 shows this dependence for different transistors having the same area but different number of fingers.

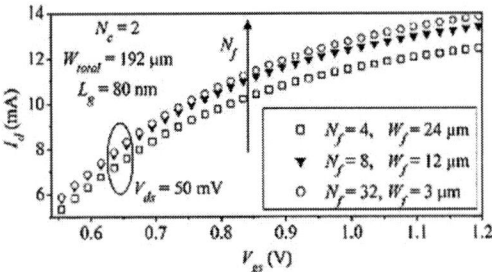

Fig. 1. Dependence of drain current on number of fingers for devices with equal gate area [2].

This effect is due to the different extrinsic source and drain resistances, which vary with the number of fingers in each cell. The variation is attributed to the number of contacts needed in each case. Fig. 2 illustrates this case.

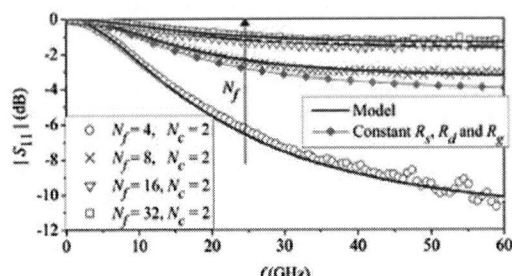

Fig. 2. Variation of S_{11} with number of fingers for equal area transistors [2].

These effects have to be considered when modeling the transistor for RF applications to obtain reliable pre-manufacturing simulations.

Another important aspect is the determination of several of the model parameters that are usually extracted using DC routines. Some of these values, however, change when the device is operating in the RF regime; clearly, the correct value has to be used. To overcome this difficulty, in [3], [4] we presented a methodology to extract all the pertinent values solely from RF measurements. As an example, Fig. 3 shows the comparison of threshold voltage extracted from RF measurements to that obtained with a conventional DC routine. The results presented in those articles include channel resistance, transconductance, subthreshold slope, and effective length, among others.

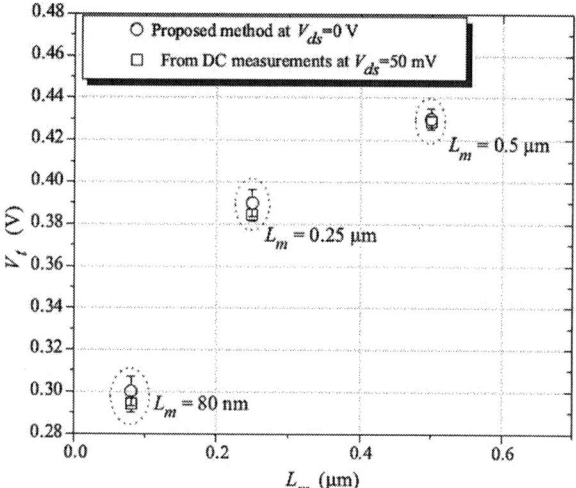

Fig. 3. Threshold voltage extracted from RF measurements and compared to that obtained with a traditional DC routine [3].

One more example deals with the unwanted, not-controllable substrate current present in MOS devices, as shown in Fig. 4.

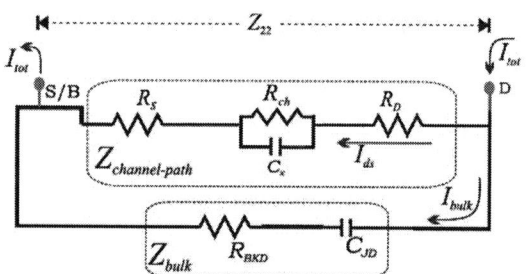

Fig. 4. Total current in a MOS device, comprised of drain current and non-controllable substrate current [3].

The magnitude of the substrate current can be considerable, severely reducing the current flowing through the channel, and thus affecting transconductance. With X_{bk} representing the fraction of substrate current to total current, I_{tot}, Fig. 5 shows its magnitude for different gate voltages as a function of frequency.

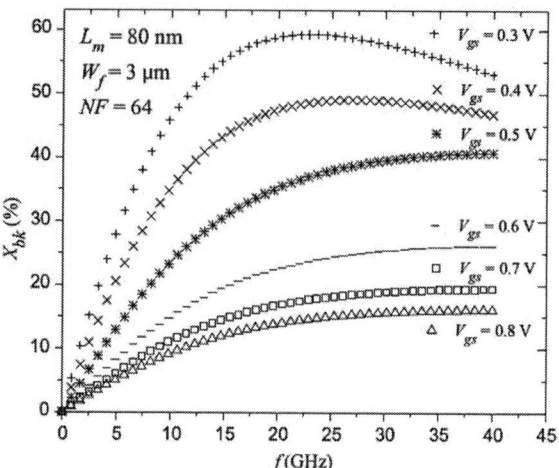

Fig. 5. Fraction of total current that flows through the substrate [3].

The graph indicates that as the channel gets inverted, more current flows through it, but there is still a considerable flow through the substrate. Take the V_{gs}=0.8V curve: almost 15% of the current will be subtracted from the desired drain current, and thus the real transconductance will be lower than the estimated one by a proportional amount.

III. INDUCTORS

Reliable integration of inductors was attained at the end of the last century. Since then, these are almost mandatory devices in RF circuits, functioning as impedance matchers, filters of all kinds, resonant circuits, and transformers. Once manufactured on-chip, the structure is present as all the other constituents of the circuit. Above the inductor's resonant frequency, it can be considered as a parasitic component affecting circuit response, and thus, it has to be characterized in the complete frequency range the circuit is designed to operate.

Moreover, the current flowing in the inductor induces an electric field in the substrate below, which in turn generates undesirable eddy currents. To reduce these currents, a ground-shield is usually included in the design. An added benefit is an increase in the Q factor. This shield can be solid or patterned, and can be made using one of the metal layers, or a polysilicon one. In [5] we present the modeling of integrated inductors with solid and patterned ground shields, shown in Fig. 6. A model based on an extended T-equivalent circuit was derived and proposed to represent their behavior up to 60 GHz, and the model was validated by measurement. This model is able to represent multiple resonances by incorporating two series of resonant circuits to account for these, which is essential as we demonstrate that their range of influence can extend several gigahertz above the resonance frequency, potentially impacting the simulated inductor response within its normal operating bandwidth.

The proposed model is shown in Fig. 7, and a comparison based on the Q factor for different geometry inductors with different ground shields is presented in Fig. 8. An important conclusion of this work is that a solid ground shield is adequate for applications for which the bandwidth is not a factor. Otherwise, a patterned ground shield is recommended.

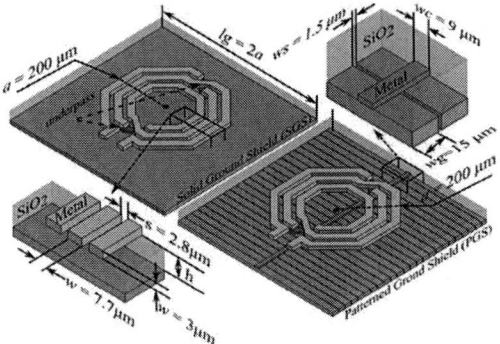

Fig. 6. Example of the studied integrated inductors [5].

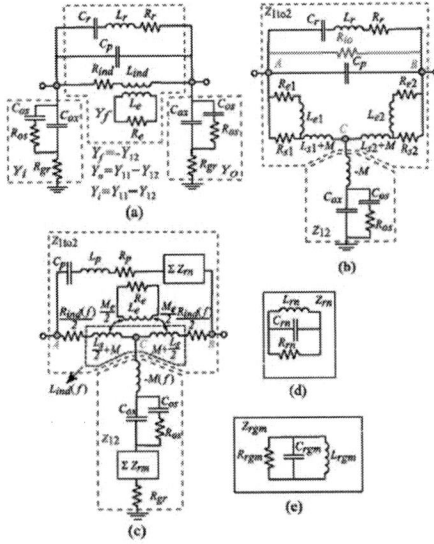

Fig. 7. Equivalent circuits for on-chip inductors. (a) Traditional p-model. (b) Conventional T-model. (c) Proposed T-model. (d) and (e) Subcircuits to represent higher order resonances [5].

IV. INTERCONNECTS

Transmission lines and coplanar waveguides are other fundamental elements of RF circuits. Stages in the same chip have to be connected to others, and eventually to the chip package. The impedance between stages has to be matched, and therefore, these structures have to be fully characterized and modeled in order to design interconnects adequately.

Furthermore, 3D integration has become common in recent years, since it is a convenient and direct form to create complex systems on-chip while reducing losses and interconnect delays. This requires conducting paths from the active part of the chip to the base of the wafer —Through Silicon Vias (TSVs)— which have to be designed and manufactured with reliability and efficiency in mind. Hence, accurate models for TSVs are also needed to design ICs for 3D integration.

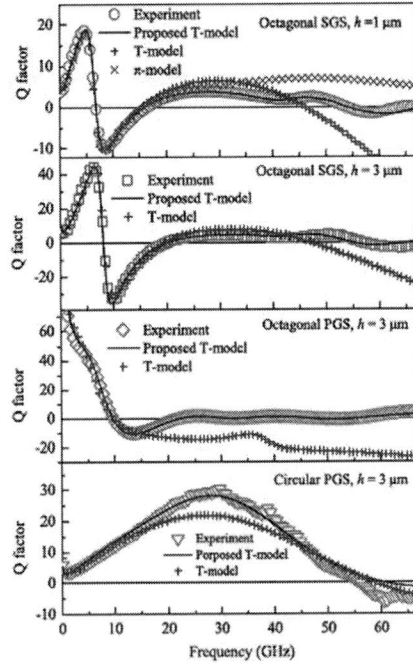

Fig. 8. Comparison of the proposed model with other commonly used ones [5].

We researched the behavior of TSVs through simulation with the goal of defining adequate designs to minimize losses. Our results were published in [6]. In this study, an equivalent circuit and parameter extraction strategy was defined to consistently represent TSVs in silicon substrates of different conductivity. Once the approach was shown to accurately represent TSVs, the concept was extended to analyze vias passing through multi-stacked chips. The assessment of TSV channels was performed in the frequency and time domains, and also considered different configurations of the ground vias used as the return path. The results show that silicon conductivity strongly limits the performance, and thus should be maintained as small as possible for microwave applications.

A SPICE compatible model for a Ground-Signal-Ground (GSG) TSV is shown in Fig. 9. A parameter extraction methodology is also presented in this reference. The model was adapted to consider stacked chips by taking into account solder bumps at the end of the vias.

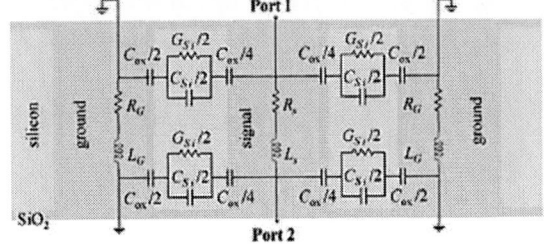

Fig. 9. SPICE compatible model for a GSG TSV [6].

In [7] we present a study of on-chip coplanar waveguides, especially focused on determining the losses caused by the ground shield. To determine them, a dedicated de-embedding routine was applied, which overcomes the limitations of the line-line method [8], [9]. Measurements also showed that port coupling is a function of frequency; the higher the frequency, the larger the current flowing from port to port through the ground shield becomes. Simulation results of this effect are shown in Fig. 10.

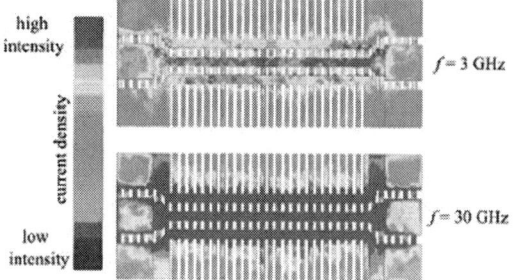

Fig. 10. Simulation of current flow through the ground shield as a function of frequency [7].

V. INTEGRATED ANTENNAS

As technology evolves, the cut-off frequency that active devices can attain is ever increasing, reaching now values of the order of several hundreds of GHz [1]. This has made the inclusion of antennas in the same IC possible, notwithstanding all the limitations that are encountered in the process. Nevertheless, research on integrated antennas has been a very active field in recent times, and it will surely continue to be so.

We have contributed to the effort throughout the years, overcoming some of the limitations [11], [12], [13], especially incorporating metamaterial properties in the design to reduce their size and improve figures of merit such as gain, efficiency and bandwidth. A recent example was the simulation and fabrication of an integrated antenna for applications in the E-band, covering the range from 72.5 to 81 GHz [13]. The antenna was manufactured in INAⁿE's Microelectronics Laboratory. The measurement results were very close to the simulation ones, proving the potential of an antenna of this kind for applications in the E and W bands. Fig. 11 shows a micrograph of the manufactured antenna. A comparison of the simulated and measured return loss (S_{11}) is shown in Fig. 12.

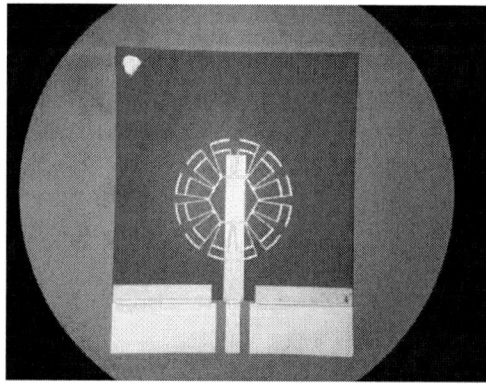

Fig. 11. Micrograph of the manufactured antenna [10].

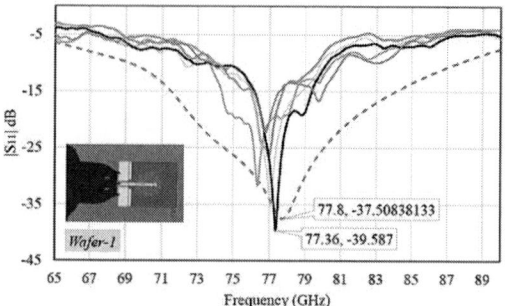

Fig. 12. Measured results for the "Flower Antenna". The red dashed line is the simulation result [10].

V. DE-EMBEDDING

To characterize and model devices for RF ICs, the measurement set-up has to be calibrated and the data must be subsequently de-embedded. As the frequency of operation increases, this is not straightforward and some additional procedures must be carried out to model the actual behavior of the device. This is because more parasitic elements become noticeable, and these can many times obscure the actual response of the device under test (DUT). Hence, new de-embedding techniques keep being published in the field's literature (a search for de-embedding or deembedding in IEEE Xplore returns 567 documents which have either word in the title. 7 correspond to 2025).

An article published in 2005 [14] presented a methodology to de-embed high-frequency measurements of a M□ST; we are proud to state that this article keeps being cited up to today. More recently, we have worked on de-embedding connectors and signal launchers, since these have a noticeable effect when modeling devices at high frequencies [15], [16].

The idea behind this study was born when noticing additional resonances evident when measuring transmission lines, both on-chip and on PCB. We attributed these resonances to the connectors used to probe the lines, and concluded they were produced by standing waves along the line due to the its influence in conjunction with the probing platforms and the connectors employed to probe them. The analyses showed that on-chip short lines are less impacted by this undesired effect when compared to lines on PCB. □n the other hand, long lines, such as those on PCB, present a considerable fluctuation which is further worsened when using transitions that exhibit a noticeable distributed nature within the measurement range.

A model to take this effect into account is shown in Fig. 13. The distributed blocks account for the transitions that introduce the oscillations that complicate the correct determination of the characteristic impedance of the lines.

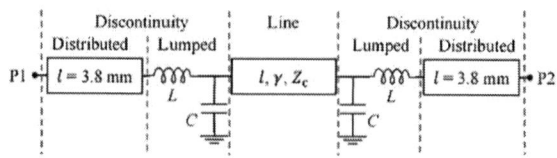

Fig. 13. Model for an IC transmission line including the transition discontinuities [16].

VI. Conclusions

The evolution of technology is trending toward circuits operating at ever higher frequencies. The associated developments will require more precise and accurate modeling of active and passive devices in order to garner reliable simulations pre-manufacture. This will also have a considerable impact in measurement techniques and characterization of devices. On the one hand, higher frequency laboratory equipment will be needed; on the other, calibration and de-embedding techniques will have to be tailored to meet the stricter interpretation of data and reduce uncertainty.

Mixed-mode circuits for RF applications include a slew of devices; of these, the ones which present the most challenges overall are integrated antennas. Much research is needed in the field in order to overcome all the associated limitations.

Acknowledgement

The authors thank the INAOE for its support of the High Frequency Laboratory through the years. They also acknowledge Mexican *Secretaría de Ciencia, Humanidades, Tecnología e Innovación (SECIHTI)* for the partial funding of the lab's project through several grants in the past 35 years.

We are also grateful to all the students that have been at the High Frequency Laboratory putting forth brilliant ideas and lots of high quality work.

References

[1] J. -C. Guo and Z. -C. Li, "Millimeter-Wave CMOS Devices Design with Mobility Enhancement and Parasitic RC Suppression for Super-350 GHz fT and fMAX in Multiring nMOSFETs", in IEEE Transactions on Electron Devices, vol. 71, no. 12, pp. 7287-7293, Dec. 2024, DOI: 10.1109/TED.2024.3487083.

[2] F. Zárate, R. Murphy, R. Torres, A. Ortiz, F. García, "Modeling the Impact of Multi-Fingering Microwave MOSFETs on the Source and Drain Resistances", IEEE Transactions on Microwave Theory and Techniques, Vol. 62, No. 12, December 2014, pp. 3255-3261. DOI 10.1109/TMTT.2014.2366105

[3] "Using S-parameter Measurements to Determine the Threshold Voltage, Gain Factor, and Mobility Degradation Factor for Microwave Bulk-MOSFETs", G. Álvarez, R. Torres, R. Murphy, Microelectronics Reliability, Vol. 51, No. 2, February 2011, pp. 342-349. DOI: 10.1016/j.microrel.2010.09.001

[4] F. Zárate, R. Torres, R. Murphy, "Consistent DC and RF MOSFET Modeling Using an S-Parameter Measurement-Based Parameter Extraction Method in the Linear Region", IEEE Transactions on Microwave Theory and Techniques, Vol. 63, No. 12, December 2015, pp. 4255-4262. DOI: 10.1109/TMTT.2015.2495363

[5] J. Valdés, R. Torres, R. Murphy, G. Álvarez, "Modeling Ground-Shielded Integrated Inductors Incorporating Frequency-Dependent Effects and Considering Multiple Resonances", IEEE Transactions on Microwave Theory and Techniques, Vol. 67, No. 4, April 2019, pp. 1370-1378. DOI: 10.1109/TMTT.2019.2895579

[6] Y. Rodríguez, R. Murphy, R. Torres, "Assessment of through-silicon-vias with different configurations of ground vias and accounting for substrate losses", International Journal of RF and Microwave Computer-Aided Engineering, July 2021, pp. 1-9. DOI: 10.1002/mmce.22811

[7] J. Valdés, R. Murphy, R. Torres, "Determination of the Contribution of the Ground-Shield Losses to the Microwave Performance of On-Chip Coplanar Waveguides", IEEE Transactions on Microwave Theory and Techniques, Vol. 69, No. 3, March 2021, pp. 1594-1601. DOI: 10.1109/TMTT.2021.3053548

[8] A. M. Mangan, S. P. Voinigescu, M.-T. Yang, and M. Tazlauanu,"Deembedding transmission line measurements for accurate modeling of IC designs," IEEE Trans. Electron Devices, vol. 53, no. 2, pp. 235–241, Feb. 2006, DOI: 10.1109/TED.2005.861726.

[9] J. A. Reynoso-Hernandez, "Unified method for determining the complex propagation constant of reflecting and nonreflecting transmission lines," IEEE Microw. Wireless Compon. Lett., vol. 13, no. 8, pp. 351–353, Aug. 2003, DOI: 10.1109/LMWC.2003.815695.

[10] K. Olan, "Antennas for High Frequency Applications", Doctoral Dissertation, INAOE, January 24, 2024.

[11] G. Rosas, R. Murphy, W. Moreno, "Small Antenna Based on MEMS and Metamaterial Properties for Reconfigurable Applications", International Journal of Antennas and Propagation, Vol. 2013, January 2013, pp. 1-10. DOI: 10.1155/2013/498176

[12] L.K. Sandoval, R. Murphy, "Development of Thick Film, CMOS Compatible Planar Millimetre-Wave Antenna for Antennas in Package Applications", Microsystem Technologies, Vol. 23, No. 7, julio 2017, pp. 2927-2930. DOI: 10.1007/s00542-016-3084-z

[13] K. Olan, R. Murphy, "A novel metamaterial-based antenna for on-chip applications for the 72.5-81 GHz frequency range", Scientific Reports, Vol. 12, February 2022, pp. 1-9. DOI: 10.1038/s41598-022-05829-0

[14] R. Torres, R. Murphy, A. Reynoso, "Analytical Model and Parameter Extraction to Account for the Pad Parasitics in RF-CMOS", IEEE Transactions on Electron Devices, Vol. 52, No. 7, julio 2005, pp. 1335-1342. DOI: 10.1109/TED.2005.850644

[15] Y. Rodríguez, R. Murphy, R. Torres, "Modeling Microwave Connectors Used as Signal Launchers for Microstrip Lines of Different Widths", IEEE Microwave and Wireless Components Letters, Vol. 32, No. 7, July 2022, pp. 1-9. DOI: 10.1109/LMWC.2022.3179927

[16] Y. Rodríguez, R. Torres, R. Murphy, "Identifying and Modeling Resonance-Related Fluctuations on the Experimental Characteristic Impedance for PCB and On-Chip Transmission Lines", Electronics, Vol. 12, 2994, July 2023, pp. 1-10. DOI: 10.3390/electronics12132994

Modern Challenges in Hardware Design

Marek Zmuda
System Security Architect
Intel Technology Poland
marek.zmuda@intel.com

Abstract—**Over the years, the technology of designing and manufacturing electronic devices has evolved dynamically, introducing new possibilities in the production of integrated circuits and systems. Despite advancements, significant challenges remain. This paper aims to highlight interesting research directions related to hardware design that are crucial from an industrial perspective. The presented problems will be supported by examples from real-world projects, including Open Hardware, Chiplets, and security issues such as supply chain attacks, side-channel vulnerabilities, and quantum computing threats.**

Keywords—**hardware design, design challenges, cybersecurity, supply chain security.**

I. Introduction

Each year, we observe increasingly rapid advancements in technologies related to the design and manufacturing of electronic circuits. We are able to effectively build complex and efficient systems composed of many components. Thanks to new possibilities, the cost of designing, implementing and validating new systems is decreasing. These new capabilities also bring new challenges that are entirely different from those we have faced in the past. This situation opens up new fields of development for scientists, researchers and engineers. This paper highlights attractive research directions related to hardware design that are crucial for industry. By examining recent trends and real-world examples from last years, the challenges related to Open Hardware, Chiplet technology, and critical security issues will be explored. The discussion will provide insights into how these problems impact design and production processes to finally consider strategies to address them.

II. Supply Chain Security

The prominent article titled "The Big Hack: How China Used a Tiny Chip to Infiltrate U.S. Companies", published by Bloomberg in 2018 [1], has sparked extensive discussion among suppliers and users of electronic equipment. It quickly became evident that considering system and data security does not solely rely on trust in software. Modern electronic systems are complex, and their creation requires using a wide range of components from various suppliers [3]. Outsourcing parts of the project to subcontractors has also become widespread. A security flaw or the deliberate introduction of a vulnerability (e.g., a backdoor) can occur at any stage of project implementation, with any supplier or subcontractor [2].

Considering the above observations, it can be concluded that security pertains to trust in the entire system: all its components, such as hardware, software, configuration, as well as its physical transport to the client, installation, and servicing. The issue has become so significant for the entire industry that the Trusted Computing Group (TCG) has established a special task force on Supply Chain Security.

Figure 1 illustrates the range of supply chain threats in various phases of project implementation. Security vulnerabilities can be introduced by an attacker even at the stage of research and development or design related to the development of new technology. These vulnerabilities may involve intellectual property theft, the introduction of backdoors through the modification of libraries or the project itself through hardware trojans [11], [12], or influencing the definition of the expected configuration. There are also known cases of attacks aimed at modifying tools (tool chain) in such a way that elements created with their help possess specific vulnerabilities [5]. The tools and libraries used at this stage mostly come from external suppliers. Therefore, the manufacturers of these elements become highly attractive targets for attacks. It also should be noted that introducing vulnerabilities at the beginning of the supply chain is highly advantageous for the attacker. The impact of their actions will be present in all projects utilizing this stage of design work. Additionally, there is a high probability that the results of the early stages of work will be directly used in other projects. The tools and libraries used at this stage mostly come from external suppliers. Therefore, the manufacturers of these elements become highly attractive targets for attacks. Subsequent phases, such as manufacturing and final assembly, are often partially or entirely carried out with the support of external partners. This phase also provides numerous opportunities for the attacker. It allows for modifications to the printed circuit board (PCB) [3], [3] and/or the addition of malicious components [1], [2]. Modern integrated circuits often incorporate intellectual property (IP) from multiple suppliers, creating an interconnected system [66]. The design and manufacturing processes involve many subcontractors, each playing a critical role. A chiplet is a small, modular integrated circuit designed to perform specific functions within a larger system-on-chip (SoC) or multi-die design. Unlike traditional monolithic chips, chiplets break down functionalities into smaller, specialized dies that can be combined within a single package. This approach allows for greater flexibility, efficiency, and scalability in chip design, enabling the integration of heterogeneous components optimized for their respective tasks. This technology offers many possibilities in terms of both performance and cost-effectiveness. However, with it arises questions regarding trust in chiplets from different manufacturers [30], [31], [35], [37] and the security of communication between them [32], [33], [34], [36], [39]. The integration phase provides the attacker with more opportunities. At this stage, it is possible to introduce malicious firmware (FW), operating system (OS), applications, or device configurations (hardware or software).

The challenges for researchers are multifaceted and demand innovative solutions. Ensuring security across the entire supply chain requires a comprehensive approach that integrates hardware, software, and human factors. Continuous efforts in developing advanced detection methods and enhancing collaboration between stakeholders are essential to mitigate the risks posed by supply chain vulnerabilities.

III. HARDWARE ROOT-OF-TRUST (RoT)

In situations where we are constructing a complex system composed of components from multiple suppliers, each manufactured in different parts of the world, managing security becomes a significant challenge [2]. However, similar to other fields of engineering, decomposition becomes the solution here as well. It turns out that dividing the system into smaller functional elements and clearly isolating the part responsible for security significantly helps address the issues. The isolated element responsible for security is called the Hardware Root-of-Trust (HW RoT) [14], [15], [16], [17], [18]. The concept assumes that if we can design and produce it in a trusted environment, it will become a trusted part of the system. When we design the system in such a way that the HW RoT is activated first and subsequently verifies the security (hardware, firmware, software configuration) of neighboring components, the verified elements will also become trusted. This creates a domino effect where the newly trusted elements verify others. In this way, after the booting process of the entire complex system is completed, we can consider it trusted.

HW RoT is becoming a crucial element not only for complex systems. The market for low-cost IoT devices, where security is not yet mature, is also addressing this aspect [19], [20]. However, this market faces additional challenges such as limited computational power and memory of devices, constrained power conditions for battery-operated devices, and the need for extremely low production costs. These factors often prevent the use of well-known solutions from the server world [21], [22], [23], [26]. Lightweight cryptography is a new direction standardized by the National Institute of Standards and Technology (NIST), which has selected the Ascon family of algorithms for this purpose [24], [28], [29]. These algorithms are designed to protect data created and transmitted by Internet of Things (IoT) devices, ensuring security despite their limited computational resources.

For several years, we have observed an intriguing trend in the industry: Open Hardware. Concept refers to physical devices whose design specifications are publicly available, allowing anyone to study, modify, create, and distribute them [7]. This approach promotes transparency and collaboration, enabling users to improve and innovate upon existing designs.

By making hardware designs accessible, Open Hardware fosters a community-driven development process similar to that of open-source software. However, it should not be forgotten that Open Hardware faces several security challenges, including vulnerabilities arising from publicly accessible design specifications, which can be exploited by malicious actors [8], [9]. Ensuring the integrity and authenticity of hardware components is difficult, as open designs may be modified and redistributed without proper verification. Additionally, the lack of standardized security measures across different open hardware projects can lead to inconsistent protection against potential threats.

In recent years, we have observed several rapidly developing HW RoT projects in the open-source format, such as OpenTI-TAN and Caliptra. These projects are increasingly becoming standards, supported to varying extents by industry consortia like the Open Compute Project (OCP). The directions discussed in this section present challenges for researchers in designing secure, attack-resistant RoT systems for both highly complex systems and small, low-cost IoT devices.

IV. SIDE-CHANNEL ATTACKS

Over the last few years, we have observed a significant increase in both the cost and scope of side-channel attacks [68]. This trend is driven by the substantially decreasing cost of tools required to execute such attacks.

Side-channel attacks on hardware exploit unintended information leakages from physical devices to gain unauthorized access to sensitive data [40], [41], [42], [43], [46], [47], [48], [50]. These attacks do not target the software or cryptographic algorithms directly but instead leverage indirect effects such as timing variations, power consumption, electromagnetic emissions, or acoustic signals. For instance, an attacker might measure the power consumption of a device during cryptographic operations to infer secret keys. The inherent physical properties of electronic systems make them susceptible to such attacks, posing significant security challenges. Mitigating side-channel attacks requires robust design practices and countermeasures to minimize information leakage and enhance hardware security. Just a few years ago, the capabilities of tools designed to carry out side-channel attacks were very limited, which translated into difficulties in carrying out attacks in a real environment. With the development of electronic systems' capabilities and a significant reduction in their cost, today these tools have become a real threat to the security of equipment. Open Source projects such as Ray-V show that even a hobbyist can now build a Laser-Fault Injection tool that successfully reads or modifies the contents of a device's RAM. Another rapidly developing attack vector in this area is the use of AI solutions to increase the probability of an attack using existing low-cost techniques [44], [45]. A good example of this is the attack on cryptographic keys stored in hardware presented by security specialists from Accenture during the DEFCON 2024 conference.

Taking into account the above trends, it can be stated that scientists are facing new challenges related to the developing

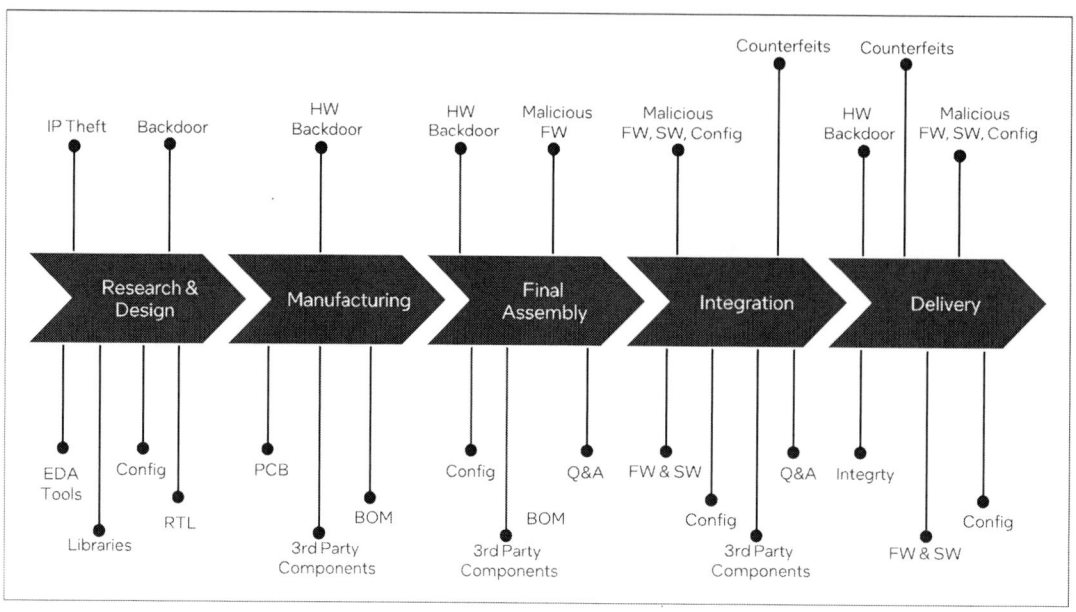

Fig. 1. Security threats related to the supply chain in various project phases

threats of site-channel attacks. These are new concepts, architectures and mechanisms that make it difficult to carry out this type of attack [51], [53]. The second important research direction is techniques for detecting and responding to this type of attack [50], [52].

V. QUANTUM COMPUTING

The capabilities of quantum computers are increasing significantly. Currently, widely known implementations of quantum computers do not allow for effective attacks on classical cryptographic algorithms. However, it is anticipated that with the advancement of quantum technology, this will change in the future [69].

Currently, we can observe a series of actions that are coming to an end, standardizing algorithms of cryptographic systems (XMSS, LMS, Kyber Dilithium, SPHINCS+) resistant to attacks using quantum computers. In the near future, we can expect a growing number of products on the market that implement them. The area of effective and secure implementation [54], [56], [58], [60], [61], [62], [63] and cryptanalytic research [55] in this area seems to be an interesting research thread from the point of view of both industry and science. Issues related to creating keys, subcodes and certificates in an effective way (especially in devices with limited hardware resources) is still a current challenge [57], [59], [64].

VI. CONCLUSION

The dynamic evolution of electronic hardware design introduces significant security challenges that demand innovative solutions from the scientific community. Ensuring trust across the entire supply chain, from initial research and design to final assembly and delivery, necessitates comprehensive strategies to combat vulnerabilities introduced at various stages. Scientists are tasked with developing robust Hardware Root-of-Trust (HW RoT) systems for both highly complex systems and resource-constrained IoT devices, alongside addressing the inherent security risks associated with the increasing adoption of Open Hardware. The escalating threat of side-channel attacks, which exploit unintended physical information leakages and are increasingly amplified by AI solutions, compels researchers to devise novel architectures and detection mechanisms. Furthermore, the anticipated advancements in quantum computing pose a long-term threat to current cryptographic standards, driving the urgent need for secure and efficient implementation of post-quantum cryptographic algorithms, particularly for devices with limited hardware resources. These multifaceted and evolving threats highlight crucial research directions in hardware design, emphasizing the continuous development of advanced security measures and enhanced collaboration across the industry.

REFERENCES

[1] R. Jordan, R. Michael, *The Big Hack: How China Used a Tiny Chip to Infiltrate U.S. Companies*, In Bloomberg Businessweek, 2018
[2] R. Jordan. R. Michael. *The Long Hack: How China Exploited a U.S. Tech Supplier*, Bloomberg 2021
[3] S. H. Russ and J. Gatlin, *Ways to hack a printed circuit board: PCB production is an underappreciated vulnerability in the global supply chain*, in IEEE Spectrum, vol. 57, no. 9, pp. 38-43, Sept. 2020
[4] I. Kabin et al., *Stealth Attacks on PCBs: An Experimental Plausibility Analysis*, 2024 IEEE International Conference on Cyber Security and Resilience (CSR), London, United Kingdom, 2024, pp. 905-912
[5] S. Sunkavilli, Z. Zhang and Q. Yu, *Analysis of Attack Surfaces and Practical Attack Examples in Open Source FPGA CAD Tools*, 2021 22nd International Symposium on Quality Electronic Design (ISQED), Santa Clara, CA, USA, 2021, pp. 504-509

[6] F. Restuccia and R. Kastner, *Towards Zero-Trust Hardware Architectures in Safety and Security Critical System-on-Chips*, 2024 IEEE 3rd Real-Time and Intelligent Edge Computing Workshop (RAGE), Hong Kong, Hong Kong, 2024, pp. 33-36

[7] J. Baehr, A. Hepp, M. Brunner, M. Malenko and G. Sigl, *Open Source Hardware Design and Hardware Reverse Engineering: A Security Analysis*, 2022 25th Euromicro Conference on Digital System Design (DSD), Maspalomas, Spain, 2022, pp. 504-512

[8] A. P. Fournaris et al., *Providing Security Assurance & Hardening for Open Source Software/Hardware: The SecOPERA approach*, 2023 IEEE 28th International Workshop on Computer Aided Modeling and Design of Communication Links and Networks (CAMAD), Edinburgh, United Kingdom, 2023, pp. 80-86

[9] T. Henkes et al., *Evaluating an Open-Source Hardware Approach from HDL to GDS for a Security Chip Design - a Review of the Final Stage of Project HEP*, 2024 Design, Automation & Test in Europe Conference & Exhibition (DATE), Valencia, Spain, 2024, pp. 1-6

[10] H. Pearce and B. Tan, *Large Language Models for Hardware Security (Invited, Short Paper)*, 2024 IEEE 6th International Conference on Trust, Privacy and Security in Intelligent Systems, and Applications (TPS-ISA), Washington, DC, USA, 2024, pp. 420-423

[11] V. T. Hayashi and W. Vicente Ruggiero, *Hardware Trojan Detection in Open-Source Hardware Designs Using Machine Learning*, in IEEE Access, vol. 13, pp. 37771-37788, 2025

[12] K. I. Gubbi et al. *Securing AI Hardware: Challenges in Detecting and Mitigating Hardware Trojans in ML Accelerators*, 2023 IEEE 66th International Midwest Symposium on Circuits and Systems (MWSCAS), Tempe, AZ, USA, 2023, pp. 821-825

[13] G. Tao et al., *Open Se Cura: First silicon results of an auditable and transparent hardware root-of-trust system using open electronic design automation in 16 nm*, in IEEE Solid-State Circuits Magazine, vol. 16, no. 2, pp. 58-66, Spring 2024

[14] S. Zhao, J. Lin, W. Li and B. Qi, *Research on Root of Trust for Embedded Devices based on On-Chip Memory*, 2021 International Conference on Computer Engineering and Application (ICCEA), Kunming, China, 2021, pp. 501-505

[15] Y. Wu, G. Skipper and A. Cui, *Uprooting Trust: Learnings from an Unpatchable Hardware Root-of-Trust Vulnerability in Siemens S7-1500 PLCs*, 2023 IEEE International Symposium on Hardware Oriented Security and Trust (HOST), San Jose, CA, USA, 2023, pp. 179-190

[16] V. Naik, T. Masur, K. Spandana, W. Sanyal and S. Kumar, *A Hardware Implementation of DICE on a RISC-V Processor*, 2024 International Conference on Circuit, Systems and Communication (ICCSC), Fes, Morocco, 2024, pp. 1-6

[17] A. Ehret, E. Del Rosario, K. Gettings and M. A. Kinsy, *A Hardware Root-of-Trust Design for Low-Power SoC Edge Devices*, 2020 IEEE High Performance Extreme Computing Conference (HPEC), Waltham, MA, USA, 2020, pp. 1-6

[18] M. S. U. I. Sami et al., *Advancing Trustworthiness in System-in-Package: A Novel Root-of-Trust Hardware Security Module for Heterogeneous Integration*, in IEEE Access, vol. 12, pp. 48081-48107, 2024

[19] W. Iqbal, H. Abbas, M. Daneshmand, B. Rauf and Y. A. Bangash, *An In-Depth Analysis of IoT Security Requirements, Challenges, and Their Countermeasures via Software-Defined Security*, in IEEE Internet of Things Journal, vol. 7, no. 10, pp. 10250-10276, Oct. 2020

[20] M. S. Sharbaf, *IoT Driving New Business Model, and IoT Security, Privacy, and Awareness Challenges*, 2022 IEEE 8th World Forum on Internet of Things (WF-IoT), Yokohama, Japan, 2022, pp. 1-4

[21] A. M. Awadelkarim Mohamed and Y. Abdallah M. Hamad, *IoT Security: Review and Future Directions for Protection Models*, 2020 International Conference on Computing and Information Technology (ICCIT-1441), Tabuk, Saudi Arabia, 2020, pp. 1-4

[22] J. Singh, G. Singh and S. Negi, *Evaluating Security Principals and Technologies to Overcome Security Threats in IoT World*, 2023 2nd International Conference on Applied Artificial Intelligence and Computing (ICAAIC), Salem, India, 2023, pp. 1405-1410

[23] A. A. Mezaal et al., *Automated IOT Security Testbed: Design and Implementation Insights*, 2024 International Conference on IoT, Communication and Automation Technology (ICICAT), Gorakhpur, India, 2024, pp. 561-567

[24] N. M. Karie, N. M. Sahri, W. Yang, C. Valli and V. R. Kebande, *A Review of Security Standards and Frameworks for IoT-Based Smart Environments*, in IEEE Access, vol. 9, pp. 121975-121995, 2021

[25] M. Khan, M. Ilyas and O. Bayat, *Enhancing IoT Security Through Hardware Security Modules (HSMs)*, 2024 International Conference on Intelligent Computing, Communication, Networking and Services (ICCNS), Dubrovnik, Croatia, 2024, pp. 278-282

[26] R. Sood and V. Sharma, *Analysis of Security Breach using IoT Devices in Smart Cities*, 2024 IEEE 4th International Conference on ICT in Business Industry & Government (ICTBIG), Indore, India, 2024, pp. 1-5

[27] S. Bansal and V. K. Tomar, *Challenges & Security Threats in IoT with Solution Architectures*, 2022 2nd International Conference on Power Electronics & IoT Applications in Renewable Energy and its Control (PARC), Mathura, India, 2022, pp. 1-5

[28] H. -S. Han, T. -h. Choi and J. -S. Yoon, *Enhancing Security in Low-Power Wide-Area (LPWA) IoT Environments: The Role of HSM, Tamper-Proof Technology, and Quantum Cryptography*, in Journal of Web Engineering, vol. 23, no. 6, pp. 787-800, September 2024

[29] A. Rakshe and N. Dongre, *Survey on Security Protocols for IoT*, 2024 IEEE 9th International Conference for Convergence in Technology (I2CT), Pune, India, 2024, pp. 1-5

[30] E. Lee, K. Park, J. Lee and D. J. Park, *Security Architecture for Heterogeneous Chiplet-Based Mobile SoC*, 2025 IEEE International Conference on Consumer Electronics (ICCE), Las Vegas, NV, USA, 2025, pp. 1-6

[31] J. Suzano, F. Abouzeid, G. Di Natale, A. Philippe and P. Roche, *On Hardware Security and Trust for Chiplet-Based 2.5D and 3D ICs: Challenges and Innovations*, in IEEE Access, vol. 12, pp. 29778-29794, 2024

[32] A. Deric, K. Mitard, S. Tajik and D. Holcomb, *Evaluating Vulnerability of Chiplet-Based Systems to Contactless Probing Techniques*, 2024 IEEE International Test Conference (ITC), San Diego, CA, USA, 2024, pp. 71-75

[33] X. Wang, Y. Wang, Y. Jiang, A. K. Singh and M. Yang, *On Task Mapping in Multi-chiplet Based Many-Core Systems to Optimize Inter- and Intra-chiplet Communications*, in IEEE Transactions on Computers, vol. 74, no. 2, pp. 510-525, Feb. 2025

[34] X. Wang, M. Xu, A. K. Singh, Y. Jiang and M. Yang, *On Optimizing Inter- and Intra-Chiplet Interconnection Topologies for Robust Multi-Chiplet Systems*, in IEEE Transactions on Computer-Aided Design of Integrated Circuits and Systems

[35] M. Sasago et al., *Next Generation Chiplet Technology Development: Focusing on Fine RDL Patterning*, 2025 International Conference on Electronics Packaging and iMAPS All Asia Conference (ICEP-IAAC), Nagano, Japan, 2025, pp. 209-210

[36] M. T. Mahmud and K. Wang, *A Chiplet-Based High-Performance and Secure Hybrid Interconnection Network Design Against DoS and Sniffing Attacks*, SoutheastCon 2025, Concord, NC, USA, 2025, pp. 874-879

[37] M. Nabeel, M. Ashraf, S. Patnaik, V. Soteriou, O. Sinanoglu and J. Knechtel, *2.5D Root of Trust: Secure System-Level Integration of Untrusted Chiplets*, in IEEE Transactions on Computers, vol. 69, no. 11, pp. 1611-1625, 1 Nov. 2020

[38] G. A. Chacon, C. Williams, J. Knechtel, O. Sinanoglu and P. V. Gratz, *Hardware Trojan Threats to Cache Coherence in Modern 2.5D Chiplet Systems*, in IEEE Computer Architecture Letters, vol. 21, no. 2, pp. 133-136, 1 July-Dec. 2022

[39] M. S. Ul Islam Sami, K. Zamiri Azar, H. M. Kamali, F. Farahmandi and M. Tehranipoor, *PQC-HI: PQC-enabled Chiplet Authentication and Key Exchange in Heterogeneous Integration*, 2024 IEEE 74th Electronic Components and Technology Conference (ECTC), Denver, CO, USA, 2024, pp. 464-471

[40] S. Mukherjee, S. k. Saikia, S. Anand, R. Chouhan and H. Das, *A Counter Measure to Prevent Timing-based Side-Channel Attack on FPGA*, 2021 6th International Conference on Communication and Electronics Systems (ICCES), Coimbatre, India, 2021, pp. 983-988

[41] H. Jeon, N. Karimian and T. Lehman, *A New Foe in GPUs: Power Side-Channel Attacks on Neural Network*, 2021 22nd International Symposium on Quality Electronic Design (ISQED), Santa Clara, CA, USA, 2021, pp. 313-313

[42] T. Mizuno, H. Nishikawa, X. Kong and H. Tomiyama, *Empirical Analysis of Side-Channel Attack Resistance of HLS-designed AES Circuits*, 2023 International Conference on Electronics, Information, and Communication (ICEIC), Singapore, 2023, pp. 1-4

[43] J. Yao, *Simulation Platform Architecture Design for Side Channel Attack*, 2024 IEEE 4th International Conference on Data Science and Computer Application (ICDSCA), Dalian, China, 2024, pp. 93-97

[44] F. Meng, Z. Li, B. Mao, W. Hu, M. Qin and Q. Zhou, *Adversarial Profiled Side-Channel Attack with Unsupervised Domain Adaptation*, 2023 9th International Conference on Computer and Communications (ICCC), Chengdu, China, 2023, pp. 974-979

[45] J. Xinchen, D. Ailing, P. Yuxi, X. Yanping and X. Jianwei, *A Two-Stage Classification Model on Detecting Cache-Based Side-Channel Attacks*, 2023 20th International Computer Conference on Wavelet Active Media Technology and Information Processing (ICCWAMTIP), Chengdu, China, 2023, pp. 1-5

[46] M. Zhao and G. E. Suh, *Remote Power Side-Channel Attacks on FPGAs*, in IEEE Design & Test, vol. 42, no. 1, pp. 13-19, Feb. 2025

[47] N. M. Rahman, U. Kamal, V. C. Krishna Chekuri, A. Singh and S. Mukhopadhyay, *Passive Lightweight On-chip Sensors for Power Side Channel Attack Detection*, 2024 IEEE International Symposium on Circuits and Systems (ISCAS), Singapore, Singapore, 2024, pp. 1-5

[48] Y. Yang, C. Ou, Y. Wei, W. Li, Y. Fan and X. Shen, *Broader but More Efficient: Broad Learning in Power Side-channel Attacks*, 2024 IEEE 23rd International Conference on Trust, Security and Privacy in Computing and Communications (TrustCom), Sanya, China, 2024, pp. 1528-1533

[49] M. J. Hettiarachchi, H. H. G. D. Sandanuwan, R. W. Balasooriya, R. Hettiarachchi, K. Y. Abeywardena and K. Yapa, *Time Analysis Side Channeling Attack in Symmetric Key Cryptography*, 2023 8th International Conference on Information Technology Research (ICITR), Colombo, Sri Lanka, 2023, pp. 1-5

[50] N. Lungu, B. B. Dash, U. Chandra De, B. B. Dash, S. Singh and S. S. Patra, *Multi-vector Monitoring, Detecting and Classifying GPU Side-Channel Attack Vectors on a Secure GPU Execution Framework*, 2024 8th International Conference on I-SMAC (IoT in Social, Mobile, Analytics and Cloud) (I-SMAC), Kirtipur, Nepal, 2024, pp. 500-505

[51] D. Galli, A. Guarisco, W. Fornaciari, M. Matteucci and D. Zoni, *The Impact of Run-Time Variability on Side-Channel Attacks Targeting FP-GAs*, 2024 31st IEEE International Conference on Electronics, Circuits and Systems (ICECS), Nancy, France, 2024, pp. 1-4

[52] W. Shen, L. Cui, G. Shen and X. Zhang, *Security Attack Detection Method Combining Side Channel with Fault Injection*, 2023 International Conference on Computer Simulation and Modeling, Information Security (CSMIS), Buenos Aires, Argentina, 2023, pp. 37-42

[53] N. Shrivastava and S. R. Sarangi, *Toward an Optimal Countermeasure for Cache Side-Channel Attacks*, in IEEE Embedded Systems Letters, vol. 15, no. 3, pp. 141-144, Sept. 2023

[54] Y. Liu et al., *Hardware Design of PQC Classic McEliece Finite Field Operations and Encryption Module*, 2023 IEEE 17th International Conference on Anti-counterfeiting, Security, and Identification (ASID), Xiamen, China, 2023, pp. 67-71

[55] L. Li, L. Wu, J. Hu, Y. Yang and X. Zhang, *Hardware Design and Security Analysis for SHA-256 in PQC Sphincs+*, 2024 9th International Conference on Integrated Circuits and Microsystems (ICICM), Wuhan, China, 2024, pp. 93-97

[56] A. Dolmeta, M. Martina and G. Masera, *Hardware architecture for CRYSTALS-Kyber post-quantum cryptographic SHA-3 primitives*, 2023 18th Conference on Ph.D Research in Microelectronics and Electronics (PRIME), Valencia, Spain, 2023, pp. 209-212

[57] Y. Tu, P. He, C. -H. Chang and J. Xie, *LTE: Lightweight and Time-Efficient Hardware Encoder for Post-Quantum Scheme HQC*, in IEEE Computer Architecture Letters, vol. 23, no. 2, pp. 187-190, July-Dec. 2024

[58] P. He, T. Bao and J. Xie, *High-Performance Instruction-Set Hardware Accelerator for Ring-Binary-LWE-Based Lightweight PQC*, in IEEE Transactions on Very Large Scale Integration (VLSI) Systems, vol. 33, no. 5, pp. 1417-1421, May 2025

[59] M. P. Çukur and E. O. Güneş, *Optimization of the SPHINCS+ PQC Algorithm with Custom Instructions and LLVM Integration on a RISC-V Processor*, 2024 32nd Signal Processing and Communications Applications Conference (SIU), Mersin, Turkiye, 2024, pp. 1-4

[60] J. Xie, W. Zhao, H. Lee, D. B. Roy and X. Zhang, *Hardware Circuits and Systems Design for Post-Quantum Cryptography—A Tutorial Brief*, in IEEE Transactions on Circuits and Systems II: Express Briefs, vol. 71, no. 3, pp. 1670-1676, March 2024

[61] Y. Cui, J. Li, J. Chen, F. Lyu, C. Wang and W. Liu, *Hardware Security Linking Everything: from Lightweight PUF to Post-Quantum Cryptography Hardware*, 2024 IEEE 17th International Conference on Solid-State & Integrated Circuit Technology (ICSICT), Zhuhai, China, 2024, pp. 1-6

[62] N. Wang, L. Wu, L. Li, M. Chinbat and X. Zhang, *Design and Implementation of a NTT Accelerator Based on RISC-V for PQC Kyber*, 2024 IEEE 18th International Conference on Anti-counterfeiting, Security, and Identification (ASID), Xiamen, China, 2024, pp. 24-27

[63] Y. Dai, Y. Song, J. Tian and Z. Wang, *High-Throughput Hardware Implementation for Haraka in SPHINCS+*, 2023 24th International Symposium on Quality Electronic Design (ISQED), San Francisco, CA, USA, 2023, pp. 1-6

[64] N. Gupta, A. Jati, A. Chattopadhyay and G. Jha, *Lightweight Hardware Accelerator for Post-Quantum Digital Signature CRYSTALS-Dilithium*, in IEEE Transactions on Circuits and Systems I: Regular Papers, vol. 70, no. 8, pp. 3234-3243, Aug. 2023

[65] H. Zhang, X. Qiao, J. Tian, S. Song and Z. Wang, *Fast Hardware Architecture With Efficient Matrix Computations for the Key Generation of Classic McEliece*, in IEEE Transactions on Circuits and Systems I: Regular Papers, vol. 72, no. 3, pp. 1321-1331, March 2025

[66] S. Maragkou, L. Rappel, H. Dettmer, T. Sauter and A. Jantsch, *The Pains of Hardware Security: An Assessment Model of Real-World Hardware Security Attacks*, IEEE Open Journal of the Industrial Electronics Society, vol. 6, pp. 603-617, 2025

[67] M. S. Sharbaf, *IoT Driving New Business Model, and IoT Security, Privacy, and Awareness Challenges*, 2022 IEEE 8th World Forum on Internet of Things (WF-IoT), Yokohama, Japan, 2022, pp. 1-4

[68] T. M. Ignatius, T. Birjit Singha and R. Paily Palathinkal *Power Side-Channel Attacks on Crypto-Core Based on RISC-V ISA for High-Security Applications*, in IEEE Access, vol. 12, pp. 150230-150248, 2024

[69] D. Bellizia et al. *Post-Quantum Cryptography: Challenges and Opportunities for Robust and Secure HW Design*, 2021 IEEE International Symposium on Defect and Fault Tolerance in VLSI and Nanotechnology Systems (DFT), Athens, Greece, 2021, pp. 1-6

Advancing FOSS Compact Modelling: From OTF Transistors to Mott Memristors

A Generic Approach for Compact Modeling of Variability and Low-Frequency Noise in Organic Thin-Film Transistors

Alexander Kloes[1], Ghader Darbandy[1], Benjamin Iniguez[2], Aristeidis Nikolaou[1,2]

[1] NanoP, THM University of Applied Sciences, 35390 Giessen, Germany
[2] DEEEA, Universitat Rovira i Virgili, Tarragona, Spain
Email: alexander.kloes@ei.thm.de

Abstract—**For organic thin-film transistors, a generic approach for modeling the drain-current variability and low-frequency noise due to carrier-number and correlated mobility fluctuations is presented. Both effects are dominated mainly by grain-boundary traps and lead to similar analytical expressions, which are ready for compact model implementation. Additionally, equations for percolative mobility fluctuation are presented, which becomes dominant for below threshold operation. Results of the model are shown to be in good agreement with measurements performed on organic thin-film transistors.**

Keywords—**noise, variability, organic, thin-film transistor, mobility fluctuation, traps, compact model.**

I. INTRODUCTION

In organic thin-film transistors (OTFT), low-frequency noise (LFN) is dominated mainly by grain-boundary traps and mobility fluctuation [1]. Accurate LFN modeling of OTFTs has been successfully demonstrated in [2] by adopting the theory of carrier-number and correlated mobility fluctuations (ΔN noise) [3], [4], [5] and the empirical Hooge approach accounting for mobility fluctuation due to percolative effects ($\Delta\mu$ noise) [6], [7]. Furthermore, OTFTs are sensitive to process variability. Charges being trapped in the channel region cause a local variation of the accumulated charge density, having impact on the threshold voltage of the device [8], [9] and reducing the effective carrier mobility in the channel by Coulomb scattering. As a result, the drain-current variability is also correlated to carrier-number and correlated mobility fluctuations and therefore allows for a similar modeling approach as for ΔN noise [9]. Based on the results published in [2], [9], we present a generic physics-based modeling approach for drain-current fluctuations by carrier-number and correlated mobility fluctuations, which leads to similar expressions for drain-current variability and ΔN noise in OTFTs.

A. Generic Approach

The model is derived from the charge-based compact model for OTFTs presented in [10]. Starting from the local fluctuation of the current in the channel, closed-form model equations

have been derived for the fluctuation of the total device current. The final expressions for drain-current variability

$$\frac{\sigma^2 I_\mathrm{D}}{I_\mathrm{D}^2} = C^* \big|_{\Delta N}\, B^*(q_\mathrm{s}, q_\mathrm{d})\,\big|_{\Delta N} \tag{1}$$

and for ΔN noise

$$\frac{S_\mathrm{ID}}{I_\mathrm{D}^2}\bigg|_{\Delta N} = C^*_\mathrm{noise}\,\big|_{\Delta N}\, B^*(q_\mathrm{s}, q_\mathrm{d})\,\big|_{\Delta N} \tag{2}$$

show the same bias dependent part $B^*(q_\mathrm{ch})$, which is given by the following equation:

$$B^*(q_\mathrm{s}, q_\mathrm{d})\,\big|_{\Delta N} =$$
$$\frac{1}{i_\mathrm{d}} \left(\frac{2(a-1)a^*\mu}{a} + 1 \right) \ln\left(\frac{1+aq_\mathrm{s}}{1+aq_\mathrm{d}} \right)$$
$$+ \frac{1}{i_\mathrm{d}} \left(\frac{1-a}{1+aq_\mathrm{s}} - \frac{1-a}{1+aq_\mathrm{d}} \right) + (a^*\mu)^2 + \frac{2}{i_\mathrm{d}}a^*\mu\,(q_\mathrm{s} - q_\mathrm{d}) \,. \tag{3}$$

Here, a is the subthreshold slope degradation factor, a^* is the normalized Coulomb scattering coefficient, and μ denotes the carrier mobility in the channel. Parameters $q_\mathrm{s/d}$ are the normalized charge densities at source and drain end of the channel, and i_d is the normalized drain current. For details on charge and current normalization please refer to [2].

An individual bias-independent prefactor C^* is obtained for variability and noise. In case of drain-current variability this factor in equation (1) depends on the density N_t of trapped charges:

$$C^* \big|_{\Delta N} = \frac{q^4 N_\mathrm{t}}{WLa^2(kT)^2 C_\mathrm{ox}'^2}\,. \tag{4}$$

For ΔN noise, parameter C^* in equation (2) is given by:

$$C^*_\mathrm{noise}\,\big|_{\Delta N} = \frac{q^4 \lambda N_\mathrm{T}}{WLa^2 kT C_\mathrm{ox}'^2 f^\mathrm{AF}}\,, \tag{5}$$

where N_T is the density of traps per energy available for capture and release of charge carriers. Further parameters are the frequency f, frequency exponent AF of the LFN model, the capacitance of the dielectric per gate area C_ox', the Boltzmann constant k, and the transistor's channel width and length W and L.

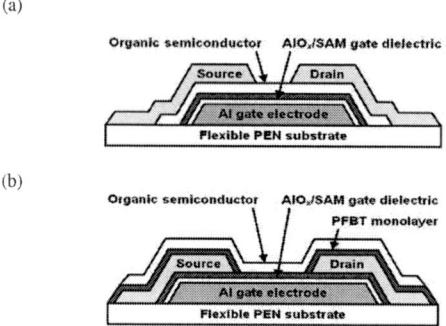

(a)

(b)

Fig. 1. Schematic cross-section of organic TFTs fabricated in the staggered (a) and coplanar (b) device structure [11].

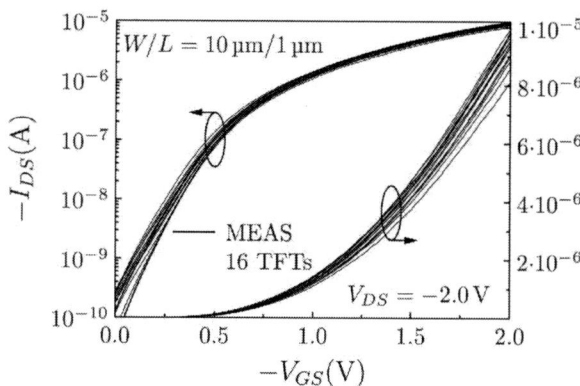

Fig. 2. Measured transfer characteristics of the 16 nominally identical p-channel organic C10-DNTT TFTs in the saturation regime.

II. RESULTS

Measurements performed on fabricated OTFTs in staggered (C10-DNTT) or coplanar (Dph-DNTT) architectures (Fig. 1) show drain-current variability (Fig. 2, [11]). In [9] it has been shown that the drain-current variability in the subthreshold regime is dominated by carrier-number fluctuations, whereas for above threshold operation the mobility-fluctuation effect by correlated Coulomb scattering comes to the fore (Fig. 3).

In case of LFN, measurements on staggered DNTT OTFTs have shown that ΔN noise alone is not sufficient to describe the noise spectra in the deep subthreshold regime of operation (refer to Fig. 4) [2]. Here, LFN due to mobility fluctuation ($\Delta \mu$ noise) must be included with the following equations:

$$\left.\frac{S_{\mathrm{ID}}}{I_{\mathrm{D}}^2}\right|_{\Delta\mu} = C_{\mathrm{noise}}^*|_{\Delta\mu} \; B_{\mathrm{noise}}^*(q_{\mathrm{s}}, q_{\mathrm{d}})|_{\Delta\mu} \qquad (6)$$

$$C_{\mathrm{noise}}^*|_{\Delta\mu} = \frac{a_{\mathrm{H}}q^2}{W\,L\,a\,kT\,C_{\mathrm{ox}}'\,f^{\mathrm{AF}}}, \qquad (7)$$

$$B_{\mathrm{noise}}^*(q_{\mathrm{s}}, q_{\mathrm{d}})|_{\Delta\mu} = \frac{1}{i_{\mathrm{d}}}\left(\ln(q_{\mathrm{s}}) - \ln(q_{\mathrm{d}}) + q_{\mathrm{s}} - q_{\mathrm{d}}\right) \qquad (8)$$

This results in an additional and different bias dependent expression with a prefactor including the empirical Hooge parameter. Finally, the total LFN is given by:

$$\frac{S_{\mathrm{ID}}}{I_{\mathrm{D}}^2} = \left.\frac{S_{\mathrm{ID}}}{I_{\mathrm{D}}^2}\right|_{\Delta N} + \left.\frac{S_{\mathrm{ID}}}{I_{\mathrm{D}}^2}\right|_{\Delta\mu}. \qquad (9)$$

III. CONCLUSIONS

In conclusion, following a generic modeling approach, the combined equations for carrier-number and correlated mobility fluctuations allow consideration drain-current variability and LFN noise in a charge-based compact model of OTFTs following similar expressions for all regions of operation. The results have been shown to be good agreement with measurements. However, for percolative mobility fluctuation dominant in operation below threshold additional expressions have to be considered for modeling LFN noise.

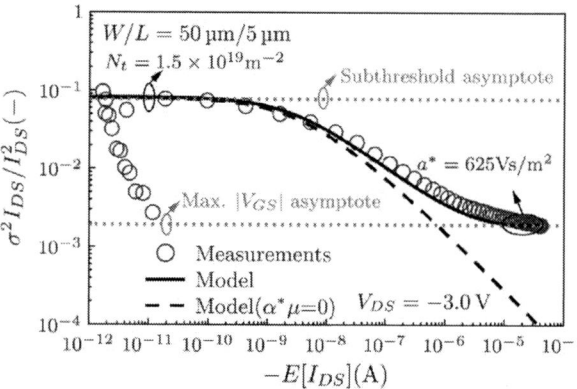

Fig. 3. Normalized drain-current variance $\sigma^2 I_{\mathrm{D}}/I_{\mathrm{D}}^2$ versus mean-value drain current $E[I_{\mathrm{DS}}]$ for OTFTs with L = 5 μm, measured at V_{DS} = -3.0 V [9]. The experimental mean values were calculated over a population of 16 nominally identical transistors. Full line: model including, dashed line: without Coulomb scattering [9].

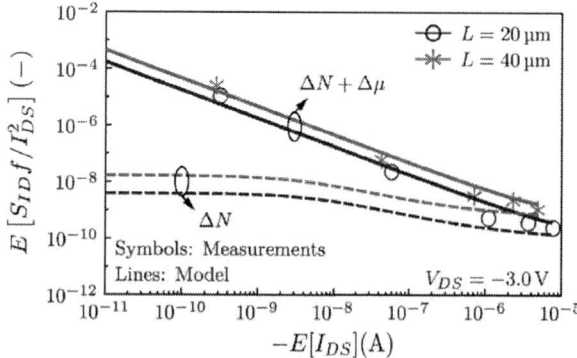

Fig. 4. Mean-value power spectral densities $E[S_{\mathrm{ID}}/I_{\mathrm{D}}^2]$ @ 1 Hz vs. drain current, measured at V_{DS} = -3.0 V [2]. Full line: model including, dashed line: without $\Delta\mu$ noise. The experimental mean values were calculated over a population of 15 nominally identical transistors [2].

ACKNOWLEDGMENT

This project was funded by the German Federal Ministry of Education and Research ("SOMOFLEX", No. 13FH015IX6), German Research Foundation (DFG) under Grant KL 1042/9-2 (SPP FFlexCom), the Spanish Ministry of Science (PRX21/00726), and the EU EIC-PATHFINDER (BAYFLEX, no 101099555). We acknowledge AdMOS GmbH, Germany, for support, and Max Planck Institute for Solid State Research Stuttgart, Germany, for fabrication of devices.

REFERENCES

[1] Lin Ke, Surani Bin Dolmanan, Lu Shen, Chellappan Vijila, Soo Jin Chua, Rui-Qi Png, Perq-Jon Chia, Lay-Lay Chua, Peter K.-H. Ho, "Low frequency noise analysis on organic thin film transistors," Journal of Applied Physics, vol. 104, no. 12, 2008, DOI: 10.1063/1.3044440

[2] Aristeidis Nikolaou, Jakob Leise, Ute Zschieschang, Hagen Klauk, Thomas Gneiting, Ghader Darbandy, Benjamin Iñiguez, Alexander Kloes, "Compact Model for the Bias-Depended Low-Frequency Noise in Organic Thin-Film Transistors Due to Carrier-Number and Mobility Fluctuation Effects," Organic Electronics, P. 106846, 2023, DOI: 10.1016/j.orgel.2023.106846.

[3] Gino Giusi, Orazio Giordano, Graziella Scandurra, Sabrina Calvi, Guglielmo Fortunato, Matteo Rapisarda, Luigi Mariucci, and Carmine Ciofi, "Evidence of correlated mobility fluctuations in p-type organic thin-film transistors," IEEE Electron Device Letters, vol. 36, no. 4, 2015, DOI: 10.1109/LED.2015.2400422.

[4] Wondwosen E. Muhea, K. Romanjek, X. Mescot, C. G. Theodorou, M. Charbonneau, F. Mohamed, G. Ghibaudo, B. Iñiguez, "1/f noise analysis in high mobility polymer-based OTFTs with non-fluorinated dielectric," Applied Physics Letters, vol. 114, no. 24, 2019, DOI: 10.1063/1.5093266.

[5] A. McWhorter, "1/f noise and germanium surface properties," Sem.Surf.Phys, RH Kingston (Univ Penn Press), 1957.

[6] B. R. Conrad, W. G. Cullen, W. Yan, E. D. Williams, "Percolative effects on noise in pentacene transistors," Applied Physics Letters, vol. 91, no. 24, 2007, DOI: 10.1063/1.2823577.

[7] F. Hooge, "1/f noise," Physica B+C, vol. 83, no. 1, 1976.

[8] D. Tu et al., "Modeling of drain current mismatch in organic thin-film transistors," J. Display Technol., vol. 11, no. 6, 2015, doi: 10.1109/JDT.2015.2419692.

[9] Aristeidis Nikolaou, Ghader Darbandy, Jakob Leise, Jakob Pruefer, James W. Borchert, Michael Geiger, Hagen Klauk, Benjmain Iniguez, Alexander Kloes, "Charge-based model for the drain-current variability in organic thin-film transistors due to carrier-number and correlated-mobility fluctuation," IEEE Trans. Electron Devices, vol. 67, no. 11, 2020, DOI: 10.1109/TED.2020.3018694.

[10] Alexander Kloes, Jakob Leise, Jakob Pruefer, Aristeidis Nikolaou, Benjamin Iniguez, Thomas Gneiting, Hagen Klauk, Ghader Darbandy, "THM-OTFT: A Complete Physics-Based Verilog-A Compact Modl for Short-Channel Organic Thin-Film Transistors," IEEE Journal of the Electron Devices Society, 2023, doi: 10.1109/JEDS.2023.3294598.

[11] A. Kloes, N. Dersch and A. Nikolaou, "Variability-Aware Circuit Design: Monte Carlo Simulation Versus Noise Analysis," 2024 IEEE Latin American Electron Devices Conference (LAEDC), Guatemala City, Guatemala, 2024, pp. 1-3, DOI: 10.1109/LAEDC61552.2024.10555681.

Extraction of Open-Access-PDK Active Inductance Parameters with FOSS Tools

Mike Brinson

Centre for Communications Technology

London Metropolitan University

UK

email: brinsonm@londonmet.ac.uk

Abstract—The choice of Verilog-A as an approved semiconductor device hardware description language for IC design has encouraged the interchange of standardised Verilog-A BJT, MOST and BiCMOS device models across commercial and FOSS circuit simulators. Recent trends have seen the release of open-access IC production development kits for digital, analogue, RF and mixed analogue/digital design, completing the "circuit concept to IC production" cycle with FOSS software tools. This paper is concerned with an extension of modelling, simulation and parameter extraction of analogue IC cells using Qucs-S/Ngspice and the IHP-G130G2 PDK. To illustrate these techniques an investigation of a CMOS analogue single ended active inductance cell is presented together with simulation output data and extracted model parameters for 130nm thin and thick oxide CMOS devices.

Keywords—Open-Access PDKs, CMOS Active inductance, compact modeling, parameter extraction, circuit simulation, Qucs-S, Ngspice, Verilog-A, OpenVAF.

I. INTRODUCTION

A recent trend in support of integrated circuit (IC) design and manufacture has been the release of open-access production development kits (PDKs) [1] [2] [3] for free use by individuals, academia and industry. This has encouraged the application of FOSS ECAD tools at all stages in the semiconductor manufacturing cycle from circuit concept to chip production [4]. This in turn under-pins a growing "Open-Hardware" movement [5] [6] particularly through the merger of FOSS circuit simulators [7] [8] with freely available open-access PDK kits. Prior to the release of open-access PDKs, Verilog-A [9] [10] compact device modelling acted as a bridge between sub-micron semiconductor models and circuit simulation. This is still true, but the addition of PDK semiconductor processing data to the open-access IC design tool chain now allows the application of FOSS tools to be extended to the characterization of manufacturable analogue circuit blocks, and mixed signal elements [11]. The main elements in the PDK/FOSS merger are semiconductor device model libraries, characterized by Verilog-A model parameters extracted from measured device data, passive component models and digital CMOS [12]. In parallel with PDK library development there has also been a steady improvement in FOSS circuit simulator modelling capabilities that link extended analysis and simulation features with post simulation graphical visualisation. These improvements allow, at an early stage in the

IC design/manufacture sequence, extraction of analogue block parameters from simulation output data.

This paper introduces a number of extended Qucs-S/Ngspice [13] circuit simulation capabilities and demonstrates their application in the analysis and design of a single ended CMOS active inductor cell designed with the IHP-SG13G2 BiCMOS technology node. This node is high performance with a 130nm CMOS process supporting devices with thin and thick oxides that allow analogue/digital, mixed signal and RF IC design and manufacture. Particular attention is given to the modelling and simulation of the fundamental two transistor CMOS active inductance based on a admittance approach that allows simple extraction of the inductor parameters from real and imaginary admittance properties and their differentiation in the frequency domain.

II. MODELLING OF A CMOS SINGLE ENDED ACTIVE INDUCTOR

The circuit schematic for a CMOS single ended active inductor is shown in Fig. 1. [14] [15] Two nMOS transistors connected in a gyrator-c configuration realize an inductance from the P_in terminal to ground, where the resulting lossy (low Q) inductance is a function of the MOS physical and instance parameters and the d.c. bias currents. In Fig. 1 bias is set by the ideal current sources $I1$ and $I2$.

Replacing the nmos transistors drawn in Fig. 1 by transconductance controlled sources with parallel $C1, go1$ at node $n1$ and $C2, go2$ at node $n2$, yields the equivalent electrical network illustrated in Fig. 2 (a). Conventional a.c. network analysis gives the input admittance, $Yin = Iin/Vin$, as

$$Yin(\omega) = go1 + j\omega C1 + \frac{1}{\left(\frac{go2}{gm1 \cdot gm2} + \frac{j\omega \cdot C2}{gm1 \cdot gm2}\right)} \quad (1)$$

Equation 1 represents a lossy inductance with the equivalent circuit drawn in Fig. 2(b), where

$$Yin(\omega) = \frac{1}{Rp} + j\omega Cp + \frac{1}{\left(Rs + j\omega L\right)} \quad (2)$$

In Equations 1 and 2 $\omega = 2\pi f$ and f is a.c. frequency in Hz. By comparing equations 1 and 2 $Rs = \frac{go2}{gm1 \cdot gm2}$ Ω, $L = \frac{C2}{gm1 \cdot gm2}$H, $Rp = \frac{1}{go1}$ Ω and $Cp = C1$F.

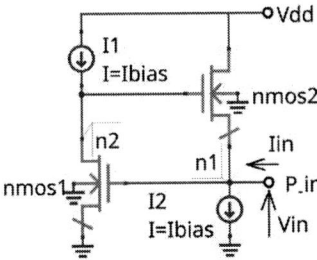

Fig. 1. A gyrator-c single ended CMOS active inductance analogue circuit block where the input admittance at terminal *P-in* is $Yin = Iin/Vin$.

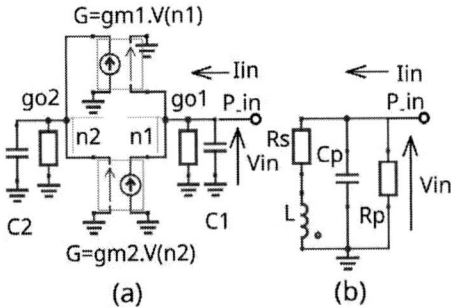

Fig. 2. A first order CMOS single ended lossy active inductor: (a) gyrator-c a.c. equivalent circuit, (b) passive component parallel model.

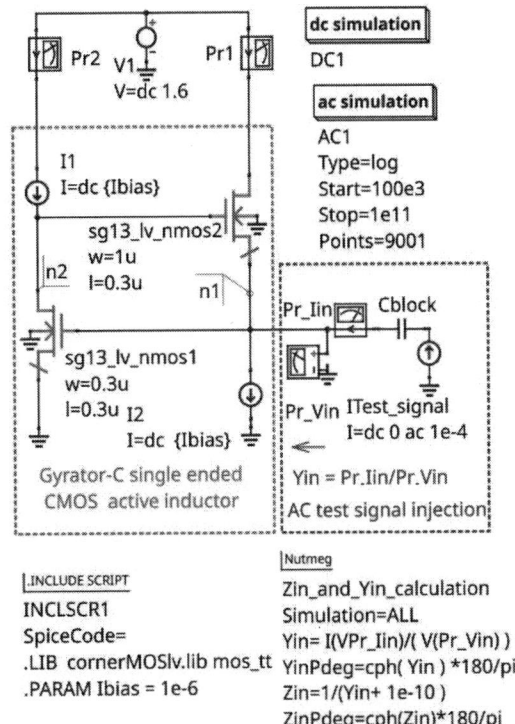

Fig. 3. A Qucs-S/Ngspice test bench for evaluating the a.c. performance of an IHP-Sg13g2 technology CMOS active inductance.

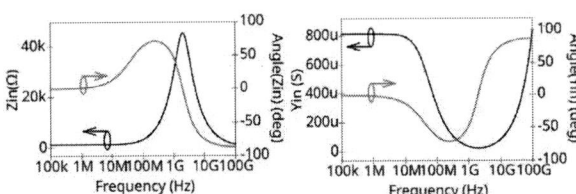

Fig. 4. Example IHP-Sg13g2 thin oxide technology CMOS active inductor Zin and Yin characteristics for $Ibias = 1\mu A$

III. SIMULATION OF AN IHP-SG13G2 OPEN-ACCESS PDK ACTIVE INDUCTOR

A Qucs-S/Ngspice test bench for simulating the admittance of a CMOS active inductance over a GHz frequency band is illustrated in Fig. 3. The lossy active inductor under test is represented by the schematic drawn inside the red dotted box. Similarly, an independent current generator ($ITest\_signal$), highlighted by the blue dotted box, injects an a.c. frequency dependent current into node $n2$, allowing Yin to be easily determined with the Ngspice nutmeg script listed in Fig. 3. Capacitor $Cblock$ d.c. isolates the $Itest\_signal$ generator from the active inductor under-test. Other schematic components set, and monitor, the nMOS bias conditions. Fig. 4 shows a set of typical Zin and Yin simulation data plots.

IV. EXTRACTION OF PDK ACTIVE INDUCTOR MODEL PARAMETERS

To design a CMOS active inductance with a specific L implies that values for W, L and d.c. bias are known for both nMOS1 and nMOS2 (Fig. 1) and that model parameters Rs, L, Rp, and Cp can be extracted from FOSS simulation output data. A well defined, and reproducible, technique for estimating Rs, L, Rp, and Cp is required. This paper suggests a new approach to active inductance model parameter estimation and extraction centred on list controlled parameter stepping and first order differentiation of admittance in the frequency domain.

Equation 2 can be written as Yin(f) = Yinr(f) + jYini(f), where

$$Yinr(f) = \frac{1}{Rp} + \frac{Rs}{Rs^2 + (2\pi fL)^2} \quad (3)$$

$$Yini(f) = 2\pi fCp - \frac{2\pi fL}{Rs^2 + (2\pi fL)^2} \quad (4)$$

$$\frac{dYinr(f)}{df} = -\frac{4\pi fRsL^2}{\left(Rs^2 + (2\pi fL)^2\right)^2} \quad (5)$$

$$\frac{dYini(f)}{df} = 2\pi Cp - \frac{2\pi L}{Rs^2 + (2\pi fL)^2}\left[1 - \frac{2(2\pi fL)^2}{Rs^2 + (2\pi fL)^2}\right] \quad (6)$$

37

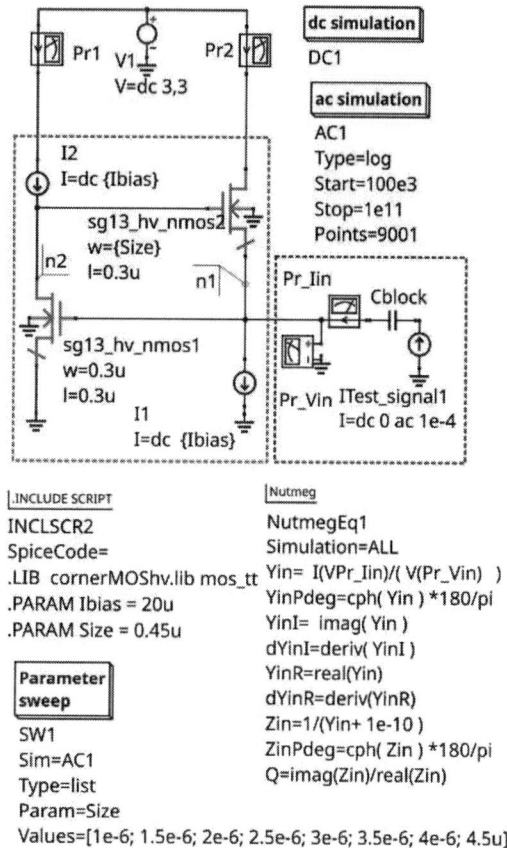

.INCLUDE SCRIPT
INCLSCR2
SpiceCode=
.LIB cornerMOShv.lib mos_tt
.PARAM Ibias = 20u
.PARAM Size = 0.45u

Parameter sweep
SW1
Sim=AC1
Type=list
Param=Size
Values=[1e-6; 1.5e-6; 2e-6; 2.5e-6; 3e-6; 3.5e-6; 4e-6; 4.5u]

Nutmeg
NutmegEq1
Simulation=ALL
Yin= I(VPr_Iin)/(V(Pr_Vin))
YinPdeg=cph(Yin) *180/pi
YinI= imag(Yin)
dYinI=deriv(YinI)
YinR=real(Yin)
dYinR=deriv(YinR)
Zin=1/(Yin+ 1e-10)
ZinPdeg=cph(Zin) *180/pi
Q=imag(Zin)/real(Zin)

Fig. 5. A Qucs-S/Ngspice IHP-Sg13g2 CMOS active inductance test bench illustrating list *Parametersweep* control of a.c. simulation and Ngspice nutmeg output data calculations: $Ibias = 1\mu A$ and thick oxide technology.

At low frequencies equation 3 predicts that when $(2\pi f L)^2 << Rs^2$ and $Rp >> Rs$

$$Yinr(f) \approx \frac{1}{Rs} \qquad (7)$$

Similarly, at high frequencies, provided $(2pifL)^2 >> Rs^2$.

$$Yinr(f) \approx \frac{1}{Rp} \qquad (8)$$

From Equation 6, again at low frequencies, provided $(2\pi f L)^2 << Rs^2$ and Cp is small

$$\frac{Yini(f)}{df} \approx -\frac{2\pi L}{Rs^2} \qquad (9)$$

Inspection of equation 6 also indicates that at high frequencies

$$\frac{Yini(f)}{df} \approx 2\pi Cp \qquad (10)$$

V. IHP-SG13G2-HV OPEN-ACCESS CMOS ACTIVE INDUCTOR PERFORMANCE

The IHP-SG13G2 technology includes nMOS and pMOS transistors with two different oxide thicknesses; firstly, a thin oxide process (target 2.85 nm for low voltage (lv) devices with $Vgs \leq 1.65$ V) that is aimed at digital logic and mixed signal IC design, and secondly, a thicker oxide process (target 7.3 nm for high voltage (hv) devices with $Vgs \leq 3.3$ V) primarily for analogue and RF design. The circuit schematic drawn in Fig. 5 shows a simulation test bench for investigating the performance of an hv single ended active inductance. The IHP-SG13G2 open-source PDK library is accessed via the Ngspice .LIB cornerMOShvlib mos_tt statement, where the mos_tt argument indicates typical PDK technology data. Similarly, the Qucs-S MOS transistor names, for example $sg13\_hv\_nmos1$, indicate that the PDK hv technology is to be used when simulating the circuit under test. Output data for the quantities defined by Equations 3 to 6 are shown plotted in Fig. 6. Three points of importance are worth noting: firstly, $sg13\_hv\_nmos1$ $W/L = 1$, with $W = L = 0.3\mu m$ constant during simulation, implies that changes in $C2$ should be minimal, secondly $sg13\_hv\_nmos2$ has a constant length, $L = 0.3\mu m$, and width W swept through a series of increasing $Size$ values, these are set by Qucs-S $Parametersweep$ $SW1$. Hence, capacitor $C1$ and transconductance $gm2$ increase in value as the simulation progresses, which in turn, changes the lossy active inductor model parameters and the frequency of maximum Q. Finally, enhancements to the Qucs-S/Ngspice software support calculation, and plotting, of first order differentiation of Yin (see equations 5 and 6) in the a.c. small signal frequency domain.

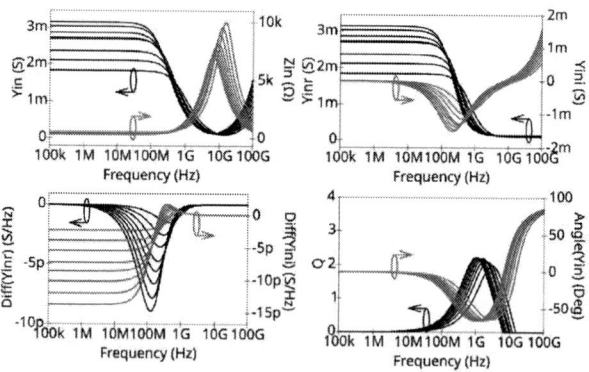

Fig. 6. Active inductor hv simulation data for the test bench drawn in Fig. 5.

Extraction of parameters Rs, L, Rp and Cp is relatively straightforward. These can be estimated by applying equations 7 to 10 to simulation output data where it is constant as frequency changes (Rs and L at low frequencies and Rp and Cp at high frequencies). Plots of extracted values are given in Fig. 7, where the direction of increasing parameter $Size$ is indicated by the insert arrow. In the case of parameters Rp and Cp the extraction frequency band is set at 20GHz to 25GHz.

This range ensures that the required parameter extraction mathematical conditions are met and does not imply that the manufactured CMOS active inductor performance will be identical to simulation data. For accurate results, above a few GHz, transistor interconnect effects, particularly those caused by metallization paths and vias, must be included in the active inductor model [16]. In saturation mode capacitance $Cp = Cds \approx \frac{2}{3} \cdot Cox \cdot W \cdot L$ which is in agreement with the simulated values shown in Fig. 7. The simulation values for Cp exhibit a level of variation indicative of electrical noise but is however, more likely to be small variations in computational accuracy that are often associated with a numerical computation noise "floor".

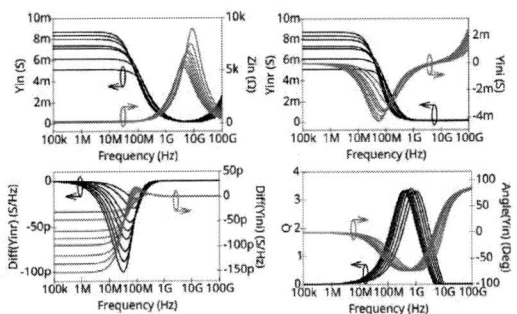

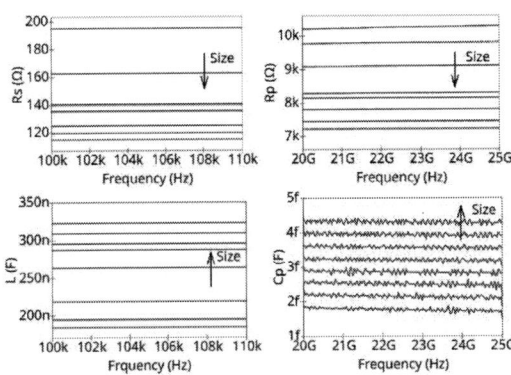

Fig. 8. Active inductor lv simulation data for the test bench drawn in Fig. 5 with reference to hv technology replaced by lv: $Ibias = 20\mu A$.

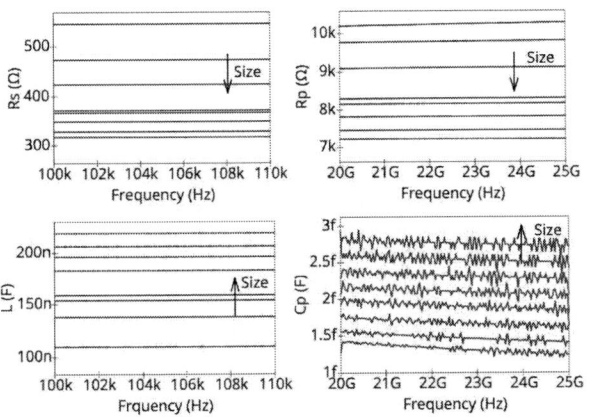

Fig. 7. Extracted IHP-Sg13g2-hv Rs, L, Rp and Cp values obtained with the test-bench given in Fig. 5.

Fig. 9. Extracted IHP-Sg13g2-lv Rs, L, Rp and Cp values obtained with the test-bench given in 5. Reference to hv technology has been replaced by lv and $Ibias = 20\mu A$.

VI. IHP-SG13G2-LV OPEN-ACCESS CMOS ACTIVE INDUCTOR PERFORMANCE

Figures 8 and 9 present the simulation output data for a IHP-Sg13g2-lv technology CMOS active inductor. These were obtained using a test-bench with the same specification as Fig. 5 except that references to hv technology were replaced by lv. Comparison of the lv and hv simulation output data indicates that the lv technology Cp and $C2$ values are higher due to the lv process thinner gate oxide, which in turn increases L, shifting Q maximum to lower frequencies.

VII. TRANSCONDUCTANCE TUNING

Active inductor parameters Rs and L are functions of the product of the nMOS transistor transconductances $gm1$ and $gm2$. By keeping $nmos1$ and $nmos2$ at a constant W/L ratio, but not necessarily the same, changes in d.c. bias currents vary $gm1$ and $gm2$ which modify L, tuning maximum Q to a specific frequency set by L, Rs and Cp. The span of maximum Q, where $Q = imag(Yin)/real(Yin)$, frequency range is therefore, to a large extent, set by the external applied d.c. bias. Moreover, in this paper the test example bias current range has been chosen to be one where the nMOS devices remain in saturation as the bias current is varied. At bias currents

above $20\mu A$, due to the lv technology limit of $Vdd = 1.66V$, this requirement is not met restricting bias current to $20\mu A$ or less. Figure 10 illustrates both hv and lv technology examples

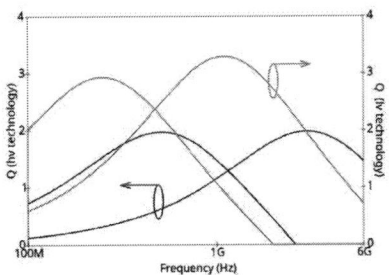

Fig. 10. Changes in active inductance Q frequency response for 1μ A and 20μ A d.c. bias current: active inductor parameters the same as previous hv and lv test benches; Black curves hv - $1\mu A$ Qmax = 1.97 at 0.5GHz, $20\mu A$ Qmax = 1.98 at 1.98GHz and red curves lv - $1\mu A$ Qmax = 2.93 at 0.25GHz, $20\mu A$ Qmax = 3.29 at 1.09GHz

for d.c. bias currents beginning at $1\mu A$ and and ending at $20\mu A$. The plotted data were extracted from Figures 11, 12, 13 and 14.

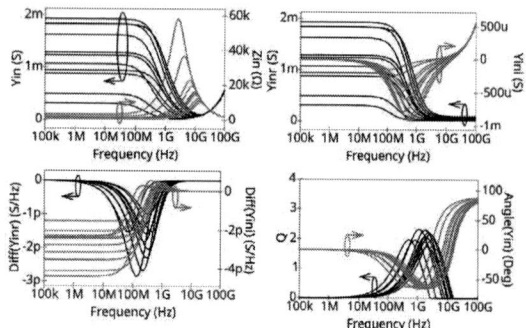

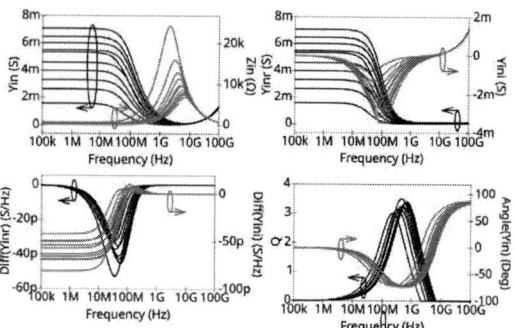

Fig. 11. Active inductor hv simulation data for the test bench drawn in Fig. 5: $sg13\_hv\_nmos1$ $W = L = 0.3$ μm, $sg13\_hv\_nmos2$ $W = 1$ μm, $L = 0.3$ μm and d.c bias parameter $Ibias$ scanned over the list 1μA, 2μA ,4μA, 6μA, 8μA, 10μA. 12μA, 14μA, 16μA. 18μA, 20μA

Fig. 13. Active inductor hv simulation data for the test bench drawn in Fig. 5: $sg13\_lv\_nmos1$ $W = L = 0.3$ μm, $sg13\_lv\_nmos2$ $W = 1$ μm, $L = 0.3$ μm and d.c bias parameter $Ibias$ scanned over the list 1μA, 2μA ,4μA, 6μA, 8μA, 10μA, 12μA, 14μA, 16μA, 18μA, 20μA.

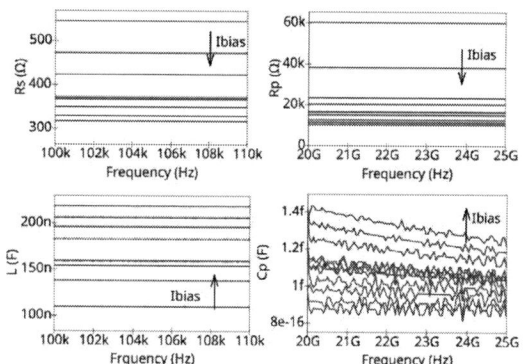

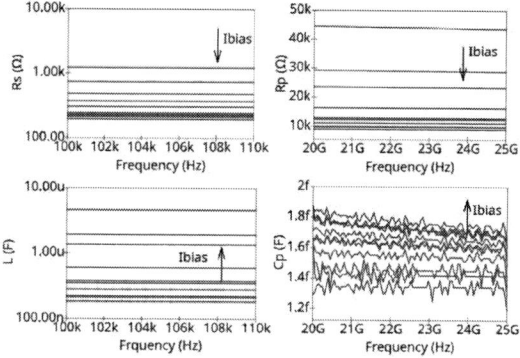

Fig. 12. Extracted IHP-Sg13g2-hv Rs, L, Rp and Cp values obtained with the test-bench given in Fig. 5: the nMOS parameters and d.c. bias test conditions are the same as given in Fig. 11.

Fig. 14. Extracted IHP-Sg13g2-lv Rs, L, Rp and Cp values obtained with the test-bench given in Fig. 5: the nMOS parameters and d.c. bias test conditions are the same as given in Fig. 11.

VIII. CONCLUSIONS

Although the single ended CMOS active inductance only requires two transistors to build a functioning gyrator-c circuit, determining values for the MOST W, L, W/L, and the bias conditions. to synthesize a specific L can be more difficult than implied by the simplicity of the circuit. The lossy active inductor model parameters Cp, Rs, Rp and L are functions of nMOS W and L and the product $gm1 \cdot gm2$, making estimating the values needed to synthesize an optimum value of inductance L, with a given frequency band width and Q, can be an involved process. Analysis of the lossy active inductor in the a.c. small signal frequency domain is often undertaken by calculating input impedance Zin. In this paper an alternative admittance analysis has been adopted and reported. This approach is simpler and allows the input admittance to be easily separated into real, $Yinr$, and imaginary, $Yini$, components. $Yinr$ and $Yini$ are also straight forward to differentiated with respect to frequency. Ongoing improvements to the Qucs-S/Ngspice FOSS circuit simulation tools support numerical differentiation and visualization in the frequency domain, which in turn, allows values for Rs, Rp,

Cp and L to be extracted directly from Qucs-S simulation output data. Earlier sections of the paper introduced examples of the extraction process for the IHP-SG13G2 hv and lv 130nm processes node. By applying established modelling techniques and model parameter extraction to analogue IC blocks, characterised by an open-access PDKs, it is possible to obtain a deeper understanding of the properties and design of analogue IC circuits.

REFERENCES

[1] Skywater, "FOSS 130nm Production PDK". [Accessed February 2025] [Online] Available https://github.com/google/skywater-pdk.
[2] Global Foundries, "GF180MCU Open Surce PDK". [Accessed February 2025] [Online] Available https://github.com/google./gf180mcu-pdk.
[3] IHP-GmbH, "IHP Open Surce PDK". [Accessed February 2025] [Online] Available https://IHP-GmbH/IHP-Open-PDK.
[4] Wladek Grabinski, Rene Scholz, Jason Verley, Eric R Keiter, Holger Vogt, Dietmar Warning. Paolo Nenzi, Francesco Lannutti, Felix Salfelder, Al Davis, Mike Brinson, Bal Virdee, Guilherme Torri, Daniel Tomaszewski, Matthias Bucher, Jean-Michel Sallese, Markus Müller, Pascal Kuthe, Mario Krattenmacher, "FOSS CAD for the Compact Verilog-A Model Standardization in Open Access PDKs", 8th IEEE Electron Devices Technology & Manufacturing Conference (EDTM). 2024. DOI: 10.1109/EDTM58488.2024.10511990.

[5] Open Source Hardware Association (OSHWA), "Opwn Source Hardware". [Accessed February 2025] [Online] Available: https://www.oshw.org.

[6] P2PF, "Open Source Hardware". [Accessed February 2025] [Online] Available: https://wikp2pfoundation.net/Open_Source_Hardware.

[7] ngspice, "for mixed-level/mixed-signal".

[8] Xyce, "Parallel electronic simulation". [Accessed February 2025] [Online]. Available: https://xyce.sandia.gov/.

[9] Qucs/ADMS, "ADMS - "An automatic device model synthesizer". [Accessed February 2025] [Online]. Available: https://github.com/Qucs/ADMS.

[10] openVAF, "A Next-Generation Verilog-A compiler that empowers the open source silicon revolution", [Accessed February 2025] [Online]. Available: https://openvaf.semimod.de/.

[11] M. Brinson, "QUCS-S: a central tool in the openPDK IC design flow" ESSERC2024: W5 MOS-AK: Compact Modeling Support for OpenPDK and FOSS IC Designs, 2024, DOI 10.5281/zenodo 14178763.

[12] IHP-GmbH, "IHP SG13G2 Open-Source Process Specification Rev. (2023-12-20)". [Accessed February 2025] [Online] Available https://IHP-GmbH/IHP-Open-PDK.

[13] Qucs-S, "Qucs with SPICE Version 25.1.0", [Accessed February 2025], [On-line]. Available: http://ra3xdh.github.io.

[14] Patel. D.P and Oza S. "CMOS active inductor: a technical review", International Journal of Applied Engineering Research, 2018, !3(11), pp. 9680-9685, ISS 0973-4562.

[15] Abodo Emad A. T. and Ahmad T. Y.,"On the design and optimization of CMOS active inductor for RF applications". Journal of Engineering Science and Technology, 2020, 15 (3), 1921-1926.

[16] IHP-GmbH, "Open—EMS for IHP SG13G2 technology", [Accessed February 2025] [Online] Available https://IHP-GmbH/IHP-Open-PDK.

42

Artificial Intelligence in Electronic Systems

Anomaly Detection on the Edge: Comparison of Reconstruction and Classification Based Approaches

Łukasz Grzymkowski, Tymoteusz Cejrowski
Arrow Electronics
Gdańsk, Poland
email: lukasz.grzymkowski@arrow.com,
tymoteusz.cejrowski@arrow.com

Tomasz P. Stefański
Faculty of Electronics, Telecommunications and Informatics
Gdańsk University of Technology
Gdańsk, Poland
email: tomstefa@pg.edu.pl

Abstract—In this work we discuss and evaluate different approaches to solving anomaly detection task when the target platform is a tiny microcontroller. We investigate modeling techniques and propose a comprehensive set of measurements to analyze performance, compute and memory requirements, and power efficiency. We run experiments to collect these measurements on platforms used in TinyML systems including Cortex-M7, Cortex-M55 and Ethos-U55 running TensorFlow Lite for Microcontrollers. The measurements are collected for an autoencoder in reconstruction-based anomaly detection and a MobileNetV2-like model trained for classification. We show which approach is more suitable depending on the system requirements and constraints. This work underscores the need for a holistic approach in selecting modeling and deployment strategies, providing empirical evidence to guide the development of efficient on-device anomaly detection systems.

Keywords—Edge computing, embedded software, tiny machine learning, anomaly detection, sound recognition.

I. INTRODUCTION

Rapid expansion of the Internet of Things (IoT) deployments generates a vast volume of data at the network edge. While this offers a significant potential, processing it in a timely and intelligent manner on-device presents significant challenges. Tiny Machine Learning (TinyML) addresses this problem by enabling machine learning (ML) models to be executed on resource-constrained microcontrollers and other low-power edge devices. This technological shift allows the devices to move beyond mere data acquisition towards autonomous, real-time decision making.

Among various TinyML applications the anomaly detection task is the main interest for many companies and industries. The identification of unusual or unexpected patterns is marked as high priority due to its cost saving potential. For instance, in industrial settings, it can facilitate predictive maintenance by recognizing subtle deviations in machinery sounds indicative of impending failures. Similarly, in healthcare, wearable devices can monitor atypical physiological readings. The capacity for on-device anomaly detection, independent of cloud connectivity, can enhance safety, operational efficiency, and system responsiveness. This allows for the development of sensors which not only perceive their environment but also interpret and react to it.

As TinyML technologies mature, with more powerful microcontrollers and specialized Artificial Intelligence (AI) ac-

celerators such as Neural Processing Units (NPUs), a fundamental question arises regarding practical capabilities of these devices. While the prospect of edge AI is compelling, rigorous constraints of limited memory, low processing power, and critical energy efficiency requirements are the source of complex optimization problems. Implementing sophisticated tasks like the anomaly detection on platforms with minimal resources necessitates innovative algorithmic design and rigorous performance evaluation.

The edge AI field has reached a point where it is insufficient to only demonstrate the feasibility of model deployment on constrained devices. There is a critical need for comprehensive assessments of the model performance due to strict business or technical requirements. There are inherent trade-offs associated with different architectural and hardware choices. This requires the understanding of the impact of model types (e.g., autoencoders vs. classifiers), data pre-processing techniques, and model compression methods on various embedded platforms.

This paper addresses the deployment of anomaly detection systems on the representative TinyML hardware. It compares distinct modeling strategies and system configurations. Through precise measurements and analysis of key metrics including latency, memory utilization, and power consumption, this work aims to provide an insight into the current capabilities and limitations of TinyML for anomaly detection.

II. RELATED WORKS

The intersection of anomaly detection and edge computing has attracted significant attention as the need for real-time on-device intelligence increases. This section reviews relevant works in TinyML benchmarking, anomaly detection approaches, and edge-based processing systems.

The evaluation of ML models on the resource-constrained devices has become increasingly important as TinyML applications mature. The MLPerf Tiny benchmark [1] establishes standardized evaluation metrics for microcontroller-based ML systems, including keyword spotting, visual wake words, image classification, and anomaly detection tasks. However, recent study [2] proposes an enhanced benchmarking methodology which directly addresses key limitations in MLPerf, namely its inability to isolate inference-specific energy and latency measurements. The authors introduce a dual-trigger signal setup to disaggregate the energy and delay contributions from pre-inference, inference, and post-inference phases, allowing for more granular analysis. Banbury et al. [3] introduce systematic evaluation methodologies for TinyML systems, emphasizing the importance of measuring latency, energy consumption, and memory utilization. Their work highlights the need for rigorous performance assessment beyond simple accuracy metrics, particularly for deployment scenarios with strict resource constraints. Work [4] surveys efficient deep learning (DL) models and optimization techniques for edge IoT, including anomaly detection. Their review underscores the viability of TinyML approaches, such as distilled CNNs and NAS-designed models, for embedded anomaly detection under hardware constraints. The authors highlight custom accelerators (e.g., ASICs, FPGAs) and memory optimization techniques, like patch-based inference, to enhance performance on edge devices.

Anomaly detection for audio signals, particularly in industrial settings, has been extensively studied in the context of the Detection and Classification of Acoustic Scenes and Events (DCASE) challenges. The DCASE Task 2 series [5], [6], [7], [8], [9] has established standard datasets, models and evaluation protocols for unsupervised anomalous sound detection in machine condition monitoring applications. While these studies have established strong foundations for anomaly detection algorithms, they often utilize complex model architectures and do not address the challenges of real-time inference on microcontrollers with severe memory and power limitations.

Although there is a great amount of work performed in the field of anomaly detection, we have not observed many studies showing the performance of models on multiple edge devices. The presented work attempts to addresses this by providing a systematic evaluation of different methodologies, which are reconstruction- and classification-based tasks, across hardware platforms including traditional microcontrollers and systems with NPUs.

III. ANOMALY DETECTION

The main challenge in the anomaly detection task is lack of anomalous samples. One usually has access to normal samples which are ubiquitous. How can we model the unknown? To address this, researchers use different techniques and formulate the task accordingly:

- Reconstruction-based - unsupervised, trained on normal samples to minimize the reconstruction (or prediction) error between the input and the reconstructed output.
- Classification - supervised, requires a balanced dataset of normal and anomalous samples, either synthetic samples or a multi-class setting, during evaluation confidence for the correct class is used.
- Embedding-based - self-supervised, a trained feature extractor generates embeddings from input samples; these are then used to learn a secondary model which, for instance, learns the data distribution or calculates the distance between samples.

A. Reconstruction

In the reconstruction-based anomaly detection, a model, typically an autoencoder, is learned using normal samples collected from the system. An autoencoder is a model built from the encoder that compresses the input sample into a latent space representations and the decoder which reconstructs the sample back into the original dimensionality. The anomaly score is usually the reconstruction error between the input and output sample. As the model is trained using only the normal samples, during evaluation when an anomalous sample is provided the model should return a high reconstruction error. Often a batch of samples is processed and the mean reconstruction error is used as the final score.

B. Classification

The classifiers are trained using supervised learning with a labeled dataset. Some of the commonly used classifier networks are MobileNetV2 [10] and ResNet [11]. In the case of anomaly detection, we only have normal samples. To address this, we may use i) the data augmentation techniques to create synthetic anomalies, ii) expand the dataset with normal samples originating from other device classes. In the latter case, during the deployment, the model is monitoring a device of the original normal class. The model under correct operating conditions should therefore return high confidence for this correct class. If it returns low probability value, then the sample is classified as an anomaly. The anomaly score is the negative logit of the correct class

$$A_\theta(X) = \frac{1}{N} \sum_{n=1}^{N} \log \frac{1 - p_\theta(\psi_{t(n)})}{p_\theta(\psi_{t(n)})} \qquad (1)$$

where ψ_t is the input sample, N is the number of samples, and p_θ is the softmax output of the model.

C. Embeddings

The anomaly detection task can also be addressed using the embedding-based approach. A feature extraction model is used to process the input samples into a lower dimensional representation which captures salient information. The embeddings are then evaluated by a secondary model to produce an anomaly score. The feature extractor is commonly a classifier network without the classification head or a model trained to generate the feature vectors. In the latter case, the model may be trained using metric learning to produce feature vectors which maximize both similarity between the samples belonging to the same class and the distance to the other classes. During the evaluation, the anomaly score is computed as the distance between an input sample and the normal samples, for instance using k-NN algorithm. The cosine distance is often used as the distance metric. Another approach is to learn GMM [12], OC-SVM [13] or similar models on normal embeddings and then evaluate the new sample fit to the learned distributions or the distance to the decision boundary.

D. Anomaly Threshold

Given the anomaly score, the next step is to select the threshold for classification as either anomalous or normal. Commonly, the area under the curve (AUC) of the receiver operating characteristic (ROC) and partial AUC (pAUC) are used [5], [6], [7], [8]. The evaluation system knows the correct labels, and the ROC curve serves as a visual representation of the performance of the model across all the thresholds, in terms of the true positive rate (TPR) and the false positive rate (FPR). An ideal system has TPR of 1.0 and FPR of 0.0. Using ROC allows for evaluating the classification model performance and is implemented in most anomaly detection systems.

IV. EXPERIMENTAL SETUP

In this section, we describe the setup used to run the experiments. This includes the models trained, the dataset used, measurements, as well as the hardware platforms used for the deployment. We refer to the system being tested as the device under test (DUT).

A. Dataset

The dataset used is the DCASE challenge task 2 dataset, the anomalous sound detection [5]. It was collected using microphones mounted on or near various types of industrial machines. It is a combination of Malfunctioning Industrial Machine Investigation and Inspection (MIMII) [14] and Toy-ADMOS [15] datasets. We used the 2020 version for the autoencoder training. The version introduced for the challenge in 2022 also includes section ids for each device type to provide more granularity. We employed this to train the classifier to distinguish between these sections. We only used the ToyCar class of the device for training and testing.

The samples in the dataset are sampled with 16 kHz. Each recording is approximately 10 seconds long. To train and evaluate the models, the raw audio samples are pre-processed into mel-spectrograms with 128 mel bands. We use a 50% overlapping, sliding window of 1024 samples (approximately 64 ms), with hop length of 512 samples.

B. Measurements

Selecting the right model and platform for the application depends on project requirements. One has to consider the trade-offs in terms of computational overhead, memory and power usage. To perform our evaluation we collect various measurements from the running system and obtained prior to the deployment:

- Inference latency - the time it takes to run a single model inference (forward path), returned as the time needed for a single inference. This may be just model execution time or may also include the time needed to load the data and run the pre- and post-processing.
- Inference energy - the amount of energy used by a single inference, either calculated as the increase of energy on top of an idle system energy or the total energy used during inference.

- Storage usage - the non-volatile memory used to store the weights of the model, the inference engine, the library code, the underlying application or the operating system code.
- Peak memory usage - the maximum amount of memory used for inference, equal to the maximum size of the activations, the weights and the intermediary buffers.
- Model parameters - the number of trainable parameters of the model. The rule of thumb is that the more parameters there are, the more complex and accurate the network is.
- Operations - the number of basic operations to run a single inference using the model.
- Duty cycle - the percentage of time the system spends inferring on the input data.
- Power consumption - the power used by the system.

To collect the latency measurements we connect the host system to the DUT via UART. We trigger the inference and run it in a loop 1000 times to calculate the mean latency. We only measure the model inference, without pre- or post-processing. The energy consumption is collected using an external current monitor, Nordic Semiconductor Power Profiler Kit II. It is connected to the platform in the current measurement mode (Ampere mode). We collect the current measurements in the idle state and during the inference by synchronizing the measurement with the UART trigger command. We take the mean current based on the inference iterations from 5% to 95%. For STM32H7, we use the total board current, as there is no header available to measure only the SoC. For the Alif kit, we monitor the current drawn by the SoC 3.3 V input voltage. The energy per inference is calculated by multiplying the inference latency with the inference mean current by the input voltage level - 3.3 V in both cases. To calculate the storage, we use ROM report tool in Zephyr to calculate the size of the binary file components. This includes primarily the model, RTOS, TFLM and CMSIS-NN. The peak memory usage is based on the model weights and the Tensor Arena size. It is a pre-allocated buffer used by TFLM for tensors and buffers. We recompile the code to find the minimal amount necessary for the tensor allocation step to succeed. The number of model parameters and operations is calculated by analyzing the model graph with the TFLite converter. The duty cycle is derived from the inference latency and the input data generation rate, based on the duration of each sample. The power consumption is measured during inference and in the idle state, and then used with the duty cycle to compute the average.

C. Models

We train the models for anomaly detection using the reconstruction error and the classification approaches. We skip training an embedding-based model, as it uses a feature extractor, which is commonly either of the other two models tested. As we focus on TinyML, we constrain the model parameter count to below 160k parameters. The rationale is that, along with the RTOS and the inference engine, the total binary size must not exceed 256 kB. We quantize the models into INT8 precision using static affine quantization. We do not perform any model pruning. The models are trained using PyTorch framework, then converted to LiteRT (previously TensorFlow Lite) format and later to the TFLM header files. For Ethos-U55 NPU, we use the Vela compiler to convert the TFLite model into the TFLM header and instruct the TFLM interpreter to use Ethos-U55.

We base the model architecture on the reference models from the DCASE challenge task 2 [5]. We modify those models to reduce the number of parameters and size to meet the test constraints.

TABLE I.
SUMMARY OF TESTED MODELS

Model	Autoencoder	TinyMobileNetV2
Task	Reconstruction	Classification
Type	Linear	Convolutional
Params [k]	137.8	120.3
Ops [M]	0.27	22.3
Model Size [kB]	156.7	149.0
Peak Memory [kB]	160.7	378.5
Test Accuracy [%]	84.57	60.0

The baseline for DCASE 2020 is an autoencoder used in the reconstruction-based anomaly detection setting. This is a fully-connected autoencoder which is used in the MLPerf Tiny benchmark [1]. The input features to the model are 5 consecutive mel-spectrogram frames with 128 mel bands each. The frames are concatenated and flattened into a 640 long vector. We use the ToyCar subset of the DCASE 2020 dataset. The baseline autoencoder uses 5 layers in the encoder and the decoder, with 8 dimensional latent space. The accuracy is evaluated as AUC for the ROC plotted for the mean anomaly score, individually for each recording in the dataset. The model achieves 86.05% accuracy, however, the parameter count is 234k and exceeds the constraint defined. We then search for a model that meets the limit while still producing sufficient accuracy. We scale down the autoencoder to 3 layers in the encoder and the decoder. As a result, we achieve 84.6% accuracy with only 156k parameters.

In the DCASE 2022 challenge, the second baseline model is introduced using the popular MobileNetV2 architecture [7], [10]. We use the updated ToyCar dataset with section ids used as labels. To test the anomaly detection accuracy, we take the output of the model for the correct class only and use the negative logit as the anomaly score. The model should return high confidence for the normal class, and low if the sample is an anomaly. Initially, we use 64 consecutive frames, each a mel spectrogram with 128 mel bands, as the input. The vanilla MobileNetV2 with this input size and structure is too large to meet the constraints on the number of parameters, although it achieves 86.5% accuracy. Then we remove layers and design a Tiny MobileNetV2. We scale it down from 17 to 7 inverted residual blocks and reduce the input size to 32 frames with 128 bands. Although the model converges to over 90% accuracy on the section classification task, we are able to achieve only 60% accuracy on the anomaly detection test. We also tested 64 frames with 64 bands, but has produced lower accuracy.

The models are summarized in Table I. The autoencoder with the same number of parameters requires significantly less dynamic memory than the classifier (160.7 kB vs. 378.5 kB), requires nearly two orders of magnitude operations. The accuracy is evaluated using ROC and AUC on the test data. The test accuracy is also higher when using an autoencoder. We note, however, that a larger classifier MobileNetV2 is able to achieve a similar level of accuracy.

D. Devices Under Test

We select the DUTs representative of the TinyML systems. We chose ST Microelectronics STM32H747XIH6 and Alif Semiconductor Ensemble E7 SoCs. The platforms are summarized in Table II. Both platforms are multi-core systems, but we only provide values for the resources used during testing. We use Zephyr RTOS and TFLM for inference.

The first SoC DUT is the STM32H747XIH6 with the ARM Cortex M7 and M4 cores. We use the STM32H747I-DISCO (Fig. 1) development kit equipped with 2 MB of flash and 1 MB of RAM, shared between the cores. We run the code on the M7 core, the high-performance core running at 480 MHz. The core features a floating point unit (FPU) and support for SIMD instructions used by the CMSIS-NN library. This platform is a general purpose system, not specifically for the edge AI.

The second platform is the Alif Semiconductor E7 SoC on the Ensemble DevKit Gen2 (Fig. 2). It is equipped with 4 cores: two Cortex-A32 application cores and two real-time Cortex-M55 cores. The real-time cores are: i) high-performance (HP) core at 400 MHz with 256 MAC wide Ethos-U55 NPU, ii) high-efficiency (HE) core at 160 MHz with 128 MAC wide Ethos-U55 NPU. ARM Ethos-U55 is an NPU specialized for acceleration of the DL models with up to 0.5 TOPS performance, with more power efficiency than the Cortex-M core. Weights and activations are fetched ahead of time using a DMA connected to system memory via AXI5 master interface. We use the HP core and the NPU. We use the ARM Vela compiler on the LiteRT model to compile the model to run it on the NPU. MRAM, a persistent RAM memory, serves both as the non-volatile storage and the dynamic memory.

TABLE II.
SUMMARY OF PLATFORMS UNDER TEST

SoC	STM32 H747XIH6	Alif Ensemble E7
Core (Used)	Cortex-M7	Cortex-M55
Core Freq.	480 MHz	480 MHz
RAM	512 kB	256 kB
ROM	1 MB	256 kB
NPU	N/A	0.5 TOPS (Ethos-U55)

V. RESULTS

A. Autoencoder

The results collected for the autoencoder model are presented in Table III. Using the Ethos-U55 NPU significantly reduces the latency. The latency is two orders of magnitude lower than with SIMD-accelerated CMSIS-NN library on the

Fig. 1. DUT1 - STM32H747I-DISCO development kit

Fig. 2. DUT2 - Alif Ensemble DevKit Gen

TABLE III.
MEASUREMENTS ON DUTs FOR AUTOENCODER MODEL

Name	STM32 H7		Alif E7		
Accel.	CPU	CPU	CPU	CPU	NPU
Uses CMSIS-NN	No	Yes	No	Yes	No
Inference					
Inf. Latency [ms/inf]	22.72	6.21	48.33	6.67	0.07
Inf. Current [mA]	163.14	164.23	31.44	32.03	30.92
Inf. Energy [mJ/inf]	12.23	3.36	5.01	0.70	0.007
Non-volatile Memory (Storage)					
Model Flash [kB]	156.7	156.7	156.7	156.7	158.8
TFLM Flash [kB]	29.5	30.2	30.2	31.7	32.4
CMSIS-NN Flash [kB]	0	17.8	0	5.8	0
Total [kB]	225.6	245.9	227.6	234.9	242.8
System Power					
Duty Cycle [%]	35.51	9.70	75.51	10.42	0.11
Infer Power [mW]	538.36	541.96	103.75	105.70	102.03
Idle Power [mW]	474.84	475.20	86.13	86.79	86.13
Average Power [mW]	497.39	481.67	99.44	88.76	86.15

CPU. We observe that using the CMSIS-NN speeds up the model inference 3.7 and 7.2 times on Cortex-M7 and Cortex-M55, respectively.

Power consumption is also lower when using an accelerator, in comparison to running the code on the CPU. The difference is not so prominent, though however, energy consumption is substantially lower with the NPU, as the computation time is shorter. Cortex-M55 SoC demonstrates more energy efficiency than the older Cortex-M7 system, although one can note that the measurement method is not ideal due to the limitation of the pins available on the development kit.

In terms of the size of code and storage requirements, adding CMSIS library increases the binary size and may not be suitable for systems with highly limited amount of flash. The bulk of the memory used is allocated to the model weights (156.7 kB), with less than 90 kB for the RTOS, inference engine, drivers and the other code.

To calculate the duty cycle, we use the latency per inference and the duration which the input represents, as well as how frequent a new sample is produced. The autoencoder uses 5 consecutive frames. Each frame is 64 ms in duration, and the sliding window is moving with hop length of 1 frame. If the model inference latency was 64 ms, the duty cycle would be equal to 100%. In all cases, the system is able to run inference with lower latency than the data rate. One should note that we only measure the model inference, without pre- or post-processing, which may also introduce significant latency.

B. Tiny MobileNetV2 Classifier

The second model tested is the Tiny MobileNetV2 classifier and the results are presented in Table IV. This model requires significantly more operations due to its convolutional nature, despite having fewer parameters than the autoencoder model. The inference latency is the lowest on the NPU with a greater improvement in this case than for the autoencoder. This is due to the NPU being specialized for the convolutional operations. As the convolution operations require intermediary buffers to store results, we believe that the latency reduction is related to improved memory management when utilizing the NPU.

Using an accelerator, either the NPU or SIMD-based CMSIS-NN, improves the power efficiency of the system. Although the energy consumption during inference is relatively similar when using NPU or CPU only, the overall time spent inferring is significantly reduced. This allows the device to remain in an idle state for longer, extending the battery life.

In terms of the storage usage, compiling with CMSIS-NN library for the H7 and for E7 adds 52.3 kB and 15.6 kB to the binary size. TFLM adds about 43-45 kBs. This increases the total binary size on the STM32 H7 to over 256 kB, more than on the Alif E7. This size increase may be prohibitive on more constrained systems indicating that, perhaps, a more streamlined implementation may be more suitable, for instance using compile-time optimization. Also, after compiling the model with ARM Vela compiler to deploy it to the NPU, its size has increased in the binary by 27 kB to 172.8 kB.

The classifier input is the concatenation of 32 consecutive frames, each representing 64 ms. We use the hop length of 8 frames with a new sample produced every 256 ms. If the model latency was 256 ms, then the duty cycle would be 100%, i.e., the platform must run inference continuously. Without the optimized kernels from the CMSIS-NN or the NPU, the latency is too high, with the duty cycle value above 100% for both DUTs (and even with CMSIS-NN for H7). This indicates inability to process all of the frames. Using the NPU, the system is able to easily process the data at a rate high enough to leave room for other tasks.

TABLE IV.
MEASUREMENTS ON DUTS FOR TINY MOBILENETV2 MODEL

Name	STM32 H7		Alif E7		
Accel.	CPU	CPU	CPU	CPU	NPU
Uses CMSIS-NN	No	Yes	No	Yes	No
Inference					
Inf. Latency [ms/inf]	2452.6	358.6	4343.5	51.5	0.08
Inf. Current [mA]	163.8	161.6	30.39	32.4	30.76
Inf. Energy [mJ/inf]	1325.9	191.3	435.6	5.5	0.008
Non-volatile Memory (Storage)					
Model Flash [kB]	145.5	145.5	145.5	145.5	172.8
TFLM Flash [kB]	43.4	43.2	45.2	44.6	13.6
CMSIS-NN Flash [kB]	0	52.3	0	15.6	0
Total [kB]	230.4	284.8	232.9	248.6	254.8
System Power					
Duty Cycle [%]	958.0	140.1	1696.7	20.1	0.03
Infer Power [mW]	540.61	533.41	100.29	106.92	101.51
Idle Power [mW]	475.83	476.69	85.64	85.60	86.13
Average Power [mW]	-	-	-	89.92	86.17

C. Discussion

Both approaches, using the autoencoder and the classifier models, can be applied to obtain accurate results for anomaly detection. The models, however, are different in architecture, inference latency and memory footprint. Without the NPU, the fully-connected autoencoder can execute and process all incoming frames without high levels of duty cycle. The classifier, on the other hand, due to the usage of convolutional layers, is characterized by higher latency and peak memory usage. We conclude that, for a more constrained system, the reconstruction-based approach with an autoencoder is more suitable.

The classifier, which is the reduced MobileNetV2 model, achieves lower accuracy than that of the autoencoder for the same number of parameters. It is possible to achieve a similar level of accuracy for the classifier by increasing the model size. This introduces more latency and increases the memory footprint. However, using the classifier's feature layer output as the feature extractor allows one to employ the embedding-based approach to the anomaly detection. These feature vectors can be used to learn a secondary model, e.g., a GMM, on the device, without having to train it prior to running the system. This allows for this model to be tuned to a specific device.

Using CMSIS-NN with SIMD instructions is recommended in every case, except when there is an insufficient amount of flash available for the increased binary. If the performance is crucial, then the NPU is the right choice, and by far outperforms the CPU-based inference. Using a SoC with NPU increases the device cost, so one must evaluate it from the business perspective as well. The deployment to the NPU is also a more difficult process, requiring more effort and integration with additional tools, introducing the increased development time.

VI. CONCLUSIONS

We present different approaches to addressing the anomaly detection task on an edge AI system. We propose a comprehensive set of measurements to provide insights into how models perform on different hardware. We test the models with standard microcontrollers, commonly used for TinyML applications. We show that using the accelerator reduces the latency and thus allows the system to remain in more efficient power state. This reduces the overall energy consumption of the system. We conclude that to select the appropriate model and platform, a holistic approach is necessary. The decision on the model used impacts the latency, memory footprint and energy. In the evaluation, we use standard models commonly applied in the research and industry. In the future work, we will focus on employing neural architecture search techniques which use the proposed measurements to find more optimal models or a model family for given design requirements.

REFERENCES

[1] C. Banbury, V. J. Reddi et al., "MLPerf Tiny Benchmark," in NeurIPS, 2021.
[2] P. Bartoli, C. Veronesi et al., "Benchmarking Energy and Latency in TinyML: A Novel Method for Resource-Constrained AI," 2025.
[3] C. R. Banbury, V. J. Reddi et al., "Benchmarking TinyML Systems: Challenges and Direction."
[4] M. Zeeshan, "Efficient Deep Learning Models for Edge IoT Devices-A Review," Authorea Preprints, 2024.
[5] Y. Koizumi, Y. Kawaguchi et al., "Description and Discussion on DCASE 2020 Challenge Task 2: Unsupervised Anomalous Sound Detection for Machine Condition Monitoring," DCASE2020, Tech. Rep., 2020.
[6] Y. Kawaguchi et al., "Description and Discussion on DCASE 2021 Challenge Task 2: Unsupervised Anomalous Sound Detection for Machine Condition Monitoring under Domain Shifted Conditions," DCASE2021, Tech. Rep., 2021.
[7] K. Dohi, K. Imoto et al., "Description and Discussion on DCASE 2022 Challenge Task 2: Unsupervised Anomalous Sound Detection for Machine Condition Monitoring Applying Domain Generalization Techniques," DCASE2022, Tech. Rep., 2022.
[8] K. Dohi, K. Imoto et al., "Description and Discussion on DCASE 2023 Challenge Task 2: First-Shot Unsupervised Anomalous Sound Detection for Machine Condition Monitoring," DCASE2023, Tech. Rep., 2023.
[9] T. Nishida, N. Harada et al., "Description and Discussion on DCASE 2024 Challenge Task 2: First-Shot Unsupervised Anomalous Sound Detection for Machine Condition Monitoring," DCASE2024, Tech. Rep., 2024.
[10] M. Sandler, A. Howard et al., "MobileNetV2: Inverted Residuals and Linear Bottlenecks," in CVPR, 2018, pp. 4510–4520.
[11] K. He, X. Zhang et al., "Deep Residual Learning for Image Recognition," in CVPR, 2016, pp. 770–778.
[12] L. Mackey, "GMMs, Expectation-Maximization," web.stanford.edu/ lmackey/stats306b, 2014.
[13] B. Schölkopf et al., "Support Vector Method for Novelty Detection," in NeurIPS, vol. 12, 1999.
[14] H. Purohit, R. Tanabe et al., "MIMII Dataset: Sound Dataset for Malfunctioning Industrial Machine Investigation and Inspection," in DCASE2019, 2019.
[15] Y. Koizumi, S. Saito et al., "ToyADMOS: A Dataset of Miniature-Machine Operating Sounds for Anomalous Sound Detection," in WAS-PAA, 2019, pp. 308–312.

Mixed Design of Integrated Circuits and Systems – MIXDES 2025

Application of Dual-Q TQWT for Atrial Fibrillation Detection with Three-Layered Neural Network

Tomasz Pander

Department of Cybernetics, Nanotechnology and Data Processing
Faculty of Automatic Control, Electronics and Computer Science
Silesian University of Technology
Akademicka Str. 16, 44-100 Gliwice, Poland
Email: tpander@polsl.pl

Abstract—Atrial fibrillation is a severe heart disease that should be detected as early as possible. In the approach presented here, an ECG signal is used, involving the detection of QRS complexes and then partitioning the ECG signal into segments containing 20 QRS complexes. In a subsequent step, this single signal segment is transformed using a Dual-Q Tunable Q-factor Wavelet Transform. On this basis, the energy distributions in the frequency sub-bands for the high and low Q-factor resonance components are calculated. This allows the generation of two fixed-length vectors characterising the analysed ECG signal segment, which are fed to the input of the three-layered neural network. The presence of atrial fibrillation in the analysed ECG signal fragment alters the energy distributions in these components. An AF database from physionet.org containing 23 long-term ECG signals was used for the study, but the database only contains 23 signals. A classifier designed on the artificial neural network was then trained and tested. The tests carried out using the 5-fold cross-validation method resulted in Sen=99.02% and Prec=99.11%, among others, which compares very well with the results of the reference methods.

Keywords—atrial fibrillation, wavelet transform, artificial neural network, DQ-TQWT

I. INTRODUCTION

Cardiovascular disease (CVD) is today the leading cause of death worldwide [1]. Therefore, researchers place a strong emphasis on heart disease research. An electrocardiogram (ECG) shows the electrical activity of the heart, which can be examined by placing electrodes on the body surface of the patient. The ECG signal plays an essential role in the monitoring of CVD. ECG analysis is a common clinical cardiac test [2]. Therefore, tools for ECG signal testing and analysis have been continuously developed for nearly four decades.

The ECG signal is quasiperiodic and consists of characteristic waveforms such as the P wave, QRS complex, and T wave. Accurate and reliable R-wave detection of the QRS complex allows medical personnel to observe and diagnose cardiovascular abnormalities [3], [4], [5], it is essential to perform automatic classification of ECG signals, for example, in the direction of the detection of arrhythmias or atrial fibrillation (AF) [5], [6], [7], [8]. Detection difficulties are mainly caused by the variability of the QRS waveform, the level of noise, and artefacts that occur during stress tests or activities in daily life. The ECG signal is usually recorded with various types of noise and artefacts. Noise can originate from a variety of sources such as power line interferences, electromagnetic waves picked up by cables, muscle activity, baseline wandering or electrode movements [9], [10]. Denoising is the first step that must be performed before starting any steps to detect the QRS complex or, for example, to classify ECG signals. There are many techniques for signal denoising.

Early atrial fibrillation automatic detection algorithms are the desired early stage of diagnosis of this kind of arrhythmia.

Various methods were employed to analyse the characteristics of atrial activity in the ECG signal. One powerful approach focuses on assessing the variations in the interval between consecutive R-waves (RR). The other is based on observing a lack or abnormal P-wave shape (replaced by rapid, irregular, and disordered fibrillatory waves) [11]. The lack of a P-wave is a powerful indicator of atrial fibrillation (AF), making it essential for accurate diagnosis and timely intervention. Understanding this key sign can significantly improve patient outcomes [12]. This makes many methods, such as wavelet transform (WT) [11] or empirical mode decomposition (EMD) [13], useful for the creation of feature sets.

The Tunable-Q Wavelet Transform (TQWT) is used to extract features from the ECG signal and then fed to support vector machine classifier [14]. In the work of Rahul et al. [15], the ECG signal and its time-frequency representation are regarded as an image, and for AF detection, the bidirectional long short-term memory network is applied.

Deep learning and machine learning methods have significantly advanced the development of atrial fibrillation (AF) detection techniques. A frequently used classifier in these methods is the support vector machine (SVM) classifier [8], [16], [17] as well as the K-nearest neighbour (KNN) classifier [18]. The model combining the convolutional and recurrent neural networks is proposed to extract high-level features from segments of RR intervals (RRIs) to classify them as AF or normal sinus rhythm in [19]. ECG features extracted via a Convolutional Neural Network (CNN) [20] and loaded to a Long Short-Term Memory (LSTM) model are proposed in [21]. The hybrid feature set, which includes characteristics of atrial activity and RR intervals, is processed using three well-known classifiers: boosted trees (BoT), random forest (RF), and linear discriminant analysis (LDA) with the random subspace method (RSM) as outlined in [22]. Rahul et al. [23] used a 1-D six-layer convolutional neural network (1-D CNN) for feature extraction from processed ECG signals. After that, bidirectional long short-term memory (Bi-LSTM) was used for different types of arrhythmia classification. A dual-domain attention cascade AF detection network (D2AFNet) is proposed in [24].

This work proposes and evaluates an improved algorithm for automatically detecting atrial fibrillation (AF). The algorithm analyses the ECG signal by dividing it into segments and examining the changes associated with atrial activity over 20 consecutive heartbeats. The first step of the proposed method involves preprocessing, which includes signal filtering and the extraction of feature vectors based on two composed signals on the base of high Q-factor and low Q-factor resonance components obtained after Dual-Q Tunable Q-factor Wavelet Transform (Dual-Q TQWT).

II. MATERIALS AND METHODS

The proposed method consists of five major stages: (i) preprocessing stage that includes denoising and R-peaks detection, (ii) data gathering as the feature vectors after the Dual-Q TQWT transform, and (iii) finally, the classification stage, which contains training (for creating a classification model) and testing. The implemented feature estimation methods are all inherently finite or possess stopping criteria to avoid latency or run-time errors. With a data window width of 20 beats, the window is marked with a pretrained classifier. The onset of an AF episode can be detected on this basis with a delay of 20 beats [22].

A. Preprocessing

ECG noise removal purpose is to reduce the noise. At first, the wandering baseline is removed from the signal. The signal $\mathbf{x}_0$ is carried out with a nonlinear median filter with window length $d_1 = \lfloor 0.5 \cdot f_s \rfloor$ as follows $\mathbf{x} = \mathbf{x}_0 - \text{median}(\mathbf{x}_0, d_1)$, where $\mathbf{x}_0$ signal to be filtered, $median(\cdot)$ the median filter and f_s is the sampling frequency [25]. Additionally, since raw ECG signals are frequently contaminated by various types of noise, such as muscle noise and 50/60 Hz power line interference, Savitzky-Golay filtering is used. An example of an ECG signal after the preprocessing step with AF episodes is presented in Figure 1.

B. Dual-Q Tunable Q-factor Wavelet Transform

Since atrial fibrillation can exhibit oscillatory characteristics, a Tunable Q-factor Wavelet Transform (TQWT) was chosen to observe both oscillatory and transient components of the ECG signal across different frequency sub-bands. The TQWT is an innovative approach in the field of wavelet transforms. This technique effectively decomposes a signal into various sub-bands, facilitating the recovery of the original signal by emphasising the sub-band with the highest energy by applying the inverse TQWT [26].

The constant Q factor characterises the conventional wavelet transform, that is, the constant ratio of its centre frequency to its bandwidth [27]. The operation of the TQWT depends on the combination of the values of the three parameters. The Q factor (Q), the redundancy (r) and the level of decomposition (J) should be known during the decomposition of TWQT. The Q factor has an impact on the oscillatory behaviour of the wavelet. Redundancy r is the total number of wavelet coefficients divided by the length of the signal to which the TQWT is used. The value of J is the level of two-channel filter banks attached to the low-pass filter output, resulting in $J+1$ subbands [26], [28]. The low-pass ($H_L^{(j)}(\omega)$) and high-pass ($H_H^{(j)}(\omega)$) filters are defined as [28]

$$
H_L^{(j)}(\omega) = \begin{cases} \prod_{m=0}^{j-1} H_L\left(\frac{\omega}{\alpha^m}\right), & \text{for} \quad |\omega| \leqslant \alpha^j \pi, \\ 0, & \text{for} \quad \alpha^j \pi < |\omega| \leqslant \pi, \end{cases} \quad (1)
$$

$$
H_H^{(j)}(\omega) = \begin{cases} H_H\left(\frac{\omega}{\alpha^{j-1}}\right) \prod_{m=0}^{j-2} H_L^{(j)}\left(\frac{\omega}{\alpha^m}\right), \\ \qquad \text{for} \quad (1-\beta)\alpha^{j-1}\pi \leqslant |\omega| \leqslant \alpha^{j-1}\pi \\ 0, \\ \qquad \text{for other} \quad \omega \in [-\pi, \pi], \end{cases}
$$
$$(2)$$

where low-pass scaling $\alpha \leqslant 1$, high-pass scaling $\beta \leqslant 1$. The parameters Q and r are defined as:

$$
r = \frac{\beta}{1-\alpha} \quad \text{and} \quad \text{Q} = \frac{f_c}{\text{BW}} = \frac{2-\beta}{\beta}, \quad (3)
$$

and BW and f_c are the bandwidth and the centre frequency, respectively. The transition bands of $H_L(\omega)$ and $H_H(\omega)$ can be constructed applying any 2π-cyclic power-complementary function. According to [27] the transition bands are given in terms of $\theta(\omega)$ by

$$
H_L(\omega) = \theta\left(\frac{\omega + (\beta - 1)\pi}{\alpha + \beta - 1}\right), \quad (4)
$$

$$
H_H(\omega) = \theta\left(\frac{\alpha\pi - \omega}{\alpha + \beta - 1}\right), \quad (5)
$$

for $(1-\beta)\pi < \omega < \alpha\pi$.

The Dual-Q TQWT application comprises the simultaneous use of the two Q-factors of the wavelet transform [27], [29]. This transformation allows decomposition of the signal $\mathbf{x}$ into two components $\mathbf{x}_1$ and $\mathbf{x}_2$, where $\mathbf{x}_1$ consists mostly of sustained oscillations and $\mathbf{x}_2$ consists mostly of non-oscillatory transients [28]. Let TQWT$_1$ and TQWT$_2$ denote the TQWT with two different Q-factors (high and low Q-factors), then the decomposition of $\mathbf{x}$ can be derived by solving an optimisation problem with constraints in sub-band-dependent regularisation form [27]

$$
\arg\min_{\mathbf{w}_1, \mathbf{w}_2} \sum_{j=1}^{J_1+1} \lambda_{1,j} ||\mathbf{w}_{1,j}||_1 + \sum_{j=1}^{J_2+1} \lambda_{2,j} ||\mathbf{w}_{2,j}||_1, \quad (6)
$$

such that

$$
\mathbf{x} = \text{TQWT}_1^{-1}(\mathbf{w}_1) + \text{TQWT}_2^{-1}(\mathbf{w}_2), \quad (7)
$$

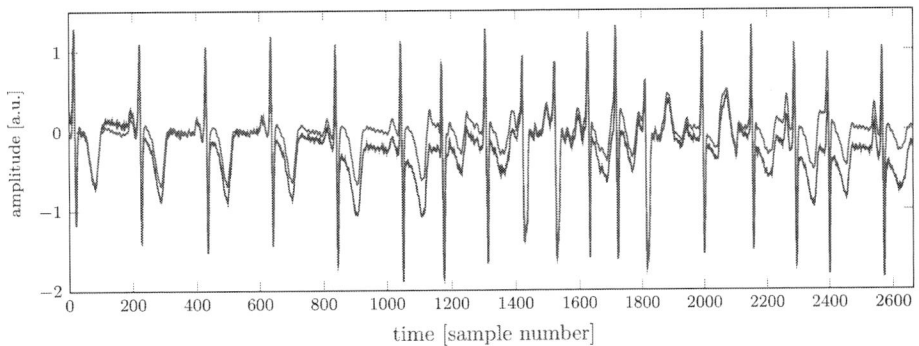

Fig. 1. An example of raw (blue line) and after denoising (red line) ECG signal segment with AF episodes.

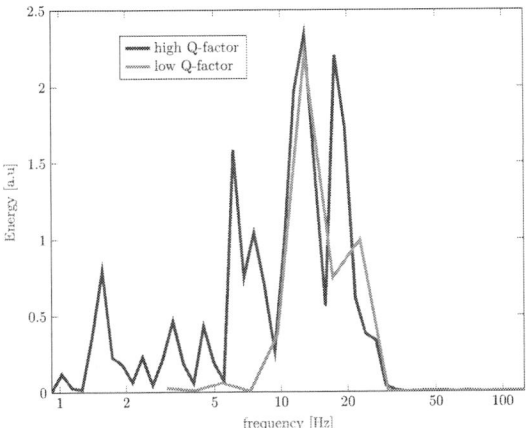

Fig. 2. Example of energy distribution within sub-bands in wavelet coefficients obtained for high and low Q-factor in the case of normal sinus rhythm (frequency axis is a log scale).

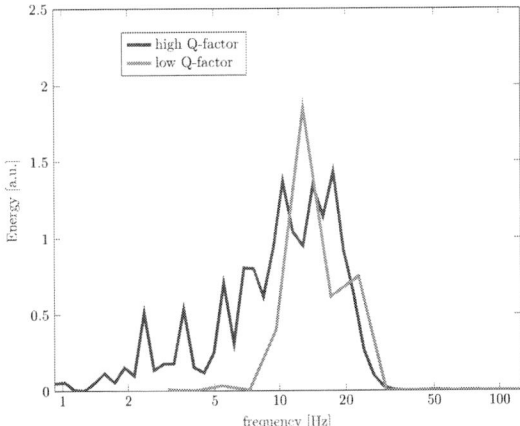

Fig. 3. Example of energy distribution within sub-bands in wavelet coefficients obtained for high and low Q-factor in the case of atrial fibrillation (frequency axis is a log scale).

where $\mathbf{w}_{i,j}$ denotes wavelet coefficients of TQWT_i for $i = 1, 2$ and TQWT^{-1} is the inverse transform. To start decomposition, the knowledge of six parameters is required: Q_1, r_1, J_1 for high Q-factor TQWT and Q_2, r_2, J_2 for low Q-factor TQWT, respectively. The high Q-factor and low Q-factor resonance components of $\mathbf{w}_1$ as well as $\mathbf{w}_2$ sub-bands are used for further analysis. The energy from each sub-band is determined on the base of $\mathbf{w}_1$ and $\mathbf{w}_2$ according to

$$E_{i,j} = \sum_{j=1}^{J_i+1} |w_{i,j}|^2, \tag{8}$$

where J_i is the decomposition level of the TQWT and $i = 1, 2$. As was shown in [17], the presence or absence of AF episodes provides changes in the energy distribution for high Q-factor and low Q-factor components decomposition $\mathbf{w}_1$ and $\mathbf{w}_2$ throughout the frequency. The energy distribution within the sub-bands for high- and low-resonance components is presented in Figure 2 and 3.

C. Three-layered neural network classifier

Artificial Neural Networks (ANNs) are fundamental to machine learning and deep learning, particularly adept at analysing complex datasets such as those obtained in AF detection. These ANNs comprise interconnected processing elements, termed nodes or neurons, organised into distinct layers. Each neuron executes specific mathematical operations, and the collective activity across these layers empowers the network to learn intricate relationships within data. This architecture allows them to capture and model the nonlinear and intricate relationships in sub-bands of energy distribution, making them ideal for predictive tasks where conventional statistical approaches may struggle.

A distinctive ANN involves an input layer, numerous hidden layers, and an output layer. The input layer collects raw data. The hidden layers, filled with numerous neurons with adjustable weights and biases, process this data. These neurons use activation functions like ReLU or sigmoid to combine nonlinearity, serving the network to detect complex relationships and interactions in the data.

TABLE I
THE DETAILED PARAMETERS OF THE NEURAL NETWORK STRUCTURE

#	Network layer type	Description
1	Feature Input	58 features with 'zscore' normalisation
2	Fully Connected	500 fully connected layer
3	Layer Normalization	Layer normalization
4	ReLU	ReLU
5	Fully Connected	300 fully connected layer
6	Layer Normalization	Layer normalization
7	ReLU	ReLU
8	Fully Connected	50 fully connected layer
9	Layer Normalization	Layer normalization
10	ReLU	ReLU
11	Fully Connected	2 fully connected layer
12	Softmax	softmax

Training an ANN implies adjusting the weights and biases to decrease variations between actual and predicted outputs through backpropagation. During this method, the network reduces a predefined loss function using optimisation algorithms like stochastic gradient descent, refining the prediction accuracy with each iteration [30].

The structure of the neural network applied in this work is presented in Table I.

D. Performance metrics and experimental setup

The performance of the AF classification is evaluated by the following quality factors: accuracy (Acc), sensitivity (Sen), specificity (Spec), positive predictivity value (PPV), and F1 score. Factors are calculated on the unseen (testing) data set using a confusion matrix. These factors are defined respectively as

$$\text{Acc} = \frac{TP + TN}{TP + FN + FP + TN}, \quad (9)$$

$$\text{Sen} = \frac{TP}{TP + FN}, \quad (10)$$

$$\text{Spec} = \frac{TN}{FP + TN}, \quad (11)$$

$$\text{PPV} = \frac{TP}{TP + FP}, \quad (12)$$

$$\text{F1} = \frac{2 \cdot \text{Sen} \cdot \text{PPV}}{\text{Sen} + \text{PPV}}, \quad (13)$$

where TP is True Positive, TN is True Negative, FP is False Positive and FN is False Negative.

The proposed method is evaluated using the publicly available MIT-BIH Atrial Fibrillation Database (AFDB) of ECGs [31]. AFDB consists of two-channel ECG signals sampled with a 12-bit resolution in the $\pm$ 10 mV range. In general, 41% of the beats are labelled AF, 57% as NSR (normal sinus rhythm), and the remaining 1.2% are from other arrhythmia [22]. The AFDB allows for the use of existing annotations of R-peak locations [31]. A sequence of 20 QRS complexes in the ECG signal is labelled as AF if the expert identifies at least half of the beats as AF; otherwise, the 20 QRS complexes

are labelled as non-AF. The percentage of AF episodes in the data, which was created by collecting 20 consecutive QRS complexes, is very similar to the percentage of AF episodes when considering QRS complexes separately, as was done in the study [17]. The overall percentage of AF cases in AFDB is about 46%. The total number of feature vectors obtained on the base of AFDB is 56415.

A Matlab 2024b environment (Mathworks, Natick, USA) was used to implement the proposed method. All signal processing as well as model training and testing for this study, were custom programmed in Windows 10 on a computer equipped with AMD(R) AMD Ryzen 7 2700X with 32 GB RAM, as well as the NVIDIA GeForce RTX1070 video card.

The cross-validation (CV) statistical method is used in this work to estimate the skill of the classifier used for AF detection method assessment. This is carried out by K-fold CV to estimate model efficiency on unobserved data. During cross-validation, the training data set is equally divided into K parts, and each part will be used as a validation/test set. For 5-fold ($K = 5$) CV, the data set is divided into 5 equal parts. The model is then trained on four of these parts, and its performance is examined on the one remaining part. This process is repeated five times until all parts are used as a test set. The average performance of the model on all test sets is then calculated. The use of K-fold CV produces an efficient model for imbalanced data. The set of parameters for DQ-TQWT transformation is the following $Q_1 = 6$, $r_1 = 4$, $J_1 = 45$, $Q_2 = 1$, $r_2 = 4$, $J_2 = 11$. It allows to prepare a feature vector that contains $(J_1 + 1) + (J_2 + 1)$ energy coefficients $E_{i,j}$ defined in eq. (8) for every ECG signal segment with 20 QRS complexes.

III. RESULTS AND DISCUSSION

The performance of the proposed algorithm was compared against several methods on the same, obtained in this approach feature vectors, as well as methods which were also evaluated on the MIT-BIH Atrial Fibrillation Database. The results for comparison are summarised in the Table II and III. The overview presented in the table presents only some of the results of work undertaken by researchers on AF detection, limited to a single MIT-BIH AFDB database, and efforts have been made to include results obtained by machine and deep learning methods.

For the methods using the same set of feature vectors developed in this paper, the best results were achieved with the Ensemble Subspace K-Nearest Neighbours method, which produced the highest specificity at 98.96% and positive predictive value (PPV) at 99.14%. In contrast, the proposed AF detection method in this paper, which utilises a three-layered artificial neural network (ANN), achieved the best results for sensitivity at 99.02%, accuracy at 98.97%, and F1 index at 99.07%. The Support Vector Machine (SVM) and K-Nearest Neighbours methods yielded slightly lower results.

Table III summarises the atrial fibrillation (AF) detection methods developed in the past 5-6 years. A notable commonality among these methods is that they have all been

TABLE II
RESULTS OF AF DETECTION WITH DIFFERENT METHODS ON THE
OBTAINED FEATURE DATABASE

Classifiers	Sen %	Spec %	PPV %	Acc %	F1 %
SVM	98.71	97.32	97.82	98.09	98.27
KNN	98.91	98.76	98.98	98.73	98.85
Ensable Subspace KNN	98.91	**98.96**	**99.14**	98.93	99.03
three-layered NN	**99.02**	98.91	99.11	**98.97**	**99.07**

TABLE III
AN OVERVIEW OF PUBLISHED RESULTS OF EXISTING AF DETECTION
METHOD USING MIT-BIH ATRIAL FIBRILLATION DATABASE IN
COMPARISON TO THE PROPOSED WORK (N/A - NOT AVAILABLE, RESULTS
ARE GIVEN IN %).

Method	Classifiers	Sen	Spec	PPV	Acc	F1
[8]	LSVM	98.94	98.80	98.39	98.86	98.66
[11]	ANN	98.7	98.9	n/a	98.8	n/a
[17]	LS-SVM	98.86	98.96	99.04	98.95	98.84
[19]	CNN-LSTM	98.98	96.95	95.76	97.8	n/a
[21]	CNN-LSTM	97.87	99.29	n/a	n/a	n/a
[22]	RF	98.00	97.4	n/a	97.6	97.1
[24]	D2AFNet	98.39	98.57	**99.19**	98.45	98.78
[32]	RF	97.7	98.5	n/a	n/a	97.7
[33]	decision tree	97.9	**99.6**	n/a	n/a	n/a
this work	three-layered NN	**99.02**	98.91	99.11	**98.97**	**99.07**

tested on the same AFDB database. This allows for effective comparison, even though the methods utilise different feature vectors and detection approaches. The method proposed in this paper demonstrates the highest F1 index, sensitivity, and accuracy. In contrast, the highest specificity value, at 99.6%, is achieved by the decision tree-based method [33]. Meanwhile, the D2AAFNet method [24] achieves the highest precision index value, with a positive predictive value (PPV) of 99.19%.

IV. CONCLUSION

Atrial fibrillation (AF) is a medical condition that can lead to serious health consequences if left undetected. Diagnosing this type of arrhythmia typically requires long-term heart monitoring, often through the recording of ECG signals. AF episodes can be brief and occur sporadically, making automatic detection methods valuable for improving the identification of AF in long-term monitoring scenarios.

This study presents and evaluates a novel and reliable algorithm for detecting AF episodes in ECG signals. The algorithm utilises a classifier based on a three-layered neural network and incorporates a feature vector that includes energy indices from both high and low Q-factor sub-bands derived from the Dual-Q Tunable Q-factor Wavelet Transform (Dual-Q TQWT). The method was trained and tested using a five-fold cross-validation approach, yielding better results than traditional reference methods. Notably, the proposed method requires an ECG signal containing 20 QRS complexes for effective AF detection.

ACKNOWLEDGMENT

This research was supported by statutory funds (BK–2024/2025) of the Department of Cybernetics, Nanotechnology and Data Processing of the Faculty of Automatic Control, Electronics and Computer Science, Silesian University of Technology, Gliwice, Poland.

REFERENCES

[1] Y. Jin, C. Qin, J. Liu, Y. Liu, Z. Li, C. Liu. "A novel deep wavelet convolutional neural network for actual ECG signal denoising". Biomed. Signal Process. Control., vol. 87, 2024, 105480.

[2] M. Elgendi, "Fast QRS detection with an optimized knowledge-based method: Evaluation on 11 standard ECG databases", PLOS ONE, vol. 8 (9), 2013, pp. 1–18.

[3] Ö. Yakut, E.D. Bolat, "A high-performance arrhythmic heartbeat classification using ensemble learning method and PSD based feature extraction approach", Biocybernetics and Biomed. Eng., vol. 42, 2022, pp. 667–680.

[4] M.B. Hossain, S.K. Bashar, A.J. Walkey, D.D. McManus, K.H. Chon, "An accurate QRS complex and P wave detection in ECG signals using complete ensemble empirical mode decomposition with adaptive noise approach", IEEE Access, vol. 7, 2019, pp. 128869–128880.

[5] J. Rahul, M. Sora, L.D. Sharma, V.K. Bohat, "An improved cardiac arrhythmia classification using an RR interval-based approach", Biocybernetics and Biomed. Eng., vol. 41 (2), 2021, pp. 656–666.

[6] R. Singh, N. Rajpal, R. Mehta, "An empiric analysis of wavelet-based feature extraction on deep learning and machine learning algorithms for arrhythmia classification", Int. Journal of Interactive Multimedia and Art. Intell., vol. 6, 2021, pp. 25-34.

[7] R. Singh, N. Rajpal, R. Mehta, "Application-specific discriminant analysis of cardiac anomalies using shift-invariant wavelet transform", Int. Journal of E-Health and Med. Comm., vol. 12(4), 2021.

[8] R. Czabanski, K. Horoba, J. Wrobel, A. Matonia, R. Martinek, T. Kupka, M. Jezewski, R. Kahankova, J. Jezewski, J.M. Leski, "Detection of atrial fibrillation episodes in long-term heart rhythm signals using a support vector machine", Sensors, vol.20 (3), 2020, 765.

[9] N. Mourad, "ECG denoising based on successive local filtering", Biomed. Sig. Proc. and Cont., vol. 73, 2022, 103431.

[10] J. Moeyersons, E. Smets, J.F. Morales, A.V. Gómez, W.D. Raedt, D. Testelmans, B. Buyse, C.V. Hoof, R. Willems, S.V. Huffel, C. Varon, "Artefact detection and quality assessment of ambulatory ECG signals", Comp. Met. and Progr. in Biomed., vol. 182, 2019, 105050.

[11] J. Wang, P. Wang, S. Wang, "Automated Detection of Atrial Fibrillation in ECG Signals Based on Wavelet Packet Transform and Correlation Function of Random Process", Biomed. Sig. Proc. and Contr., vol. 55, 2020, 101662,

[12] A. Rizwan, A. Zoha, I.B. Mabrouk, H.M. Sabbour, A.S. Al-Sumaiti, A. Alomainy, M.A. Imran, Q.H. Abbasi, "A Review on the State of the Art in Atrial Fibrillation Detection Enabled by Machine Learning", IEEE Reviews in Biomed. Eng., vol. 14, 2020, pp. 219–239.

[13] S. Pal, U. Maji, M. Mitra, "Characterizing Atrial Fibrillation in Empirical Mode Decomposition Domain", J. Med. Biol. Eng., vol. 36, 2016, pp. 693–703,

[14] C.K. Jha, M.H. Kolekar, "Cardiac arrhythmia classification using tunable-Q wavelet transform based features and support vector machine classifier", Biomed. Sig. Proc. and Contr., vol. 59, 2020, 101875.

[15] J. Rahul, L.D. Sharma, "Artificial intelligence-based approach for atrial fibrillation detection using normalised and short-duration time-frequency ECG", Biomed. Sig. Proc. and Contr., vol. 71, 2022, 103270.

[16] R.S. Andersen, E.S. Poulsen, S. Puthusserypady, "A novel approach for automatic detection of Atrial Fibrillation based on Inter Beat Intervals and Support Vector Machine", 39th Annual Int. Conf. of the IEEE Eng. in Med. and Bio. Soc. (EMBC) 2017, pp. 2039—2042.

[17] T. Pander, "An Improved Approach for Atrial Fibrillation Detection in Long-Term ECG Using Decomposition Transforms and Least-Squares Support Vector Machine", Appl. Sci., vol. 13, 2023, 12187.

[18] K. Padmavathi, K.S. Ramakrishna KS, "Classification of ECG signal during atrial fibrillation using autoregressive modeling", Procedia Computer Sci., vol. 46, 2015, pp.53—59.

[19] R.S. Andersen, A. Peimankar, S. Puthusserypady, "A deep learning approach for real-time detection of atrial fibrillation", Expert Syst. with App., vol. 115, 2019, pp. 465—473.

[20] X. Chen, Z. Cheng, S. Wang, G. Lu, G. Xv, Q. Liu, X. Zh, "Atrial fibrillation detection based on multi-feature extraction and convolutional neural network for processing ECG signals", Comp. Meth. and Progr. in Biomed., vol. 202, 2021, 106009.

[21] G. Petmezas, K. Haris, L. Stefanopoulos, et al, "Automated Atrial Fibrillation Detection using a Hybrid CNN-LSTM Network on Imbalanced ECG Datasets", Biomed. Sig. Proc. and Contr., vol. 63, 2021, 102194.

[22] G. Hirsch, S.H. Jensen, E.S. Poulsen, S.K. Puthusserypady, "Atrial fibrillation detection using heart rate variability and atrial activity: A hybrid approach". Expert Sys with App., vol. 169, 2021, 114452.

[23] J. Rahul, L.D. Sharma, "Automatic cardiac arrhythmia classification based on hybrid 1-D CNN and Bi-LSTM model", Biocyb. and Biomed. Eng., vol. 42, 2022, pp. 312–324.

[24] P. Zhang, C. Ma, F. Song, Y. Sun, Y. Feng, Y. He, T. Zhang, G. Zhang, "D2AFNet: A dual-domain attention cascade network for accurate and interpretable atrial fibrillation detection", Biomed. Sig. Proc. and Contr., vol. 82, 2023, 104615.

[25] J. Rahul, M. Sora, L.D. Sharma, "Dynamic thresholding based efficient QRS complex detection with low computational overhead", Biomed. Sig. Proces. and Contr. vol. 67, 2021, 102519.

[26] A. Kumar, A. Prakash, R. Kumar, "Tunable Q-Factor Wavelet Transform for Extraction of Weak Bursts in the Vibration Signal of an Angular Contact Bearing", Procedia Technol., vol. 25, 2016, pp. 838–845.

[27] I.W. Selesnick, "Wavelet Transform With Tunable Q-Factor", IEEE Trans. Signal Process., vol. 59, 2011, pp. 3560–3575.

[28] J. Liu, C. Zhang, Y. Zhu, T. Ristaniemi, T. Parviainen, F. Cong, "Automated Detection and Localization System of Myocardial Infarction in Single-Beat ECG Using Dual-Q TQWT and Wavelet Packet Tensor Decomposition", Comp. Meth. and Progr. in Biomed., vol. 184, 2020, 105120.

[29] I.W. Selesnick, "Resonance-Based Signal Decomposition: A New Sparsity-Enabled Signal Analysis Method", Sig. Proces., vol. 91, 2011, pp. 2793–2809.

[30] R. Setiono, H. Liu. "Neural-network feature selector", IEEE Trans. on Neu. Net., vol. 8(3), 1997, pp. 654-62.

[31] G.B. Moody, R.G. Mark, "A new method for detecting atrial fibrillation using R-R intervals", Computers in Cardiology, vol. 10, 1983, pp. 227-230.

[32] V. Kalidas, L.S. Tamil, "Detection of atrial fibrillation using discrete-state Markov models and Random Forests", Comp. in Biol. and Med, vol. 113, 2019, 103386.

[33] Y. Hu, Y. Zhao, J. Liu, J. Pang, C. Zhang, P. Li, "An Effective Frequency-Domain Feature of Atrial Fibrillation Based on Time–Frequency Analysis", BMC Med Inform Decis. Mak., vol. 20, 2020, 308.

Application of Modified Particle Swarm Optimization Algorithm in FIR Filter Design

Kamil Pipka[1], Tomasz Talaska[1,2], Rafał Długosz[2,3], Witold Pedrycz[4]

[1] Gdańsk University of Technology, Faculty of Electronics Telecommunications and Informatics
Gabriela Narutowicza 11/12, 80-233, Gdańsk, Poland
[2] Bydgoszcz University of Science and Technology,
Faculty of Telecommunication, Computer Science and Electrical Engineering,
Kaliskiego 7, 85-796 Bydgoszcz, Poland
[3] Aptiv Services Poland, ul. Podgórki Tynieckie 2, 30-399, Kraków, Poland
[4] University of Alberta, Department of Electrical and Computer Engineering,
11th Floor, Donadeo Innovation Centre for Engineering Edmonton, Alberta, Canada T6G 1H9

Abstract—In this work we present an application of Particle Swarm Optimization (PSO) algorithm as a support in the design of Finite Impulse Response (FIR) filters. The conventional PSO algorithm was not sufficient to obtain desired filter parameters for filters longer than 20 coefficients. For this reason, we used an adaptive PSO algorithm that allows for adjustment of several key parameters during the optimization process of the swarm. In comparison to existing adaptive algorithms, in which for example only the inertia coefficient was subject to change, in our approach the possibility of changing several parameters simultaneously has been introduced. In our approach we additionally modify the social and cognitive coefficients in parallel. As a result, it was possible to obtain satisfactory results for FIR filters of lengths exceeding 50. In this work, we focused in particular on examining the effect of the swarm population size on the algorithm convergence. It turned out that for FIR filters of lengths around 40-50, satisfactory results are obtained with the number of particles in the swarm at the level of 100-150.

Keywords—PSO algorithm, PSO adaptive algorithm, FIR filter design

I. INTRODUCTION

The PSO algorithm was proposed about 30 years ago by James Kennedy and Russell C. Eberhart [1]. Its key advantages include computational simplicity and high efficiency, but the latter is largely dependent on the number of particles used in the swarm. For this reason, it affects the computational complexity to some extent. The algorithm has found applications in many engineering fields, in very different optimization tasks. One such application is finding such values of the FIR filter coefficients that the filter frequency response meets given requirements [2], [3], [4]. In this case, the algorithm can be considered as an alternative solution to classical FIR filter design methods, but with some limitations. The most popular conventional approaches to FIR filter design include window, least mean square, and Parks-McClellan methods. We will not discuss these methods in detail here. It is worth mentioning, however, that they can be considered universal, i.e. suitable for very different filter lengths and other parameters. For the comparison, in the case of using the PSO algorithm, the filter length is usually limited to several dozen. Such filter lengths are sufficient in many applications. However, further research in this area may consist in searching for solutions that would

increase the lengths of FIR filters that can be designed using the PSO algorithm. This is one of the goal of the presented project.

Over the years, many modifications have been made to the conventional PSO algorithm to adapt it to different applications [5], [6], [7]. The aim of these modifications was to increase the efficiency or improve the convergence of the optimization process. In next Section, we briefly presents possible variants of this algorithm. The application of the PSO algorithm in the design of the FIR filters also requires some modifications to the original algorithm. The problem can be formulated as how to enable the design of filters with larger lengths and at what computational cost. In the presented project, the goal was to obtain satisfactory results of the swarm optimization process with a relatively small number of particles, so as to minimize the execution time and computational cost. This is important in the case of a hardware implementation of this algorithm. The ability to obtain good optimization results with impulse response lengths above 40-50 is not a simple task, because increasing the filter length linearly increases the dimension of the search space. To minimize this problem, we introduced the possibility of simultaneous modification of several key algorithm parameters during the optimization process.

II. OVERVIEW OF THE PSO ALGORITHM

The PSO algorithm is an iterative optimization method that involves modifying the position of a specific group of particles (agents) in such a way as to find an optimal solution, which is understood as the minimum of the so-called objective function. This function can be defined in various ways. For example, artificial functions of this type, often described by complex mathematical formulas, are used as a benchmark to evaluate the performance of the swarm algorithms in a comparative way. In real solutions, such functions result from a specific problem to solved by the PSO algorithm. In this work, the objective function is defined in such a way as to support the optimization of several key parameters of FIR filters, such as the linearity of the filter phase response, the cutoff frequencies of the pass and stop bands, and the distortions in the filter pass band. The search space is determined by the filter coefficients in such a way that the dimension is equal to half the the length of the impulse response. This results from the assumed symmetry of the filter transmittance, which is important from the point of view of the linearity of the phase response of the filter. Each particle in the swarm has its own velocity and position, which are updated in subsequent iterations. The particle position x_i, velocity v_i, the best position so far p_i and the global best position p_d are defined as follows [8]:

$$
\begin{aligned}
\boldsymbol{x}_i &= [x_{i1}, x_{i2}, \ldots, x_{id}], \\
\boldsymbol{v}_i &= [v_{i1}, v_{i2}, \ldots, v_{id}], \\
\boldsymbol{p}_i &= [p_{i1}, p_{i2}, \ldots, p_{id}], \\
\boldsymbol{p}_d &= [p_{d1}, p_{d2}, \ldots, p_{dd}],
\end{aligned}
\tag{1}
$$

where:
- d – size of the search space
- x_i – position of the ith particle in the search space. Each element of x_{ij} represents a coordinate value in the corresponding dimension j.
- v_i is the velocity vector of the ith particle that determines the direction and speed of the particle's motion in each dimension.
- p_i is the personal best position of the ith particle, i.e. the position where the particle has achieved the best result so far (personal best).
- p_d is the global best position in the overall swarm, i.e. the best solution found by all particles (global best).

Each particle updates its velocity according to following equation:

$$
\begin{aligned}
\boldsymbol{v}_i(k) = \ & \alpha \boldsymbol{v}_i(k-1) + \\
& c_1 r_1(\boldsymbol{p}_i(k-1) - \boldsymbol{x}_i(k-1)) + \\
& c_2 r_2(\boldsymbol{p}_d(k-1) - \boldsymbol{x}_i(k-1))
\end{aligned}
\tag{2}
$$

Then, the particle's position is updated according to formula:

$$
\boldsymbol{x}_i(k) = \boldsymbol{x}_i(k-1) + \boldsymbol{v}_i(k)
\tag{3}
$$

where:
- α – inertia coefficient that determines an impact of the previous velocity on the new velocity of the particle. Higher values of α favors exploration, while a lower value favors exploitation.
- c_1 – attraction coefficient for the best personal solution. It is responsible for the particle's trust in its own previous experience.
- c_2 – attraction coefficient for the best global solution in the swarm. It is responsible for the particle's trust in the experience of other particles in the swarm.
- r_1, r_2 – random numbers in the range $[0, 1]$, which introduce an element of stochasticity, allowing for a more dynamic search for optimal solutions.
- k – current iteration of the algorithm.

After each update of $x_i(k)$ it is checked whether its value does not go outside the range $< -1, 1 >$. The filter coefficients are normalized to such values. In case these limits are exceeded by a given coefficient, its value is set to the corresponding limit.

The influence of the cognitive and social components on the particle velocity is controlled by the acceleration factors c_1 and c_2 together with the random numbers r_1 and r_2. A proper balance between the values of c_1 and c_2 [9] is required, since their improper selection can lead to divergent or cyclical behavior of the swarm. In general, there is usually a need to find a proper balance between mentioned coefficients. In our approach this balance as well as the values of these parameters depend on the course of the optimization process.

A. Modifications of conventional PSO algorithm

As mentioned earlier, various versions of the PSO algorithm have been proposed in the literature. They include hybrid algorithms that combine the operation of the PSO algorithm with, for example, evolutionary computation techniques, such as Genetic Algorithms (GA) or Differential Evolution (DE). The GA-PS version allows for an increase in population diversity through selection and mutation mechanisms. This version allows for large changes in the position of particles through mutation. This can accelerate the convergence process. DE-PSO allows for avoiding local minima and accelerating convergence in difficult search spaces. In this approach, the classic PSO algorithm is extended with a mutation operator, which increases the diversity of the particle population. While these approaches are efficient in terme of algorithm performance, they are complex and difficult to implement in hardware.

From the point of view of this work, more interesting are the adaptive versions of the PSO algorithm, in which mechanisms for modifying the algorithm parameters based on the current optimization results are introduced. Such algorithms are generally simpler from the implementation side. An example is the reduction of the inertia weights as the results improve or along with the course of the optimization process. This prevents the phenomenon of particle stagnation and allows for a more accurate search for optimal solutions. In some implementations, fuzzy logic is used to adjust parameters depending on the population state. In our work, we also used solutions from the area of multiobjective PSO (MO-PSO), which aims to optimize a particle swarm based on several different criteria simultaneously [10]. The simplest approach to solving multi-objective problems in MOPSO is to aggregate all optimization criteria into a single objective function with specified weights. This approach is characterized by the difficulty of finding appropriate weights, which is why in most cases it is based on the concept of Pareto front (non-dominated objectives), which is the best compromise between different objectives.

III. PSO ALGORITHM IN FIR FILTER DESIGN

Designing FIR filters with the use of the PSO algorithm is an alternative approach to the ones presented in previous Section. It requires defining an appropriate fitness function, which allows for the evaluation of differences between the theoretical and desired frequency responses. One of the assumptions of this work was to achieve a linear phase response. The design criteria may also include the passband and stopband boundaries, as well as minimizing signal distortion in the passband. The objective function can be defined in such a way to minimize the sum of absolute errors in the passband and stopband [11]. Optimization of this function leads to better attenuation of signals in the stopband, minimal ripples in the passband, and optimal transition width between these bands.

In this work we used the objective function that is a weighted sum of two criteria related to the accurate reconstruction of the passband and stopband of an ideal filter. Both bands are defined using cutoff frequencies. Additionally, a parameter associated with the transition band width is introduced, which masks error accumulation within specific frequency ranges. This is because the dynamics in this band largely depends on the filter size and could mislead the algorithm.

The implementation of the PSO algorithm for FIR filter design consists of several steps presented below, the aim of which is to optimize the filter coefficients. In this work, the focus is on the example of a low-pass filter. It is worth adding, however, that this is a typical approach, because the high-pass filter can be easily obtained from the previously designed low-pass filter [12].

1) Filter parameter definition: Determination of the FIR filter parameters, such as the cutoff frequencies of the passband ω_p and stopband (ω_s), ripples in both bands (δ_p) and (δ_s), the transition bandwidth and the filter order $N = M - 1$.

2) Particle Initialization PSO: Random initialization of particles representing potential solutions (sets of $h(m)$ coefficients) in the search space.

3) Objective function calculation: For each particle, the objective function is calculated as the sum of errors between the filter frequency response and the ideal response, which allows for minimizing ripples and achieving appropriate damping. The function we applied in our approach is given by Eq. 4.

4) Particle velocity and position update: Particle velocities and positions are updated based on the personal best and global best solutions towards optimal filter coefficients.

5) Termination criterion: The process is repeated until the minimum objective function value is reached or until the number of iterations is exhausted, which terminates the optimization.

$$I = G_{\mathrm{p}} \cdot \sum_{\omega \in \omega_{\mathrm{p}}} \begin{cases} 0 & \text{if cond. 1} \\ \mathrm{abs}(\mathrm{abs}(H(\omega)) - 1) & \text{otherwise} \end{cases} + \\ + G_{\mathrm{s}} \cdot \sum_{\omega \in \omega_{\mathrm{s}}} \begin{cases} 0 & \text{if cond. 2} \\ \mathrm{abs}(\mathrm{abs}(H(\omega)) - 1) & \text{otherwise} \end{cases} \quad (4)$$

cond. 1 $\rightarrow \mathrm{abs}(\mathrm{abs}(H(\omega)) - 1) \leq \delta_{\mathrm{p}}$
cond. 2 $\rightarrow \mathrm{abs}(\mathrm{abs}(H(\omega)) - 1) \leq \delta_{\mathrm{s}}$

In Eq. 4 the δ_{p} and δ_{s} factors are maximum ripples (deflections from desired values) in the passband and stopband, respectively. G_{p} and G_{s} are possible to apply gains (playing the role of weights) in the passband and stoppand of the frequency response. In many practical application suppressing ripples in the passband may be more important than in the stopband. In this case G_{p} should be larger than G_{s}.

For the initial tests, the values of the PSO parameters and the size of the searched filter defined in the example shown in [13] were assumed. They are presented in the Table I.

TABLE I.
EXAMPLE ADOPTED VALUES OF THE PARAMETERS
OF THE DESIGNED FIR FILTER

Parameter	Value
Population size (N)	25
Number of iterations (T)	40
Local and global influence coefficients (c_1 and c_2)	2.05
Maximum inertia ($\alpha_{\max}$)	1.0
Minimum inertia ($\alpha_{\min}$)	0.4
Number of filter coefficients (M)	21

With the adopted parameters, the PSO algorithm was implemented to determine the values of the coefficients of the low-pass filter, which is supposed to suppress frequencies above the cut-off frequency f_c. With the fixed values of the parameters, the PSO algorithm very rarely finds satisfactory solutions and stops at local optima.

In order to increase the exploratory capabilities of the algorithm, modifications were introduced, compared to the previous parameters. For this purpose, the particle population was increased to 100 and the number of iterations to 200. A dynamic change of the coefficients c_1 and c_2 was also applied in a way analogous to the dynamics of the inertia. The cognitive component at the beginning of the algorithm had a greater influence than the social component in such a situation, while during the final iterations the c_2 coefficient had a greater emphasis.

The magnitude response of the filter in linear scale with coefficients with parameters adjusted for greater exploration is shown in Fig. 1 (a). It was compared with the obtained course of the filter determined by the window method with an identical number of coefficients.

Based on the shown results, it can be concluded that the filter determined using the PSO algorithm performs better in damping oscillations in the passband. However, it is characterized by a wider transition band than the classical window method. The comparative analysis of the filter is also presented on a logarithmic scale in Fig. 1 (b).

The filter designed using the swarm algorithm offers a similar attenuation to the one obtained using the classical design method, and even more strongly attenuates interference at the beginning of the stopband. Similarly to the linear diagram, a wider transition band is visible, which means that a larger range of frequencies near the cutoff frequency will be less attenuated.

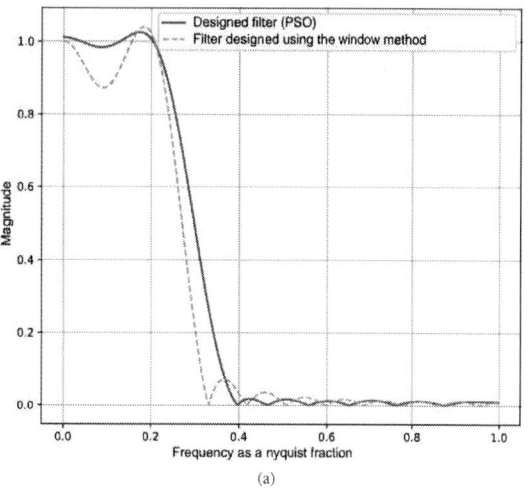 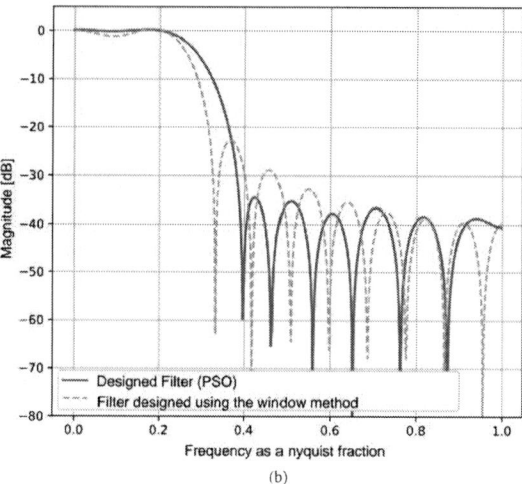

Fig. 1. Magnitude response of the low-pass FIR filter obtained using the PSO algorithm with improved parameters, for (a) linear and (b) logarithmic scale of the value axis.

Fig. 2 presents the phase response of the obtained filter to ensure that it meets all the assumptions related to maintaining the shape of the signal with different frequency components. The unwound phase was used in the Figure to more clearly present its linear character. The characteristic maintains linearity in the passband region, i.e. in the range up to the cutoff frequency $f_c = 0.275$ (fraction of the Nyquist frequency). This results from the forced symmetry of the filter coefficients, which simultaneously allowed to reduce the dimension of the search space by half.

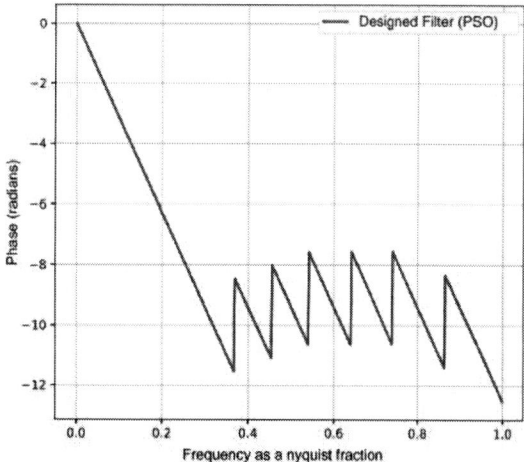

Fig. 2. Phase response of the low-pass FIR filter obtained using the PSO algorithm with improved parameters.

IV. EXPERIMENTAL RESULTS

The aim of the research presented in this paper was to evaluate the effectiveness and efficiency of the PSO algorithm in the process of designing FIR filters for different numbers of particles in the swarm. In this paper, we present the results of filter optimization with lengths exceeding 40 coefficients.

The population size in the PSO algorithm significantly affects the FIR filter optimization process. The population defines the number of particles exploring the solution space, which directly affects the quality and convergence speed of the design process. This parameter has also a direct influence on the implementation complexity. In case of a parallel implementation of such a solution, the hardware complexity almost linearly depends on the number of particles in the swarm.

The analysis considered bandpass FIR filters with previously defined default parameters, where $f_1 = 0.4$ and $f_2 = 0.66$. Different population sizes $N = 10, 50, 100, 150, 200$ were assumed. Each configuration was tested in 10 experiments to take into account the variability of results related to the random nature of the algorithm. The obtained results in the form of the best value of the objective function are presented in Table II. The objective function for bandpass filters takes a twice higher value (despite meeting the criteria) due to unavoidable errors in two transition bands.

TABLE II.
OBJECTIVE FUNCTION VALUES FOR DIFFERENT POPULATION SIZES

Population size	10.0	50.0	100.0	150.0	200.0
Experiment 1	632.54	241.03	424.44	231.46	423.78
Experiment 2	575.12	425.71	231.57	424.88	231.31
Experiment 3	515.15	423.63	235.78	231.2	422.95
Experiment 4	425.56	430.98	423.67	231.21	230.96
Experiment 5	875.31	428.99	424.27	231.49	231.02
Experiment 6	915.51	425.46	231.65	230.96	231.19
Experiment 7	628.94	242.07	423.33	426.63	230.91
Experiment 8	730.73	446.19	425.44	231.94	426.04
Experiment 9	589.46	432.24	428.16	231.93	233.21
Experiment 10	895.23	424.18	230.98	423.75	231.01
Average result	678.35	392.04	347.92	289.54	289.24
Best result	425.56	241.03	230.98	230.96	230.91

The results show the relationship between the population size and the algorithm's efficiency in finding the solution. Data analysis shows that at the smallest sizes, such as 10, the algorithm has difficulty finding the optimal solution. This is evidenced by the high average value and the large divergence of results between samples. As the population size increases, the results improve significantly. The differences between the best values obtained for populations of sizes 100, 150 and 200 are negligible, which means that further population growth does not allow for finding better solutions. However, the average values of the objective function are improved.

Fig. 3 presents the best courses of the objective function obtained in each group with different population sizes. The graph clearly shows the differences in the dynamics of the algorithm's convergence depending on the population size. For the smallest populations, we observe a slower decrease in the objective function value, which indicates the algorithm's difficulty in exploring the solution space and greater instability in the optimization process. With the increase in the number of particles, the decrease in the objective function value becomes faster and more systematic, which indicates a better balance between exploration and exploitation of the solution space.

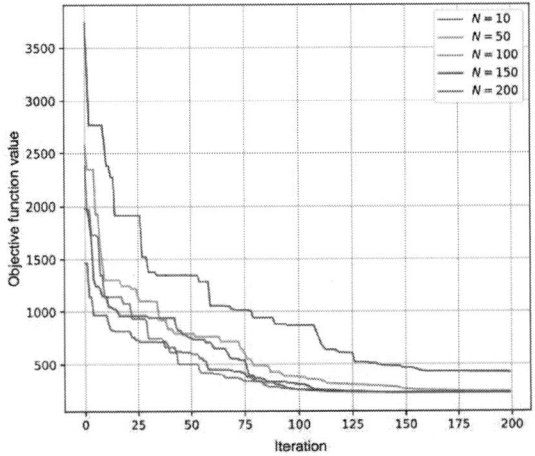

Fig. 3. Objective function values of the best solutions in each population group.

It is also worth noting that larger populations lead to an improvement in the initial value of the objective function, which may be due to the greater diversity in the generated initial solutions. The results for populations of sizes $N = 150$ and $N = 200$ stabilize in a similar number of iterations, while the differences in the final values of the objective function are negligible.

In summary, a population of size 150 seems to be an optimal compromise, combining fast convergence, stability of results, and moderate computational complexity. On the other hand, smaller populations require more iterations to achieve satisfactory results, while increasing the number of particles to 200 does not bring significant benefits, generating only higher computational costs.

V. CONCLUSIONS

The paper presents an application of the PSO algorithm in FIR filter design. The work is part of wider research aimed at developing an adaptive PSO algorithm capable of optimizing longer filters of this type. In this paper, we focus on the size of the particle swarm, as this parameter directly affects the implementation complexity of such a solution. The analyses carried out show that the size of the population plays an important role in balancing the exploration and exploitation of the solution space. Larger population sizes lead to a more systematic search for optimal solutions.

In following stages of the project, it is planned to split filters with long impulse responses into shorter sections connected in series, and then optimize them separately. Optimizing each of such shorter sections separately would be easier due to the significantly smaller dimensions of the search space.

REFERENCES

[1] Kennedy J., Eberhart R. C., Shi Y.: Swarm Intelligence. A volume in The Morgan Kaufmann Series in Artificial Intelligence, 2001
[2] Sarangi, Archana et al.:Design of Linear Phase FIR High Pass Filter Using PSO with Gaussian Mutation. Swarm, Evolutionary, and Memetic Computing. vol. 8947. Switzerland: Springer International Publishing AG, 2015
[3] Boudjelaba, Kamal, Frédéric Ros, and Djamel Chikouche: Potential of Particle Swarm Optimization and Genetic Algorithms for FIR Filter Design.Circuits, systems, and signal processing, Vol. 33(10), 2014
[4] Ababneh, Jehad I, and Mohammad H Bataineh.: Linear Phase FIR Filter Design Using Particle Swarm Optimization and Genetic Algorithms. Digital signal processing, Elsevier, Vol. 18(4), 2008
[5] Del Valle Y., Venayagamoorthy G.K., Mohagheghi S., Hernandez J.C., Harley R.G.: Particle Swarm Optimization: Basic Concepts, Variants and Applications in Power Systems. IEEE Transactions on Evolutionary Computation, Vol. 12, No. 2, 2008
[6] Eberhart R.C., Shi Y.: Particle Swarm Optimization: Developments, Applications and Resources. Congress on Evolutionary Computation, 2001
[7] Shi Y., Eberhart R.C.: A Modified Particle Swarm Optimizer. IEEE International Conference on Evolutionary Computation Proceedings. IEEE World Congress on Computational Intelligence, USA, 1998
[8] Rostami M., Berahmand K., Nasiri E., Forouzandeh S.: Review of swarm intelligence-based feature selection methods, Engineering Applications of Artificial Intelligence, Vol. 100, 104210, 2021
[9] Jain M., Saihjpal V., Singh N., Singh S.B.: An Overview of Variants and Advancements of PSO Algorithm. Applied Sciences, 12(17), 2022
[10] Mohd Zain, Mohamad Zihin bin, et al.: A Multi-Objective Particle Swarm Optimization Algorithm Based on Dynamic Boundary Search for Constrained Optimization. Applied soft computing, Vol. 70, Elsevier, 2018
[11] Neha, Singh A. P.: Design of Linear Phase Low Pass FIR Filter using Particle Swarm Optimization Algorithm. International Journal of Computer Applications, Vol. 98, No. 3, 2014
[12] Smith S.,: Digital Signal Processing: A Practical Guide for Engineers and Scientists, Newnes, 2002
[13] Praneeth A., Shah P.K.: Design of FIR Filter Using Particle Swarm Optimization. International Advanced Research Journal in Science, Engineering and Technology, Vo. 3, Iss. 5, 2016

Design Flow for AI-driven Medical Systems Demonstrated through an Example in Dental Imaging Analysis

Marek Fechner, Konrad Śniatała,
Paweł Śniatała
Poznan University of Technology
Poznan, Poland
marek.fechner@put.poznan.pl

Shangping Ren
San Diego State University
San Diego, CA, USA
sren@sdsu.edu

Renata Śniatała, Tamara Pawlaczyk
Poznan University of Medical Sciences
Poznan, Poland
rsniatala@ump.edu.pl

Abstract—**The use of Machine Learning (ML) mechanisms for image analysis in dentistry may potentially bring many benefits. At the same time, potential challenges may include the availability of dental images, the availability of specialist imaging equipment, and the selection of an appropriate ML model. The main goal of this paper is to propose a procedure for designing an ML-based module of a dental system for analyzing dental images, using the example of a tooth detection algorithm within images obtained from popular devices such as a camera or a smartphone. As part of this paper, a review of open access dental image datasets with assumed characteristics was conducted, followed by a review of ML models used for tooth detection. Next, a proposal for the architecture of a tooth detection module was developed along with a prototype implementation version. The works carried out also constituted an extension of the DentIO system, enabling, among other things, the generation of dental diagrams based on voice commands. Further research directions may include expert verification of the results presented based on dental knowledge, as well as extension of the format of the analyzed data to include dental video analysis.**

Keywords—**image analysis, tooth detection, dental system, camera photos, smartphone photos, Machine Learning, Deep Learning**

I. INTRODUCTION

The use of technology in dentistry has been of interest for years both in dentists and in industry. Examples include hardware applications, such as the production of digitally assisted dental materials, and software applications, such as the digitization of dental data through electronic dental record (EDR) [1]. In addition, in recent years, special attention has been paid to the possibilities of using Artificial Intelligence (AI) in dentistry, such as the classification and identification of diseases, automatic interpretation and correction of dental images, and modeling [2]. Despite many potential benefits, it is not easy to predict the potential consequences of using these technologies in the daily work of dentists at this point. On the one hand, both dentists and patients might have some hopes for AI applications, however, on the other hand, some concerns may also emerge in this regard [3]. Among the indications of

the advantages and challenges, in the literature, for example, the following can be found [4], [5], [6], [7]:

- the possibility of automation of daily tasks,
- diagnostics and treatment support,
- the possibility of procedures standardization,
- potential system complexity,
- system availability for Dentists and Patients,
- potential need for Dentists and Patients training,
- the need to collect the appropriate amount of data,
- potential inexplicability of the results achieved by the model (depending on the algorithm).

In this paper, in the context of the challenges and opportunities characterized above, a fragment (presented in Fig. 1) of the system that supports dental practice is analyzed using the automatic analysis mechanism of dental images for tooth detection, its efficiency, complexity, and potential availability for dentists and patients. In particular, in terms of the availability of the algorithm, one of the assumptions made is to limit the input data only to images from relatively common devices, such as a camera or a smartphone. Such an approach might potentially increase the range of recipients due to the lack of the need to use specialist dental imaging equipment, however, on the other hand, it might leave the question of the potential algorithmic efficiency. An additional challenge associated with this approach might be the availability of public dental image datasets with such characteristics on the basis of which AI models could be trained. In the longer term, beyond the scope of the current article, the suggested algorithm could be one of the elements of the method developed to support the diagnosis of Temporomandibular Joint Disorder (TMD). TMD is described as a set of factors that cause pain and dysfunction of the mandibular joint and muscles controlling jaw movements, and is considered one of the most common pain syndromes in the oral-facial area [8]. The origin of the disease and the diagnostic path are not always clear [9], however, some publications indicate a connection between the path of movement and mobility of the jaw and the appearance

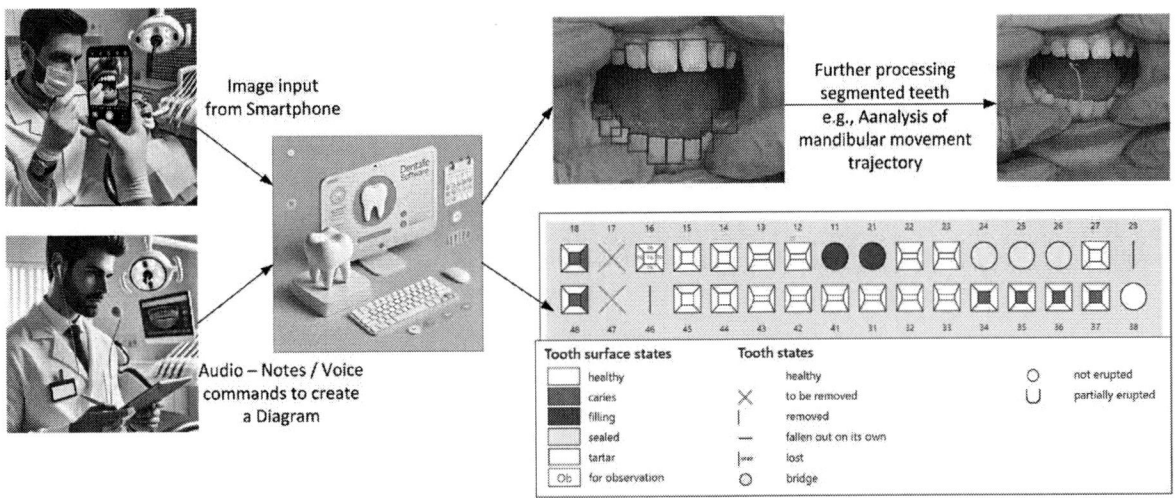

Fig. 1. Example functionality of the DentIO system

of TMD [10]. The suggested tooth detection mechanism might potentially be used in the analysis of jaw movement tracking based on the recorded video image.

The following research questions are addressed in this paper.

RQ1 What publicly available dental image collections containing camera images could be found in the literature?

RQ2 What classes of algorithms have been suggested in the literature for tooth detection from dental images?

The paper has the following structure. In the next part, work on similar topics will be briefly presented. Section III briefly characterized the solutions chosen. Then, in Section IV, we briefly describe our original *DentIO* system developed on the basis of selected solutions. The paper will be summarized with a brief conclusion.

II. SIMILAR WORKS

The automatic analysis of medical images is a research area that has been interesting to scientists for a long time. Both studies that review this topic in a more general sense (for example, [16], [17], [18]), as well as in more domain-specific contexts (for example, [19], [20], [21]), can be found. In particular, in the context of this paper, among others, the following studies could be mentioned: [22], [23], [24], [25], [26], [27]. In the paper [22], the authors develop a tooth segmentation mechanism for occlusal views obtained from photos, comparing four network architectures and reporting an accuracy of more than 95% for the most effective model. The study [23] also addresses the problem of image-based segmentation, focusing on the segmentation of four regions around the mouth: lips, teeth, tongue, and the whole mouth, during relaxation exercises and speech therapy. The authors, in the study mentioned [23], report an average precision (AP) for the detection of 99% for the lips and about 80% overall, while the Dice segmentation index exceeds 0.83 for each articulator, reaching the highest result, 0.95, for the whole mouth. The study [24] explores the possibility of developing

a diagnostic tool to aid in the detection of early caries and cavities based on smartphone images. In the paper mentioned above [24], four Deep Learning (DL) models are analyzed, among which YOLOv3 and Faster R-CNN are reported as the ones with the highest sensitivity, at 87.4% and 71.4%, respectively. The problem of detecting dental caries and fissure sealants is discussed in the study [25]. The DL model suggested in the study [25], ToothNet, achieved a reported efficiency of AUC of 0.925 for caries detection and AUC of 0.902 for sealant detection. The authors of the study [26] also focus on the problem of caries detection, at the same time introducing the element of tooth number recognition, using Region-Based Deep Convolutional Neural Network (R-CNN) model. For the problem of tooth number recognition, the reported average mean Average Precision (mAP) of the model is around 88%, while for caries detection it is around 77%, with individual scores ranging from 69.5% to 89.3% [26]. The article [27] describes an experiment using the YOLOv8 [28] model, trained on the public DentalAI [15] dataset, to detect caries.

In the context of reviewing the available collections of dental images, the study [11] might be mentioned. The authors [11] reviewed 16 publicly available datasets containing dental images obtained using various techniques, such as intraoral photos, scans, radiographs, etc., with particular emphasis on panoramic radiographs, consisting of 58. 8% of the reviewed datasets.

This paper contribution might be as follows. In the literature, data sets of dental images collected by authors are often used. This paper focuses on the review and use of publicly available dental image datasets with the characteristics assumed in the section I, that is, obtained using commonly available devices such as cameras or smartphones. In addition, this paper provides a brief review of the classes of algorithms used in tooth detection for dental images of different modalities, followed by a short experiment of using a selected class of

TABLE I
DENTAL IMAGE DATASETS BASED ON CAMERA PHOTOS

Reference	Dataset	Main area of dataset	Images	Labeled
[11], [12]	Oral Cancer (Lips and Tongue) images	Oral pathology - binary classification	131	Yes
[11], [13]	tooth-marked-tongue	Oral pathology - binary classification	1250	Yes
[14]	Teeth or Dental image dataset	Identification of teeth, cavity detection and dental malalignment	9562	No
[15]	Dentalai Computer Vision Project	Instance segmentation, semantic segmentation, and object detection	2495	Yes

algorithms for an input dataset based on a selected publicly available dental image dataset, containing images obtained using commonly available devices such as cameras or smartphones. In addition, the paper presents a proposal for the system architecture and a proposal for a Machine Learning (ML) pipeline enabling the detection of teeth based on images with assumed characteristics in a theoretically selected model based on the conducted review.

III. THE CHOSEN SOLUTIONS

In the following subsections, the results of the shortened review can be found, in relation to the research questions formulated in section I.

A. Datasets

In the context of considering the applications of AI in image analysis, publicly available image collections might be a very valuable resource. In particular, such datasets allow, especially at the stage of initial experiments, to relatively quickly evaluate the effectiveness of the algorithms considered and their potential applications for clinical solutions. In the field of dental images, such collections are not very common in general, and the need to take into account image modalities appropriate for the assumed applications further limits the obtained results. For example, in the study [11], a review can be found containing sixteen datasets obtained from publicly available sources, each of the datasets containing at least fifty images. The most common image modality in the presented review [11] is panoramic radiographs (nine datasets), followed by intraoral photos (two datasets), CBCT, cephalometric radiographs, 3D intraoral scans, periapical radiographs (one data set each) and one data set containing a combination of CBCT and panoramic radiographs.

In the context of this paper, dental image datasets based on photos or videos taken with a camera or smartphone might be of potential interest. A short summary of selected data sets of this type can be found in table I.

B. Algorithms

Among the main classes of algorithms indicated in the literature and used in the automatic analysis of dental images, those belonging to Deep Learning (DL) and Machine Learning (ML) might be indicated. For example, in study [2], the following classes of algorithms are characterized, among others:

- Convolutional Neural Networks (CNNs) in applications related to disease classification, identification and dental image segmentation,

- Generative Adversarial Networks (GANs) in applications related to enhancing images,
- Random Forest (RF) in applications related to dental image segmentation.

In particular, among the suggested network architectures, used for tooth detection and segmentation, the following can be found in the literature [2], [3], [29]:

- U-Net [30],
- V-Net [31],
- O-CNN with modifications [32],
- Visual Geometry Group (VGG) [33], [34], [35],
- Residual Network (ResNet) [36],
- Faster R-CNN [37], [38], [39],
- You only look once (YOLO) [27], [40], [41], [42], [43], [44],
- Ant Lion-based Convolution Neural Model [45],
- DenseNet [36],
- Inception [38],
- Detectron2 [46], [47].

Other classes of algorithms can also be found in the literature. For example, the study [2] also mentions fuzzy logic (FL) in applications related to color analysis.

This paper presents our results achieved with the application of the YOLO model [48]. We have implemented it in software developed to support dentists in their dental practice.

IV. DENTIO SYSTEM

A. High-level view

The proposed solution was implemented as a part of a DentIO system. The system enables the dentist to schedule visits, manage visits, store patient clinical data and generate a dental diagram based on voice commands via Web Speech API [49]. The speech recognition module allows doctors to automatically fill in a dental diagram based on information (commands) spoken into a microphone. As part of the system expansion, it was planned to prepare an additional module within the system, enabling the detection of teeth based on dental images placed in the system by Dentist. The functionalities of the DentIO system discussed in this paper are illustrated in Fig. 1 An important assumption for the aforementioned tooth detection module was to prepare the algorithm in such a way that it would be possible to analyze images taken without professional cameras but using, for example, a smartphone. Having a photo of the patient's teeth, taken even with a non-specialized camera (such as a smartphone), it is possible to make a preliminary diagnosis of some dental conditions.

Based on the image, orthodontic problems (such as a tooth outside the arch) and problems related to tooth discoloration can be diagnosed. Sometimes, also, it may be possible to diagnose caries. These are just some possible diagnoses. With not only a static photo but also a video sequence, it may be possible to detect other symptoms, such as an abnormal trajectory of mandibular movement. In our work, we used this observation to develop a computer-assisted interpretation of the clinical picture of temporomandibular joint involvement in patients with juvenile idiopathic arthritis [50].

As we described in Section III, it may be important to choose the right AI tool for the job. As a result of our analysis and comparisons, to segment teeth based on photos and videos, we considered the use of YOLO [48] or Detectron2 [51] models. These are object detection models that allow you to tag an object in an image using what is defined in the data set. In this article, we present the results obtained using the YOLO model [48], which we have trained with a collection of tooth images taken from an open source database [15].

The goal was to perform tooth detection, which allowed us to perform more analysis, such as determining the trajectory of mandibular movement. The mandibular motion trajectory was determined by tracking the motion of a point located between the lower central incisors (teeth labeled 31 and 41) and a fixed point that was located between the upper central incisors (teeth 11 and 21). We have checked the quality of the selected model by calculating the following characteristics: Recall-Confidence Curve, Precision-Confidence Curve, F1-Confidence Curve. These characteristics confirm the performance of the selected model in terms of tooth segmentation.

Fig. 2 presents a Confusion Matrix, which summarizes the performance of classification. The main diagonal (from top left to bottom right) shows the number of correct predictions made for each class. The off-diagonal display will show the errors made. It can be seen that tooth segmentation (just finding the tooth in the image) works very well - the normalized coefficient is equal to 0.98 (max=1.0).

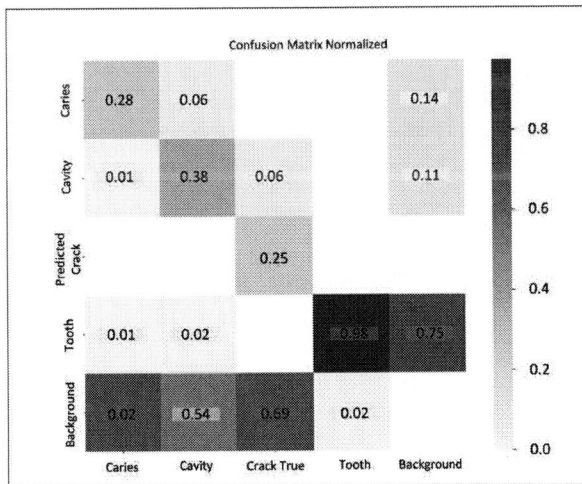

Fig. 2. Normalized Confusion Matrix of the YOLO model

B. System container diagrams and tech stack

The system's technology stack is as follows:

- Typescript [52] and Angular [53] in the frontend part (web application),
- Node.js [54] and Express [55] in the backend part (REST API),
- MongoDB [56] for data storage,
- Firebase [57], with Firebase Admin Node.js SDK [58] and Firebase Admin Python SDK [59] for file storage and handling,
- Web Speech API [49] for voice commands recognition,
- Python [60] for handling ML pipeline.

The container diagram [61] of the system, in the scope of planning the visit, filling in patient data, and providing dental images and videos, is shown in Fig. 3.

The suggested architecture may allow for separating the system mechanisms related to handling and storing Patient data from the system mechanisms related to using the ML model for tooth detection. Such separation might allow the addition of subsequent system modules without the need for significant interference in existing system elements, and the proposed shape of the system, together with the tooth detection module, may, in the long term, constitute the foundation for a potential system that supports the diagnosis of Temporomandibular Joint Disorder (TMD), the outline of which is also presented in this article.

V. Conclusions

The paper discussed a design approach that is typically used to choose AI tools that are suitable for a given task. A brief review of the ML models used for tooth detection for images of different modalities was presented. As an example, we have discussed an AI tool which would help segment (recognized) teeth from images. The YOLO model was chosen for this application. Next, we present the possible open data set that can be used to train a model. The described case needed a teeth dataset, which was used to train the YOLO model. We have presented the parameters obtained from the trained model. As an example, the practical application of the presented solutions, DentIO, a system that supports dental work, was briefly described. The elaborated module was used successfully for teeth segmentation. This result is an initial point for other algorithms that can be used to improve dental diagnosis, e.g., analysis of mandibular movement trajectory. The elaborated DentIO system is currently under preliminary tests in the dental clinic of the Poznan University of Medical Science.

Acknowledgements

This work was supported by a PhDBoost Program at the Poznan University of Technology.

Conflict of Interest

To the best of our knowledge, the named authors have no conflict of interest, financial or otherwise.

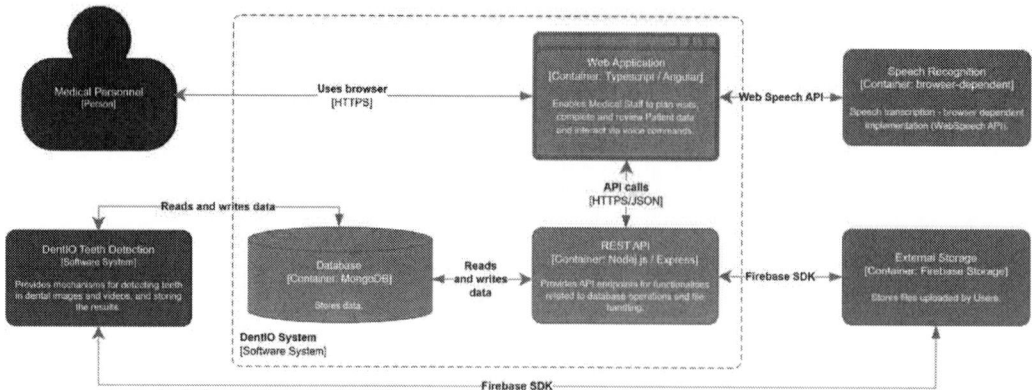

Fig. 3. DentIO System container diagram, created with [62]

REFERENCES

[1] O. Schierz, C. Hirsch, K.-F. Krey, C. Ganss, P. W. Kämmerer, and M. A. Schlenz, "DIGITAL DENTISTRY AND ITS IMPACT ON ORAL HEALTH-RELATED QUALITY OF LIFE," *Journal of Evidence-Based Dental Practice*, vol. 24, no. 1, p. 101946, Jan. 2024.

[2] F. Carrillo-Perez, O. E. Pecho, J. C. Morales, R. D. Paravina, A. Della Bona, R. Ghinea, R. Pulgar, M. D. M. Pérez, and L. J. Herrera, "Applications of artificial intelligence in dentistry: A comprehensive review," *Journal of Esthetic and Restorative Dentistry*, vol. 34, no. 1, pp. 259–280, Jan. 2022.

[3] R. Vashisht, A. Sharma, T. Kiran, S. S. Jolly, P. K. Brar, and J. V. Puri, "Artificial intelligence in dentistry — A scoping review," *Journal of Oral and Maxillofacial Surgery, Medicine, and Pathology*, vol. 36, no. 4, pp. 579–592, Jul. 2024.

[4] R. Mitra and G. Tarnach, "Artificial intelligence - A boon for dentistry," *International Dental Journal of Student's Research*, vol. 10, no. 2, pp. 37–42, Jul. 2022.

[5] S. Chakravorty, B. K. Aulakh, M. Shil, M. Nepale, R. Puthenkandathil, and W. Syed, "Role of Artificial Intelligence (AI) in Dentistry: A Literature Review," *Journal of Pharmacy & Bioallied Sciences*, vol. 16, no. Suppl 1, pp. S14–S16, Feb. 2024.

[6] J. Ma, L. Schneider, S. Lapuschkin, R. Achtibat, M. Duchrau, J. Krois, F. Schwendicke, and W. Samek, "Towards Trustworthy AI in Dentistry," *Journal of Dental Research*, vol. 101, no. 11, pp. 1263–1268, Oct. 2022.

[7] F. Schwendicke and J. Krois, "Data Dentistry: How Data Are Changing Clinical Care and Research," *Journal of Dental Research*, vol. 101, no. 1, pp. 21–29, Jan. 2022.

[8] T. List and R. H. Jensen, "Temporomandibular disorders: Old ideas and new concepts," *Cephalalgia*, vol. 37, no. 7, pp. 692–704, Jun. 2017.

[9] The Johns Hopkins University, The Johns Hopkins Hospital, and Johns Hopkins Health System, "Temporomandibular Disorder (TMD)." https://www.hopkinsmedicine.org/health/conditions-and-diseases/temporomandibular-disorder-tmd, Aug. 2021.

[10] D. V. Da Cunha, V. V. Degan, M. Vedovello Filho, D. P. Bellomo, M. R. Silva, D. A. Furtado, A. O. Andrade, S. T. Milagre, and A. A. Pereira, "Real-time three-dimensional jaw tracking in temporomandibular disorders," *Journal of Oral Rehabilitation*, vol. 44, no. 8, pp. 580–588, Aug. 2017.

[11] S. Uribe, J. Issa, F. Sohrabniya, A. Denny, N. Kim, A. Dayo, A. Chaurasia, A. Sofi-Mahmudi, M. Büttner, and F. Schwendicke, "Publicly Available Dental Image Datasets for Artificial Intelligence," *Journal of Dental Research*, vol. 103, no. 13, pp. 1365–1374, Dec. 2024.

[12] "Oral Cancer (Lips and Tongue) images," https://www.kaggle.com/datasets/shivam17299/oral-cancer-lips-and-tongue-images.

[13] "Tooth-marked-tongue," https://www.kaggle.com/datasets/clearhanhui/biyesheji.

[14] S. Chaudhary, P. Shah, P. Paygude, S. Chiwhane, P. Mahajan, P. Chavan, and M. Kasar, "Varying views of maxillary and mandibular aspects of teeth: A dataset," *Data in Brief*, vol. 56, p. 110772, Oct. 2024.

[15] "Dentalai computer vision project," https://datasetninja.com/dentalai.

[16] X. Li, L. Zhang, J. Yang, and F. Teng, "Role of Artificial Intelligence in Medical Image Analysis: A Review of Current Trends and Future Directions," *Journal of Medical and Biological Engineering*, vol. 44, no. 2, pp. 231–243, Apr. 2024.

[17] S. Saratkar, R. Raut, T. Thute, A. Chaudhari, and G. Thakre, "Review of Machine Learning and Deep Learning Techniques for Medical Image Analysis," in *2024 Second International Conference on Intelligent Cyber Physical Systems and Internet of Things (ICoICI)*. Coimbatore, India: IEEE, Aug. 2024, pp. 1437–1443.

[18] O. A. M. F. Alnaggar, B. N. Jagadale, M. A. N. Saif, O. A. M. Ghaleb, A. A. Q. Ahmed, H. A. A. Aqlan, and H. D. E. Al-Ariki, "Efficient artificial intelligence approaches for medical image processing in healthcare: Comprehensive review, taxonomy, and analysis," *Artificial Intelligence Review*, vol. 57, no. 8, p. 221, Jul. 2024.

[19] A. I. Dumachi and C. Buiu, "Applications of Machine Learning in Cancer Imaging: A Review of Diagnostic Methods for Six Major Cancer Types," *Electronics*, vol. 13, no. 23, p. 4697, Jan. 2024.

[20] K. Kryszan, A. Wylęgała, M. Kijonka, P. Potrawa, M. Walasz, E. Wylęgała, and B. Orzechowska-Wylęgała, "Artificial-Intelligence-Enhanced Analysis of In Vivo Confocal Microscopy in Corneal Diseases: A Review," *Diagnostics*, vol. 14, no. 7, p. 694, Jan. 2024.

[21] P. K. Verma and J. Kaur, "Systematic Review of Retinal Blood Vessels Segmentation Based on AI-driven Technique," *Journal of Imaging Informatics in Medicine*, vol. 37, no. 4, pp. 1783–1799, Mar. 2024.

[22] A. R. El Bsat, E. Shammas, D. Asmar, G. E. Sakr, K. G. Zeno, A. T. Macari, and J. G. Ghafari, "Semantic Segmentation of Maxillary Teeth and Palatal Rugae in Two-Dimensional Images," *Diagnostics*, vol. 12, no. 9, p. 2176, Sep. 2022.

[23] A. Sage and P. Badura, "Detection and Segmentation of Mouth Region in Stereo Stream Using YOLOv6 and DeepLab v3+ Models for Computer-Aided Speech Diagnosis in Children," *Applied Sciences*, vol. 14, no. 16, p. 7146, Jan. 2024.

[24] M. T. G. Thanh, N. Van Toan, V. T. N. Ngoc, N. T. Tra, C. N. Giap, and D. M. Nguyen, "Deep Learning Application in Dental Caries Detection Using Intraoral Photos Taken by Smartphones," *Applied Sciences*, vol. 12, no. 11, p. 5504, Jun. 2022.

[25] Y. Xiong, H. Zhang, S. Zhou, M. Lu, J. Huang, Q. Huang, B. Huang, and J. Ding, "Simultaneous detection of dental caries and fissure sealant in intraoral photos by deep learning: A pilot study," *BMC Oral Health*, vol. 24, no. 1, p. 553, May 2024.

[26] K. Yoon, H.-M. Jeong, J.-W. Kim, J.-H. Park, and J. Choi, "AI-based dental caries and tooth number detection in intraoral photos: Model development and performance evaluation," *Journal of Dentistry*, vol. 141, p. 104821, Feb. 2024.

[27] A. Germanov, "Teeth caries detection using YOLOv8 neural network," https://dev.to/andreygermanov/teeth-caries-detection-using-yolov8-neural-network-3fap, Feb. 2024.

[28] R. Varghese and S. M., "YOLOv8: A Novel Object Detection Algorithm with Enhanced Performance and Robustness," in *2024 International Conference on Advances in Data Engineering and Intelligent Computing Systems (ADICS)*, Apr. 2024, pp. 1–6.

[29] T. Bonny, W. Al Nassan, K. Obaideen, T. Rabie, M. N. AlMallahi, and S. Gupta, "Primary Methods and Algorithms in Artificial-Intelligence-Based Dental Image Analysis: A Systematic Review," *Algorithms*, vol. 17, no. 12, p. 567, Dec. 2024.

[30] M. Xu, Y. Wu, Z. Xu, P. Ding, H. Bai, and X. Deng, "Robust automated teeth identification from dental radiographs using deep learning," *Journal of Dentistry*, vol. 136, p. 104607, Sep. 2023.

[31] Y. Chen, H. Du, Z. Yun, S. Yang, Z. Dai, L. Zhong, Q. Feng, and W. Yang, "Automatic Segmentation of Individual Tooth in Dental CBCT Images From Tooth Surface Map by a Multi-Task FCN," *IEEE Access*, vol. 8, pp. 97 296–97 309, 2020.

[32] S. Tian, N. Dai, B. Zhang, F. Yuan, Q. Yu, and X. Cheng, "Automatic Classification and Segmentation of Teeth on 3D Dental Model Using Hierarchical Deep Learning Networks," *IEEE Access*, vol. 7, pp. 84 817–84 828, 2019.

[33] M. K. Alam, T. Haque, F. Akhter, H. N. Albagieh, A. B. Nabhan, M. A. Alsenani, A. Natesan, N. R. Ramanujam, and S. Islam, "Teeth segmentation by optical radiographic images using vgg-16 deep learning convolution architecture with r-cnn network approach for biomedical sensing applications," *Optical and quantum electronics*, vol. 55, no. 9, 2023.

[34] K. Zhang, J. Wu, H. Chen, and P. Lyu, "An effective teeth recognition method using label tree with cascade network structure," *Computerized Medical Imaging and Graphics*, vol. 68, pp. 61–70, Sep. 2018.

[35] Y. Mine, Y. Iwamoto, S. Okazaki, K. Nakamura, S. Takeda, T.-Y. Peng, C. Mitsuhata, N. Kakimoto, K. Kozai, and T. Murayama, "Detecting the presence of supernumerary teeth during the early mixed dentition stage using deep learning algorithms: A pilot study," *International Journal of Paediatric Dentistry*, vol. 32, no. 5, pp. 678–685, 2022.

[36] M. Al-Sarem, M. Al-Asali, A. Y. Alqutaibi, and F. Saeed, "Enhanced Tooth Region Detection Using Pretrained Deep Learning Models," *International Journal of Environmental Research and Public Health*, vol. 19, no. 22, p. 15414, Jan. 2022.

[37] D. M. Alalharith, H. M. Alharthi, W. M. Alghamdi, Y. M. Alsenbel, N. Aslam, I. U. Khan, S. Y. Shahin, S. Dianišková, M. S. Alhareky, and K. K. Barouch, "A Deep Learning-Based Approach for the Detection of Early Signs of Gingivitis in Orthodontic Patients Using Faster Region-Based Convolutional Neural Networks," *International Journal of Environmental Research and Public Health*, vol. 17, no. 22, p. 8447, Nov. 2020.

[38] E. Bilgir, İ. Ş. Bayrakdar, Ö. Çelik, K. Orhan, F. Akkoca, H. Sağlam, A. Odabaş, A. F. Aslan, C. Ozcetin, M. Kılıı, and I. Rozylo-Kalinowska, "An artificial intelligence approach to automatic tooth detection and numbering in panoramic radiographs," *BMC Medical Imaging*, vol. 21, no. 1, p. 124, Dec. 2021.

[39] D. V. Tuzoff, L. N. Tuzova, M. M. Bornstein, A. S. Krasnov, M. A. Kharchenko, S. I. Nikolenko, M. M. Sveshnikov, and G. B. Bednenko, "Tooth detection and numbering in panoramic radiographs using convolutional neural networks," *Dentomaxillofacial Radiology*, vol. 48, no. 4, p. 20180051, May 2019.

[40] E. Kaya, H. G. Gunec, S. S. Gokyay, S. Kutal, S. Gulum, and H. F. Ates, "Proposing a CNN Method for Primary and Permanent Tooth Detection and Enumeration on Pediatric Dental Radiographs," *Journal of Clinical Pediatric Dentistry*, vol. 46, no. 4, p. 293, 2022.

[41] I. D. S. Chen, C.-M. Yang, M.-J. Chen, M.-C. Chen, R.-M. Weng, and C.-H. Yeh, "Deep Learning-Based Recognition of Periodontitis and Dental Caries in Dental X-ray Images," *Bioengineering*, vol. 10, no. 8, p. 911, Aug. 2023.

[42] S. Gülüm, S. Kutal, K. Cesur Aydin, G. Akgün, and A. Akdağ, "Effect of data size on tooth numbering performance via artificial intelligence using panoramic radiographs," *Oral Radiology*, vol. 39, no. 4, pp. 715–721, Oct. 2023.

[43] H. A. Hasan, F. H. Saad, S. Ahmed, N. Mohammed, T. H. Farook, and J. Dudley, "Experimental validation of computer-vision methods for the successful detection of endodontic treatment obturation and progression from noisy radiographs," *Oral Radiology*, vol. 39, no. 4, pp. 683–698, Oct. 2023.

[44] T. H. Bui, K. Hamamoto, and M. P. Paing, "Automated Caries Screening Using Ensemble Deep Learning on Panoramic Radiographs," *Entropy*, vol. 24, no. 10, p. 1358, Oct. 2022.

[45] P. Jaiswal and D. Bhirud, "An intelligent deep network for dental medical image processing system," *Biomedical Signal Processing and Control*, vol. 84, p. 104708, Jul. 2023.

[46] J. H. Dewan, B. Sonare, H. Date, A. Gavali, V. Ghulaxe, and A. Pathan, "Comprehensive Analysis and Automation of Dental Treatment Outcomes: A Multi-Faceted Approach," in *2024 8th International Conference on Computing, Communication, Control and Automation (ICCUBEA)*, Aug. 2024, pp. 1–6.

[47] S. Sadr, H. Mohammad-Rahimi, M. S. Ghorbanimehr, R. Rokhshad, Z. Abbasi, P. Soltani, A. Moaddabi, S. Shahab, and M. H. Rohban, "Deep learning for tooth identification and enumeration in panoramic radiographs," *Dental Research Journal*, vol. 20, no. 1, p. 118, Nov. 2023.

[48] G. Jocher, A. Chaurasia, and J. Qiu, "Ultralytics YOLOv8." [Online]. Available: https://github.com/ultralytics/ultralytics

[49] "Web Speech API - Web APIs | MDN," https://developer.mozilla.org/en-US/docs/Web/API/Web_Speech_API, Feb. 2023.

[50] T. Pawlaczyk-Kamieńska, P. Śniatala, and T. Kulczyk, "Computer aided analysis of the clinical image of temporomandibular joint involvement in juvenile idiopathic arthritis patient."

[51] "Facebookresearch/detectron2," Meta Research, Mar. 2025.

[52] "JavaScript With Syntax For Types." https://www.typescriptlang.org/.

[53] "Angular," https://angular.dev/.

[54] "Node.js — Run JavaScript Everywhere," https://nodejs.org/en.

[55] "Express - Node.js web application framework," https://expressjs.com/.

[56] "MongoDB: The Developer Data Platform," https://www.mongodb.com/.

[57] "Cloud Storage for Firebase," https://firebase.google.com/docs/storage.

[58] "Firebase/firebase-admin-node," Firebase, Mar. 2025.

[59] "Firebase/firebase-admin-python," Firebase, Mar. 2025.

[60] "Welcome to Python.org," https://www.python.org/, Mar. 2025.

[61] "The C4 Model for Software Architecture," https://www.infoq.com/articles/C4-architecture-model/.

[62] "Draw.io," https://www.drawio.com/.

Edge Computing of Human Poselet

Tymoteusz Byrwa, Jakub Kłopotek Główczewski, Michał Czubenko

Gdańsk University of Technology
Faculty of Electronics Telecommunications and Informatics
Department of Decision Systems and Robotics
Narutowicza 11/12 80-233 Gdańsk, Poland
email: micczube@pg.edu.pl

Abstract—**The study investigated the computational capabilities of the NVIDIA Jetson Orin Nano – one of the most popular edge computing platforms – in the domain of Human Pose Estimation (HPE). A comparative evaluation was conducted across several widely adopted pose estimation architectures, assessing performance in GPU execution mode. Key performance indicators included inference latency, power consumption, memory footprint, and processor utilization. The goal was to determine the feasibility of deploying real-time pose estimation models on resource-constrained edge devices. Experimental results highlight the trade-offs between model complexity and system efficiency, offering practical insights for the edge-based deployment of deep learning models.**

Keywords—**pose estimation, edge computing, computational capability**

I. INTRODUCTION

The rise of Industry 4.0 has accelerated the integration of advanced digital technologies into manufacturing and industrial environments, enabling smarter, more autonomous systems. Industrial approaches need computation performed close to data sources – such as sensors, cameras, and robotic systems – rather than relying solely on centralized cloud infrastructure. On the other hand, the rapid advancement of artificial intelligence (AI) gives new opportunities, especially from the computer vision branch, such as defect detection, predictive maintenance, object tracking, and human-machine interaction. Thus there is an escalating demand for powerful computing devices capable of executing modern AI algorithms. This trend is driving the rise of „smart" devices, with applications ranging from home automation systems through complex smart city infrastructures to industrial environment – all dependent on real-time AI inference at the edge of the network.

Modern edge computing devices have evolved significantly, offering solutions for running artificial intelligence algorithms. Among the existing solutions, we can distinguish platforms like the NVIDIA Jetsons, Google Coral, ASUSU Thinker Edge, Intel Neural Compute Stick, and Raspberry Pi (of Revolution PI in case of Industrial solution) [22]. With AI accelerators based on Graphics Processing Units (GPU), Tensor Processing Units (TPU), or even Neural Processing Units (NPUs), they can support compute-intensive applications such as computer vision, speech recognition, and predictive maintenance. The latest devices of this kind, the Jetson Orin family, exemplify this trend, supporting major AI frameworks – e.g., TensorFlow, PyTorch, ONNX, while maintaining quite low power consumption. These platforms are used in Industry 4.0, autonomous robotics, smart cities, and healthcare systems.

Human Pose Estimation (HPE) is one of the core problems in computer vision. It involves detecting and localizing key points of the human body (sometimes they correspond to human joints) of individuals in 2D or 3D space from images or videos. Such points, concerning an individual, connected in space create so-called *poselet*. The method serves as an enabling technology for a wide array of applications such as safety, efficiency, and human-machine collaboration. HPE enables real-time monitoring for fall detection and ensures compliance with personal protective equipment (PPE) requirements by analysing body posture and keypoint data [1], [21]. In smart manufacturing, HPE supports collaborative robotics by enabling safe interaction zones and gesture-based control systems, allowing machines to dynamically respond to human presence and actions [7].

A. Contribution

This study explores the performance of the NVIDIA Jetson Orin Nano with JetPack SDK 6.2 – a recent and significant upgrade from the widely adopted Jetson platform – in the domain of human pose estimation. We evaluate multiple state-of-the-art, lightweight pose estimation models, optimized for accuracy and reduced inference latency on the Jetson Orin Nano in GPU execution mode. Our analysis emphasizes critical performance metrics, including inference latency, memory utilization, GPU and CPU processor load, and overall power consumption. An additional focus is placed on thermal behaviour and cooling efficiency, as operating conditions in real-world environments can vary significantly and impact sustained performance.

II. STATE OF THE ART

Recent advances in HPE research have been primarily driven by deep learning techniques. In particular, models that leverage large-scale convolutional neural networks have demonstrated substantial performance improvements and currently represent the primary path of development within the field [8]. These models can be categorized based on the source type – 2- or 3-dimensional. Two-dimensional HPE involves the identification and spatial localization of poselets from images. The three-dimensional case aims to reconstruct human body positions in 3D space. It offers greater potential for applications, particularly in domains such as film production and virtual reality. 3D HPE may be based on RGBD cameras, LiDAR, other additional sensors, or depth estimation [29], which of course requires much more computing power. Hence, our study focuses exclusively on the two-dimensional approach to human pose estimation.

Approaches to 2D HPE can be further distinguished as single or multiple poselet predictions. From a methodology point of view, we can point out regression-based techniques or heatmap-based representations. Regression-based approaches directly predict poselet coordinates using deep learning architectures. Commonly used backbone networks include AlexNet [15], as in OpenPose [4], or ResNet (Residual Network) [11], as used in the compositional pose regression method. Another example of such a method, which uses shared feature representations among keypoints detectors may be found in [16]. In contrast, heatmap-based methods generate a set of heatmaps, each corresponding to a specific keypoint. These heatmaps typically take the form of two-dimensional Gaussian distributions, where each pixel represents the probability of a certain keypoint being located at that position [29].

In case of multi-person pose estimation, the algorithm must not only detect the poselet, but also identify the number of individuals present and localize each person within the image. There are also two commonly used methodologies. Top-down approach detects bounding boxes of human silhouettes and further threats them as Region of Interest (ROI) for single-person HPE. A key limitation of this method – particularly relevant in the context of edge computing – is that computational complexity scales with the number of individuals in the image. Furthermore, this approach introduces an additional source of error associated with the person detector, especially in complex scenes where individuals may be partially occluded by other objects [10]. The bottom-up approach inverts the process: it first detects individual keypoints based on local image features and then assembles them into coherent human skeletons. This method requires precise localization of keypoints, often necessitating higher-resolution input data. Models, such as DETR (object DEtection with TRansformers) [5] and HRNet (High Resolution Network) [24], achieve highly competitive results. However, their complexity makes them less suitable for deployment on resource-constrained edge devices.

Numerous 2D HPE architectures achieve high levels of performance [2]–[4], [9], [14], [18], [23], [25], [27], [28]. However, the majority of these models are computationally demanding, making them unsuitable for deployment on edge devices. Furthermore, many pre-trained models are implemented in frameworks that are either incompatible with the ARM architecture or cannot be compiled for JetPack – the framework used in this study. For instance, while BlazePose is available through the official MediaPipe library, it can only be executed on the CPU when deployed on our Jetson device. Thus, our study focuses on evaluating architectures that offer an optimal trade-off between computational efficiency and prediction accuracy. We selected three models designed for 2D multi-person pose estimation using the top-down approach described below.

A. YOLO (You Only Look Once)

YOLOv8-Pose and YOLO11-Pose from Ultralytics are single-stage, anchor-free pose estimation extensions of the YOLO family [20]. YOLOv8-Pose builds on a CSP-based (Cross Stage Partial Network) backbone and PANet (Path Aggregation Network) neck, then uses an anchor-free split Ultralytics head that decouples classification, box regression, and pose branches. The pose branch directly regresses 17 2D keypoints per detected person predicting coordinates as offsets from the bounding box for maximal inference speed. YOLO11-Pose retains this efficient head, but swaps in an improved backbone and neck architecture from YOLO11, yielding richer multi-scale features and tighter pose estimates. Both models predict the standard COCO keypoints along with confidence scores. Training combines CIoU loss for boxes, BCE for classification, and L1/MSE losses for keypoint offsets, with loss weights configurable. In our experiments, we used medium and and extreme large models: YOLOv8-Pose M (81 GFLOPs), YOLO11-Pose M (71.7 GFLOPs), YOLOv8-Pose X (263.2 GFLOPs), YOLO11-Pose X (203.3 GFLOPs).

B. RTMPose (Real-Time Multi-Person Pose Estimation)

RTMPose [13] is a two-stage, top-down multi-person pose estimation framework built on the MMPose toolbox. The architecture decouples the workflow: an off-the-shelf detector first localizes person bounding boxes, then a lightweight pose network refines each ROI to predict keypoints. RTMPose adopts the CSPNeXt (modified CSP) network as its backbone, striking an optimal balance between representational capacity and inference efficiency. In general, CSPNeXt is deployment-friendly and achieves real-time throughput on CPUs and mobile chips. Keypoint localization leverages a simple coordinate classification based paradigm [17]: two parallel classifiers predict horizontal and vertical coordinates via sub-pixel binning. The head comprises a simple 7 x 7 convolution, two fully connected layers, and a Gated Attention Unit (GAU) to fuse global and local context before classification. RTMPose is advertised as an ultra-fast pose estimation architecture, mostly optimized for CPU and edge devices. In our experiments, we used RTMPose S (0.31 GFLOPs for detection and 0.68 GFLOPs for pose estimation) and RTMPose T (0.31 GFLOPs and 0.36 GFLOPs respectively).

C. TRT_pose (Tensor RT)

NVIDIA TRT_pose is a TensorRT-accelerated HPE framework that converts PyTorch keypoint models into highly optimized inference engines for NVIDIA Jetson platforms. TRT_pose focuses solely on keypoint regression. It uses ResNet-18 and DenseNet-121 as backbones. The ResNet-18 (1.8 GFLOPs) is a lightweight, residual convolutional neural network originally designed for image classification [11]. It is characterized by its residual block architecture, which uses skip connections to mitigate the vanishing gradient problem and stabilize training for deeper networks. This design allows each residual block to learn only the difference (residual) between the input and the target feature representation, effectively acting as a dynamic feature enhancer.

The DenseNet-121 (2.8 GFLOPs) backbone used in TRT_pose is a deeper, densely connected convolutional network known for its efficient feature reuse and parameter efficiency [12]. Unlike traditional convolutional networks like ResNet, which stack layers with isolated residual connections, DenseNet introduces dense connectivity, where each layer receives feature maps from all preceding layers. This dense connectivity pattern significantly improves gradient flow, mitigates the vanishing gradient problem, and promotes feature reuse, leading to a more compact yet highly expressive representation. Based on the architecture and repository assumptions, DenseNet should be more accurate than ResNet. Further we will use only names as TRT_pose(ResNet-18) and TRT_pose(DenseNet-121) for clarity.

D. Datasets

There are several pose estimation datasets, but we are going to briefly describe the ones used in model training and evaluation. In particular, the YOLO-Pose and TRT_pose models were primarily trained on the COCO (Common Objects in Context) dataset [19], while RTMPose leverages both COCO and the AIC (Artificial Intelligence Challenger) Human Keypoint Detection dataset [26] to improve generalization across diverse human poses.

The COCO dataset is one of the most widely used benchmarks, also for human pose estimation. It contains over 200,000 images with 250,000 person instances, each annotated with 17 keypoints representing human poselet (e.g., nose, eyes, shoulders, elbows, wrists, hips, knees, ankles) showed in Figure 1. COCO presents significant challenges, including extreme scale variation, occlusions, and diverse poses, making it a critical benchmark for evaluating both accuracy and robustness in pose models. In addition to keypoints, the dataset includes object bounding boxes, segmentation masks, and multi-label annotations, supporting models that rely on rich, multi-scale context.

In contrast, the AIC Human Keypoint Detection dataset focuses on high-throughput keypoints detection in dense, real-world human activity scenes. It contains approximately 300,000 images with 1.8 million labeled human figures, each annotated with 14 keypoints. Unlike COCO, which emphasizes fine-grained full-body annotation, AIC simplifies the poselet structure by omitting less frequently visible points (e.g., ears, ankles) and concentrating on critical skeletal landmarks. This streamlined keypoint set is particularly suited to applications requiring fast, efficient posture analysis and action recognition.

III. METHODOLOGY

Our work focuses on evaluating off-the-shelf models without applying fine-tuning or multi-task learning techniques. Furthermore, we intentionally refrained from employing optimization strategies such as quantization or pruning to enable a direct comparison of models within their respective standard frameworks.

In the case of the YOLO models (both v8 and v11), we employed the Ultralytics framework, which handles all necessary preprocessing internally, including resizing to the native 640 x 640 resolution. In contrast, the RTMPose models require preprocessing – the input image should be scaled to a fixed 256 x 192 resolution. Similarly, the DenseNet and ResNet backbones in the TRT_pose models also require image resizing to their respective input resolutions 256 x 256 for DenseNet-121 and 224 x 224 for ResNet-18.

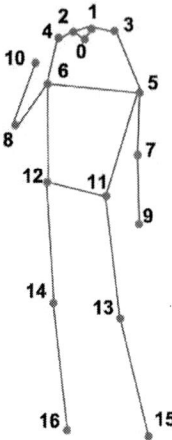

Fig. 1. 2D HPE COCO keypoints [6].

To evaluate and compare the selected architectures, we employed a subset of the COCO dataset – the 2017 validation set (COCO val2017), which contains 2,693 images depicting a small number of humans. In assessing model inference times, we also included preprocessing steps such as image resizing and data transfer to the GPU. In all cases, the time required for these operations did not exceed 10% of the model's total inference time and was therefore considered negligible. The following performance parameters have been recorded by the tegrastats tool, throughout the evaluation: CPU usage, GPU usage, power consumption, CPU temperature, GPU temperature, average inference time per image, Frames Per Second (FPS) (showed in Tab. I).

To ensure consistency and eliminate interference from background processes, the experiments have been conducted in an isolated software environment. Upon completion of each measurement sequence, the results have been saved, and standard evaluation metrics, such as Average Precision (AP) and Average Recall (AR) have been calculated. All tests were conducted in a room with an ambient temperature of approximately 20°C and minimal airflow to ensure a consistent and reproducible environment.

IV. RESULTS AND DISCUSSION

Tab. I summarizes the average precision of the models in relation to their inference speed and resource. The RTMPose models, particularly the S variant, demonstrate the highest

TABLE I
COMPARISON OF POSE ESTIMATION MODELS. NOTE THAT, INFERENCE TIME (INF.) IS AN AVERAGE, AS WELL AS CPU AND GPU USAGES. WHILE THE TEMPERATURE SHOW THE HIGHEST ONES.

Model	GFLOPs	AP	AR	FPS	Inf. [ms]	CPU [%]	CPU [°C]	GPU [%]	GPU [°C]
YOLOv8 M	81	0.536	0.619	18.60	0.054	14.500	62.187	66.145	65.281
YOLOv8 X	263.2	0.583	0.664	9.34	0.107	10.833	46.468	81.444	69.906
YOLOv11 M	71.7	0.537	0.621	17.32	0.058	15.833	61.781	67.876	64.312
YOLOv11 X	203.3	0.583	0.667	9.48	0.106	11.167	63.062	82.844	66.250
RTMpose T	0.67	0.626	0.695	5.65	0.177	16.333	57.843	65.248	60.093
RTMpose S	0.99	0.667	0.728	5.49	0.182	16.167	58.218	60.716	60.718
TRT_Pose (Resnet-18)	5	0.117	0.165	34.48	0.029	13.833	52.437	69.628	53.843
TRT_Pose (DenseNet-121)	7.5	0.148	0.210	15.53	0.064	15.167	53.375	59.818	54.843

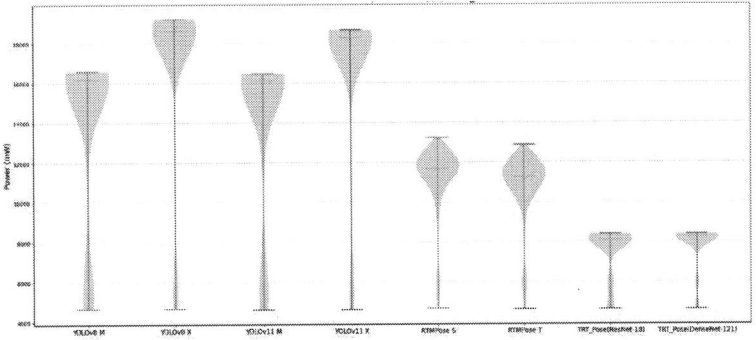

Fig. 2. Average power consumption (VDD_IN) for all models. Orange mark stands for median value.

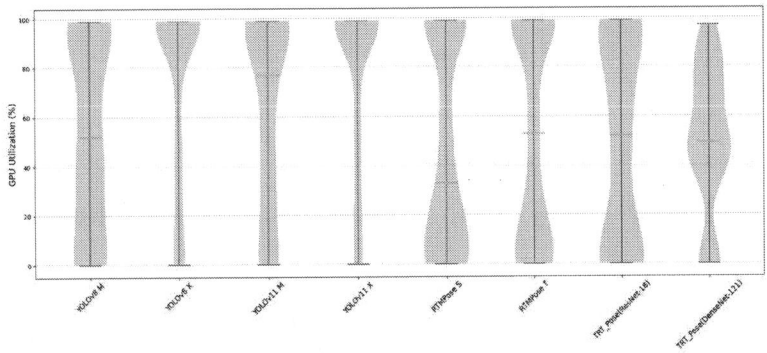

Fig. 3. GPU utilization distribution plot. Orange stands for median values.

precision among the evaluated architectures, achieving an AP of 0.667 and an AR of 0.728. However, this superior accuracy comes at the expense of inference speed, with the model reaching only around 5 FPS, which is insufficient for many real-time applications. It is worth noting that the S variant enhances prediction accuracy while incurring only a minor reduction in inference speed. Additionally, RTMPose achieves a notably higher Average Recall (AR) compared to the YOLO-based models, with an improvement averaging between 9% (yolo X models) and 17% (yolo M models). This indicates that RTMPose is significantly more effective at detecting and localizing poses comprehensively, ensuring that more

true poses are retrieved across varying evaluation thresholds. YOLO models, on the other hand, excel in terms of inference speed. YOLO-X models are approximately 67% faster than both RTMPose variants, while the M models are around 235% faster. This substantial improvement in processing speed does not come at the cost of a massive decrease in model accuracy. It is also worth pointing out the significant difference between the two YOLO architectures, as the X models are substantially slower than the M models. TRT_Pose based on ResNet18 and Densenet121 are relatively fast, but they fall significantly behind in accuracy, making them less suitable for precise pose estimation tasks.

The results for average CPU usage and maximum CPU temperature, shown in Tab. I, indicate relatively minor differences across the models. In all cases, CPU temperature stabilized around 60°C during testing. The YOLO architectures in the X versions exhibited slightly lower CPU loads, suggesting good algorithmic optimization. As with the CPU, GPU temperatures stabilized near 60°C. RTMPose models showed lower overall temperatures for both the CPU and GPU, indicating that they are particularly well-suited for deployment on edge devices. The TRT-Pose models, despite their significantly faster inference speeds, achieved the lowest temperatures among all evaluated models, maintaining CPU and GPU temperatures around 53°C. This could indicate that these models are optimized for older, lower-power Jetson devices, aligning with the fact that the TRT-Pose library was released in 2020 and has seen minimal updates since. While this efficiency is advantageous for extended deployments in power-constrained environments, it also suggests that these models may lack the algorithmic optimizations present in newer architectures like RTMPose and YOLO, potentially limiting their applicability in high-performance edge AI scenarios.

The variation in model size between the YOLO versions is evident in their GPU utilization, with YOLO architectures requiring approximately 15% more GPU resources on average. This increased demand for VRAM also results in longer inference times. Moreover, a reduction of about 5% in average RAM consumption between the two RTMPose versions further highlights the optimization improvements in the bigger model.

Regarding power consumption, as shown in Fig. 2, it is evident that the YOLO X models are the most power-consuming, drawing slightly above 18 Watts each. Interestingly, the newer version of YOLO is slightly less power-hungry while achieving better precision and recall, as demonstrated in Tab. I, indicating progress in the transition from CNN to transformer-based architectures. The M models, on the other hand, exhibit virtually no difference in power consumption.

For the RTMPose models, we observe that they require the most time for inference, consistent with their slower processing speeds, and their power usage shows more pronounced oscillations. This may suggest implementation inefficiencies, especially considering that the only official implementation is based on the MMPose toolbox, which may introduce additional overhead. This behaviour is further supported by Fig. 2, where the power distribution for RTMPose is noticeably broader compared to the YOLO models, which behave mostly stable. Regarding TRT_pose models we can see significantly lower power consumption of both models averaging only a fraction above 8 Watts which is not even a double of idle Power Consumption of used Jetson Nano model.

The vast majority of images in the dataset contain fewer than ten individuals. As a result, no significant correlation was observed between inference time and the number of people present in an image. This outcome is expected given the relatively low person count across samples. It is well established that top-down architectures tend to underperform compared to bottom-up approaches when processing scenes

with a large number of individuals. However, confirming such a relationship would require evaluation on a dataset specifically designed to represent densely populated scenes.

V. CONCLUSION

The evaluation demonstrated that the YOLO-based models (YOLOv8 and YOLOv11) significantly outperform the RTM-Pose and TRT_Pose models (ResNet-18 and DenseNet-121) in terms of overall performance, making them more suitable for real-time edge deployment. However, this speed advantage comes at the cost of relatively high power consumption and slightly reduced accuracy compared to RTMPose, which achieved the highest precision but with considerable latency.

The TRT_Pose pose models, based on ResNet-18 and DenseNet-121, offer mixed results. ResNet-18 stands out for its high inference speed (34 FPS) but struggles with weak prediction accuracy, potentially limiting its use to applications where rough, rapid estimates are sufficient. DenseNet-121 provides a modest accuracy improvement but suffers from significantly higher inference times, reducing its appeal for real-time tasks. On the other hand, YOLO models achieve superior performance not merely because of their lighter architecture, but because they eliminate the need for separate heatmap generation and decoding stages — streamlining the pose estimation process into a single forward pass optimized for GPU execution. Overall, YOLO models offer the most balanced approach, combining fast inference speeds, reasonable accuracy, and straightforward deployment, making them a practical choice for most edge computing scenarios.

Beyond raw performance metrics, the results offer practical insights into model suitability for real-world edge applications. RTMPose, for example, demonstrates outstanding precision and recall, but its reliance on the MMPose framework and multi-stage processing results in high latency and less consistent power consumption. This suggests that, while academically impressive, RTMPose may face integration challenges in production environments that demand stable, predictable performance. Meanwhile, YOLOv8-M and YOLOv11-M strike a compelling balance, offering frame rates above 17 FPS with moderate accuracy, making them ideal for real-time use cases such as gesture recognition or pedestrian tracking. These results underscore a critical trade-off in edge AI deployment: peak model accuracy is often less valuable than the consistency, efficiency, and responsiveness required by practical applications.

REFERENCES

[1] Ekram Alam, Abu Sufian, Paramartha Dutta, and Marco Leo. Real-time human fall detection using a lightweight pose estimation technique. In *International Conference on Computational Intelligence in Communications and Business Analytics*, pages 30–40. Springer, 2023.

[2] Valentin Bazarevsky, Ivan Grishchenko, Karthik Raveendran, Tyler Zhu, Fan Zhang, and Matthias Grundmann. Blazepose: On-device real-time body pose tracking, 2020.

[3] Gedas Bertasius, Christoph Feichtenhofer, Du Tran, Jianbo Shi, and Lorenzo Torresani. Learning temporal pose estimation from sparsely-labeled videos, 2019.

[4] Zhe Cao, Gines Hidalgo, Tomas Simon, Shih-En Wei, and Yaser Sheikh. Openpose: Realtime multi-person 2d pose estimation using part affinity fields, 2019.

[5] Nicolas Carion, Francisco Massa, Gabriel Synnaeve, Nicolas Usunier, Alexander Kirillov, and Sergey Zagoruyko. End-to-end object detection with transformers, 2020.

[6] MMPose Contributors. Openmmlab pose estimation toolbox and benchmark. https://github.com/open-mmlab/mmpose, 2020.

[7] Xinjian Deng, Jianhua Liu, Honghui Gong, Hao Gong, and Jiayu Huang. A human–robot collaboration method using a pose estimation network for robot learning of assembly manipulation trajectories from demonstration videos. *IEEE transactions on industrial informatics*, 19(5):7160–7168, 2022.

[8] Gaetano Dibenedetto, Stefanos Sotiropoulos, Marco Polignano, Giuseppe Cavallo, and Pasquale Lops. Comparing human pose estimation through deep learning approaches: An overview. *Computer Vision and Image Understanding*, 252:104297, 2025.

[9] Hao-Shu Fang, Jiefeng Li, Hongyang Tang, Chao Xu, Haoyi Zhu, Yuliang Xiu, Yong-Lu Li, and Cewu Lu. Alphapose: Whole-body regional multi-person pose estimation and tracking in real-time, 2022.

[10] Hao-Shu Fang, Shuqin Xie, Yu-Wing Tai, and Cewu Lu. Rmpe: Regional multi-person pose estimation. In *Proceedings of the IEEE international conference on computer vision*, pages 2334–2343, 2017.

[11] Kaiming He, Xiangyu Zhang, Shaoqing Ren, and Jian Sun. Deep residual learning for image recognition, 2015.

[12] Gao Huang, Zhuang Liu, Laurens van der Maaten, and Kilian Q. Weinberger. Densely connected convolutional networks, 2018.

[13] Tao Jiang, Peng Lu, Li Zhang, Ningsheng Ma, Rui Han, Chengqi Lyu, Yining Li, and Kai Chen. Rtmpose: Real-time multi-person pose estimation based on mmpose, 2023.

[14] Sven Kreiss, Lorenzo Bertoni, and Alexandre Alahi. Openpifpaf: Composite fields for semantic keypoint detection and spatio-temporal association, 2021.

[15] Alex Krizhevsky, Ilya Sutskever, and Geoffrey E Hinton. Imagenet classification with deep convolutional neural networks. In F. Pereira, C.J. Burges, L. Bottou, and K.Q. Weinberger, editors, *Advances in Neural Information Processing Systems*, volume 25. Curran Associates, Inc., 2012.

[16] Sijin Li, Zhi-Qiang Liu, and Antoni B. Chan. Heterogeneous multi-task learning for human pose estimation with deep convolutional neural network, 2014.

[17] Yanjie Li, Sen Yang, Peidong Liu, Shoukui Zhang, Yunxiao Wang, Zhicheng Wang, Wankou Yang, and Shu-Tao Xia. Simcc: A simple coordinate classification perspective for human pose estimation. In *European conference on computer vision*, pages 89–106. Springer, 2022.

[18] Yanjie Li, Shoukui Zhang, Zhicheng Wang, Sen Yang. Wankou Yang, Shu-Tao Xia, and Erjin Zhou. Tokenpose: Learning keypoint tokens for human pose estimation, 2021.

[19] Tsung-Yi Lin, Michael Maire, Serge Belongie, Lubomir Bourdev, Ross Girshick, James Hays, Pietro Perona, Deva Ramanan, C. Lawrence Zitnick, and Piotr Dollár. Microsoft coco: Common objects in context, 2015.

[20] Debapriya Maji, Soyeb Nagori, Manu Mathew, and Deepak Poddar. Yolo-pose: Enhancing yolo for multi person pose estimation using object keypoint similarity loss, 2022.

[21] Dawid Masłowski and Michał Czubenko. System bezpieczeństwa dla współpracującego robota przemysłowego na bazie kamer głębi. *Pomiary Automatyka Robotyka*, 23, 2019.

[22] Rafal Tobiasz, G Wilczyński, Piotr Graszka, Nikodem Czechowski, and Sebastian Łuczak. Edge devices inference performance comparison. *arXiv preprint arXiv:2306.12093*, 2023.

[23] Alexander Toshev and Christian Szegedy. Deeppose: Human pose estimation via deep neural networks. In *2014 IEEE Conference on Computer Vision and Pattern Recognition*, page 1653–1660. IEEE, June 2014.

[24] Jingdong Wang, Ke Sun, Tianheng Cheng, Borui Jiang, Chaorui Deng, Yang Zhao, Dong Liu, Yadong Mu, Mingkui Tan, Xinggang Wang, Wenyu Liu, and Bin Xiao. Deep high-resolution representation learning for visual recognition, 2020.

[25] Shih-En Wei, Varun Ramakrishna, Takeo Kanade, and Yaser Sheikh. Convolutional pose machines, 2016.

[26] Jiahong Wu, He Zheng, Bo Zhao, Yixin Li, Baoming Yan, Rui Liang, Wenjia Wang, Shipei Zhou, Guosen Lin, Yanwei Fu, Yizhou Wang, and Yonggang Wang. Large-scale datasets for going deeper in image understanding. In *2019 IEEE International Conference on Multimedia and Expo (ICME)*. IEEE, July 2019.

[27] Bin Xiao, Haiping Wu, and Yichen Wei. Simple baselines for human pose estimation and tracking, 2018.

[28] Yufei Xu, Jing Zhang, Qiming Zhang, and Dacheng Tao. Vitpose: Simple vision transformer baselines for human pose estimation, 2022.

[29] Ce Zheng, Wenhan Wu, Chen Chen, Taojiannan Yang, Sijie Zhu, Ju Shen, Nasser Kehtarnavaz, and Mubarak Shah. Deep learning-based human pose estimation: A survey, 2023.

Evaluating Device Variability in RRAM-Based Single- and Multi-Layer Perceptrons

Alan Blumenstein[1,2], Eduardo Pérez[3,4], Christian Wenger[3,4], Nadine Dersch[1,2], Alexander Kloes[1], Benjamín Iñíguez[2], Mike Schwarz[1]

[1] NanoP, THM, Giessen, Germany
[2] DEEEA, Universitat Rovira i Virgili, Tarragona, Spain
[3] IHP - Leibniz Institute for High Performance Microelectronics, Frankfurt (Oder), Germany
[4] BTU Cottbus-Senftenberg, Cottbus, Germany

Abstract—This work investigates the impact of stochastic weight variations in hardware implementations of artificial neural networks, focusing on a Single-Layer Perceptron and Multi-Layer Perceptrons. A variable neural network model is introduced, applying Gaussian variability to synaptic weights based on an adjustment rate, which controls the proportion of affected weights. By studying how stochastic variations affect accuracy, simulations under device-to-device and cycle-to-cycle variation conditions demonstrate that Single-Layer Perceptrons are more sensitive to weight variations, while Multi-Layer perceptrons show greater robustness. Additionally, stochastic quantization improves the performance of Multi-Layer Perceptrons but has minimal effect on Single-Layer Perceptrons.

Keywords—Single-Layer Perceptron, Multi-Layer Perceptrons, Device-to-device, Cycle-to-cycle, Variability, Simulation

I. Introduction

Artificial neural networks (ANNs) have gained widespread success in machine learning due to their capacity to model complex patterns in data [1]. However, when implementing ANNs on hardware platforms, such as **Resistive Random Access Memory** (RRAM)-based architectures, challenges arise due to **Cycle-to-Cycle** (C2C) and **Device-to-Device** (D2D) variation [2]. These variations induce stochastic fluctuations in synaptic weights, which can negatively impact the accuracy and stability of the network. Understanding how ANNs respond to these variations is critical for developing accurate simulation models and advancing hardware-aware learning techniques.

Previous studies have investigated the impact of C2C variation of RRAM devices performance [3]. A recent analysis further examined the influence of C2C variation on ANNs, revealing that stochastic fluctuations, especially in weight values, lead to significant performance degradation. These findings highlight the critical importance of incorporating such variations into hardware-aware neural network models [4].

Building upon this foundation, this study further explores the concept of **Adjustment Rate** (AR) [4], a parameter designed to quantify the cumulative effect of stochastic weight variations on network accuracy. To model D2D variability, a **Variable Neural Network** (VNN) is employed, incorporating controlled stochastic changes in synaptic weights using a Gaussian distribution with standard deviation (σ). Unlike traditional models where σ is applied uniformly across the network, the VNN adjusts σ at the individual neuron level, influenced by the AR. This approach enables a more nuanced investigation of how varying levels of stochasticity affect network performance. Moreover, the VNN framework provides a versatile platform to study the behavior of ANNs with respect to robustness and stability, as it can be easily adapted to various ANN topologies, ranging from simple perceptrons to more complex **Multi-Layer Perceptron (MLP)** architectures.

The experimental data utilized in this study is derived from previous works [5], [6]. N. Dersch et al. provide a model of a **Single-Layer Perceptron** (SLP), which simulates the behavior of SLP at the hardware level [5]. V. Milo et al., on the other hand, uses a MLP to evaluate the impact of C2C and D2D variation on ANN performance [6]. These experimental values form the foundation for the analysis presented here, enabling a comprehensive investigation of the relationship between weight variation and network accuracy.

To evaluate the effects of these stochastic variations at the system level, the framework is applied to two different architectures: a SLP based on [5] and a more complex MLP with one hidden layer based on [6]. The role of *stochastic quantization* is also investigated, which introduces randomness into the rounding process to mitigate precision loss. The findings indicate that the SLP exhibits a more unstable accuracy decline as AR increases, while the MLP shows a smoother degradation. Furthermore, stochastic quantization improves MLP performance but has minimal effect on the SLP, suggesting that network complexity plays a key role in determining the effectiveness of quantization techniques.

II. Methodology

A. Neural Network Architectures

This study applies the proposed framework to two neural network architectures trained on the MNIST dataset with its standard split of 60,000 training images and 10,000 testing images use to train ANNs [7].

The first architecture is a hardware-level simulation of a **SLP**, following the model presented in [5] and illustrated in Fig. 1a. It consists of 784 input neurons and 10 output neurons, with a *linear* activation function. This network achieves a software accuracy of 84.49%, which drops to 82.6% after quantization with 21 levels, ranging from [-1, 1] and a phase of 0.1. With 9 levels, using the same range and a phase of 0.25, the accuracy further decreases to 48.63%.

The second architecture is a more complex **MLP** with one hidden layer, based on the work in [6] and depicted in Fig. 1b. The network processes downsampled 14x14 MNIST images (196 input neurons), with 20 hidden neurons and 10 output neurons. It uses a *sigmoid* activation function between the input and hidden layer. The software accuracy reaches 92.70%, which decreases to 88.12% after quantization with the same conditions as the SLP, and further drops to 82.79% under the same settings.

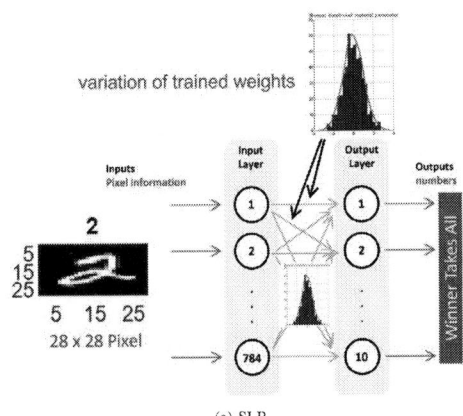

(a) SLP

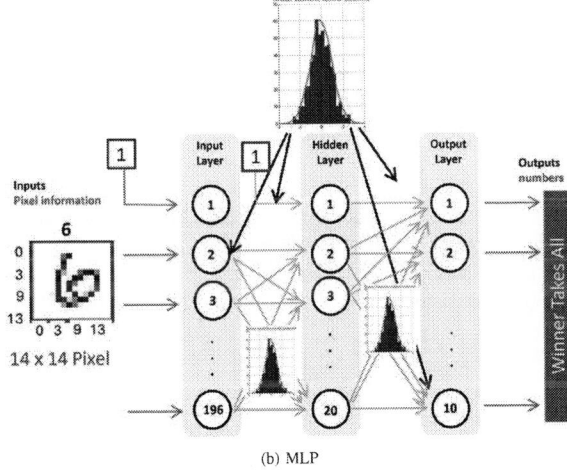

(b) MLP

Fig. 1. (a) SLP is modeled as an MLP without a hidden layer. (b) The MLP includes one hidden layer. In both cases, variability is applied to a proportional number of weights depending on AR. This variability is uniform across all affected weights in C2C mode or differs for each weight in D2D mode.

B. VNN Model

The VNN is used in this study to simulate the impact of stochastic weight variations on network accuracy. The VNN introduces controlled randomness into the synaptic weights by applying Gaussian noise with a with a standard deviation σ. The magnitude and scope of this perturbation are governed by the AR, which defines the proportion of weights in the network that are affected. Specifically, the number of weights to be modified is directly proportional to the AR; for example, an AR value of 1 implies that all weights within the ANN are perturbed, while an AR of 0 indicates no stochastic variation. Intermediate values affect a proportional subset of the weights.

Two operational modes are considered for modeling variability:

- **C2C mode**: C2C variation describes the fluctuations in the conductance of a RRAM device across repeated programming cycles, caused by stochastic differences in filament formation and dissolution [3]. In this mode, these variations are modeled by applying an uniform σ across all affected weights, with the mean values mapped to the respective synaptic weights. The Monte Carlo sampling procedure results in random variations that reflect C2C behavior. However, the magnitude of these fluctuations is generally smaller than the D2D variation, consistent with experimental observations.

- **D2D mode**: D2D variation represents the differences in conductance between multiple devices, caused by variations in filament caused by local process variations. In this mode, each synaptic weight is assigned a distinct σ, reflecting realistic D2D fluctuations. These variations are mapped onto individual synaptic weights with different mean values, resulting in a diverse spread of conductance values that captures intrinsic hardware variability. The D2D mode thus accounts for broader, more pronounced stochastic deviations compared to C2C, significantly impacting the overall network behavior.

This flexible modeling framework allows for systematic analysis of ANN robustness and stability under varying degrees of stochastic disturbance and can be easily adapted to different ANN architectures, including various forms of MLPs.

C. Stochastic Quantization

Quantization reduces the precision of weights to match hardware constraints. This study employs **stochastic quantization**, where rounding decisions are probabilistic rather than deterministic as is also implement in [8]. This technique has been shown to mitigate precision loss in deep networks. We apply quantization to both architectures and evaluate its impact on accuracy. The quantization process is implemented as follows:

$$\Delta_{i,j} = |a_i - l_j|, \quad \text{where } a_i \in A, \ l_j \in L \quad (1)$$

$$\lambda = \max\left(0.005, \min\left(0.05, \frac{\sigma(A)}{\mu(|A|)}\right)\right) \quad (2)$$

$$p_j = e^{-\frac{\Delta_{i,j}}{\lambda \cdot \sigma(\Delta)}} \quad (3)$$

$$\tilde{p}_j = \frac{p_j}{\sum_{j=1}^{M} p_j} \quad (4)$$

$$\hat{a}_i = l_j \quad \text{if} \quad C_{j-1} < r \le C_j \quad (5)$$

Here, $\Delta_{i,j}$ (Equation 1) represents the absolute difference between each element in the input array (A) and the quantization levels (L). The λ (Equation 2) is dynamically adjusted based on the σ of A values and their mean. The p_j (Equation 3) represents the probabilistic distribution over quantization levels, which is normalized in Equation 4 to ensure a valid probability distribution. Finally, each element in A is quantized by randomly selecting one of the levels (r) according to the computed probability distribution, as shown in Equation 5, where the selection is based on the cumulative distribution C_j and a randomly sampled value r that determines which quantization level l_j is chosen.

D. Relative Range

The Relative Range (RR) quantifies the variation in accuracy by measuring the spread of values relative to their median. It is defined as:

$$RR = \frac{\max(X) - \min(X)}{\text{median}(X)} \tag{6}$$

where $\max(X)$ and $\min(X)$ denote the maximum and minimum observed accuracy values, respectively, and $\text{median}(X)$ represents the average accuracy over a given range. A higher RR indicates greater variability, while a lower RR suggests more stability in accuracy measurements.

E. Experimental Setup

To evaluate the influence of the Adjustment Rate (AR) on accuracy, a Monte Carlo Simulation (MC) with 10 independent runs per configuration was performed.

The AR values were varied as follows:

- For the SLP: $AR = 0, 0.1, 0.2, \ldots, 1.0$
- For the MLP: $AR = 0, 0.2, 0.4, \ldots, 1.0$

The following four scenarios were evaluated:

1) **C2C scenario:** The MLP was evaluated under C2C variation with a fixed $\sigma = 0.04$ and no quantization applied. This σ is chosen to be lower than the values shown in Table I.
2) **D2D scenario (no quantization):** The MLP was analyzed under D2D variation without quantization, using σ values from Table I. The values are applied according to the theoretical quantization level.
3) **D2D scenario with 9-level quantization:** Both SLP and MLP architectures were tested under D2D variation, applying 9 quantization levels. The σ used correspond to the values in Table I.
4) **D2D scenario with 21-level quantization:** D2D variation applied to both networks with 21 quantization levels, using σ values estimated from the ranges presented in Table I.

This comprehensive setup allows for assessing the acurracy of both ANN models to AR under different stochastic regimes and quantization conditions.

TABLE I.
QUANTIZATION LEVELS AND σ FOR BOTH NETWORKS

Quantization Level	SLP σ	MLP σ
$-1.00, 1.00$	0.04062	0.05700
$-0.75, 0.75$	0.04255	0.05858
$-0.50, 0.50$	0.04321	0.05938
$-0.25, 0.25$	0.04075	0.06020
0.00	0.05245	0.06364

Values for the SLP are taken from [5], and values for the MLP are derived from [6], with mapped normalized values based on RRAM hardware data.

III. RESULTS AND DISCUSSION

A. C2C vs. D2D MLP Simulation Without Quantization

The impact of AR on accuracy is analyzed for both C2C and D2D simulations of the MLP without quantization. In C2C, increasing AR reduces accuracy from 92.7% to a median of 91.55%, while the stability of σ suggests that the configuration remains predictable despite this decline. In contrast, D2D exhibits a smaller accuracy drop, reaching a median of 89.72%. Both trends are shown in Fig. 2, where C2C is labeled in blue and D2D in red. The impact of AR on SLP is studied in [4].

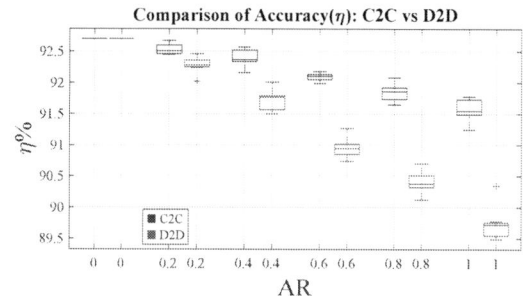

Fig. 2. MLP performance with C2C vs D2D

In D2D, accuracy also decreases with AR but remains higher than in C2C, dropping to a median of 89.72% (Fig. 2). Table II shows that RR decreases with AR, indicating more constrained accuracy fluctuations. Additionally, RR values for C2C are consistently lower than for D2D, suggesting greater stability.

TABLE II.
RR FOR MLP

AR	C2C	D2D	21 Levels	9 Levels
0.2	0.0024	0.0048	0.0045	0.0072
0.4	0.0044	0.0056	0.0049	0.0122
0.6	0.0021	0.0058	0.0121	0.0079
0.8	0.0047	0.0064	0.0090	0.0091
1.0	0.0058	0.0096	0.0094	0.0159

B. D2D Simulation of MLP at 21 and 9 Quantization Levels

The impact of AR on accuracy is analyzed for the D2D simulation of the MLP with quantization at two different levels: 21 and 9 levels. Increasing the quantization levels from 21 to 9 results in an accuracy improvement, from 88.12% to 82.79%. However, as AR increases, the accuracy decreases, reaching a median of 83.66% for 21 levels and 77.87% for 9 levels. As shown in Fig. 3, the 21-level configuration (blue) shows more stable accuracy than the 9-level configuration (red).

Table II presents the RR values, indicating that the RR for 21 levels is consistently lower than for 9 levels at all AR levels, suggesting greater stability with higher quantization. This trend aligns with the accuracy behavior observed in the figure.

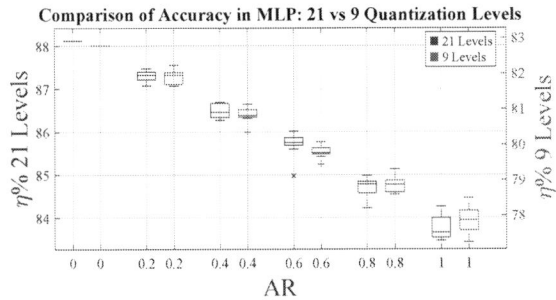

Fig. 3. MLP performance with 21 Quantization levels vs 9 Quantizatio levels

C. D2D Simulation of SLP at 21 and 9 Quantization Levels

In the D2D simulation results for the SLP, shown in Fig. 4, increasing the quantization levels from 21 to 9 results in a significant improvement in accuracy, from 82.6% to 48.63%. However, as AR increases, the accuracy decreases to a median value of 75.41% in the case of 21 levels and to a median of 46.85% with 9 levels. Additionally, the RR values presented in Table III show that the RR for 9 levels is always higher than for 21 levels, further emphasizing the higher stability of accuracy at the 21-level configuration.

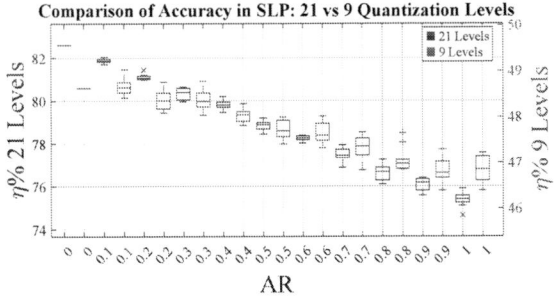

Fig. 4. SLP performance with 21 Quantization levels vs 9 Quantizatio levels

IV. CONCLUSION AND FUTURE WORK

This approach helps to identify the impact of synaptic weight variation and optimal quantization levels for hardware by analyzing their impact on accuracy, robustness, and AR. It enables selecting an optimized network architechture and

TABLE III.
RR FOR SLP

AR	21 Levels	9 Levels
0.1	0.0043	0.0125
0.2	0.0058	0.0138
0.3	0.0085	0.0153
0.4	0.0093	0.0098
0.5	0.0096	0.0124
0.6	0.0045	0.0145
0.7	0.0138	0.0175
0.8	0.0150	0.0170
0.9	0.0108	0.0192
1.0	0.0164	0.0177

practical quantization levels that balance performance and resource efficiency. Understanding AR's influence aids in hardware design, ensuring performance goals are met with minimal overhead, improving efficiency, robustness and cost-effectiveness.

Future work could focus on optimizing AR and quantization levels for different architectures, exploring their effects in complex models like U-Net, and assessing their impact on hardware implementations for neuromorphic computing applications.

ACKNOWLEDGMENT

The authors would like to thank the German Research Foundation (DFG) for funding this work under grant 546680029.

REFERENCES

[1] R. Raman, V. Kumar, B. G. Pillai, A. Verma, S. Rastogi, and R. Meenakshi, "Advances in multi-spectral image categorization through convolutional neural networks with an optimized activation function," in *2024 International Conference on Data Science and Network Security (ICDSNS)*. IEEE, Jul. 2024, pp. 1–6.

[2] E. Perez. M. K. Mahadevaiah, E. Perez-Bosch Quesada, and C. Wenger, "In-depth characterization of switching dynamics in amorphous hfo2 memristive arrays for the implementation of synaptic updating rules," *Japanese Journal of Applied Physics*, vol. 61, no. SM, p. SM1007, Jun. 2022.

[3] A. Kloes, C. Bischoff, J. Leise, E. Perez-Bosch Quesada, C. Wenger, and E. Perez, "Stochastic switching of memristors and consideration in circuit simulation," *Solid-State Electronics*, vol. 201, p. 108606, Mar. 2023.

[4] A. Blumenstein, E. Pérez, C. Wenger, N. Dersch, A. Kloes, B. Iñíguez, and M. Schwarz, "Exploring variability and quantization effects in neuronal networks using the MNIST dataset," *accepted to 11th Joint EuroSOI Workshop and International Conference on Ultimate Integration on Silicon (EuroSOI-ULIS 2025)*, 2025.

[5] N. Dersch, E. Perez-Bosch Quesada, E. Perez. C. Wenger, C. Roemer, M. Schwarz, and A. Kloes, "Efficient circuit simulation of a memristive crossbar array with synaptic weight variability," *Solid-State Electronics*, vol. 209. p. 108760, Nov. 2023.

[6] V. Milo, F. Anzalone, C. Zambelli, E. Perez, M. K. Mahadevaiah, O. G. Ossorio, P. Olivo, C. Wenger, and D. Ielmini, "Optimized programming algorithms for multilevel rram in hardware neural networks," in *2021 IEEE International Reliability Physics Symposium (IRPS)*. IEEE, Mar. 2021, pp. 1–6.

[7] L. Deng, "The MNIST database of handwritten digit images for machine learning research [best of the web]," *IEEE Signal Processing Magazine*, vol. 29, no. 6, pp. 141–142, Nov. 2012.

[8] D. Wu, Y. Wang, Y. Fei, and G. Gao, "A novel mixed-precision quantization approach for cnns," *IEEE Access*, vol. 13, pp. 49 309–49 319, 2025.

Design of Integrated Circuits and Microsystems

Mixed Design of Integrated Circuits and Systems – MIXDES 2025

CMOS OTA for Detector Readout Electronics Integrator in the ALICE FIT Project

Jakub Miszczyński[1], Piotr Otfinowski[1], Andrzej Laczewski[1], Michał Grzegorzek[1], Ireneusz Brzozowski[1],
Cezary Worek[1], Piotr Wiącek[1], Paweł Russek[1], Jacek Kitowski[1], Jacek Otwinowski[2]

[1] AGH University of Krakow, 30-059 Cracow, Poland

[2] Institute of Nuclear Physics, 31-342 Cracow, Poland

miszczynski@student.agh.edu.pl, potfin@agh.edu.pl, alaczewski@agh.edu.pl, mgrzegorzek@agh.edu.pl, brzoza@agh.edu.pl,
worek@agh.edu.pl, wiacek@agh.edu.pl, russek@agh.edu.pl, kito@agh.edu.pl, Jacek.Otwinowski@ifj.edu.pl

Abstract—This article discusses the Operational Transconductance Amplifier (OTA), which is a part of the integrator circuit for the new read-out electronics for the FIT detector in the ALICE experiment in CERN. The circuit is implemented using 180-nm CMOS technology within Cadence Virtuoso, and powered by a 3.3 V supply. This design features a two-stage, single-ended CMOS operational transconductance amplifier (OTA) incorporating cascodes, along with a load capacitance of 50 pF. This solution allows for the processing of nanosecond pulses across a broad dynamic range.

Keywords—OTA, integrator, read-out electronics, CMOS

I. INTRODUCTION

The FIT detector, part of the ALICE experiment at CERN, aims to explore the phenomena that occur at small exit angles of particles generated in heavy-ion collisions. For this, a collection of three FIT subdetectors (FT0, FV0, and FDD) is used to capture these events. In these scenarios, it was essential to develop a suitable circuit for reading and processing the data from these subdetectors. The electronics involved need to handle operations on a nanosecond analog pulse within a brief time frame (tens to hundreds of nanoseconds) frequently in an environment with high levels of noise. Circuits similar to ours are also being created for the needs of X-ray photon correlation spectroscopy or Gas Electron Multipliers (see Table I). The current read-out electronics used in the FIT-ALICE experiment consist of discrete integrated circuits (see Fig. 1) which result in large PCB area and substantial power consumption. This paper presents the design of an integrated, power-efficient and area-efficient circuit that allows for the integration of nanosecond pulses from photomultipliers. It is part of a new ASIC that enables more accurate and reliable measurements.

TABLE I
COMPARISON OF SIMILAR SOLUTIONS

Chip	SPHIRD [1]	GEMINI [2]	ABCDC1 [3]	ALPIDE [4]
Technology [nm]	40	180	250	180
Size [mm^2]	11.52	6.89	16	450
Power/chanel [mW]	0.0176	2.7	1.5	–

This work is co-financed in part by the Ministry of Science and Higher Education (Agreement Nr 2023/WK/07) and subvention for AGH

II. PRESENTLY USED INTEGRATOR

The present integrator circuit utilizes an operational amplifier configured for integration. Such a circuit is characterized by very high precision and linear integration. However, the problem here is the operational amplifier itself. It is responsible for the operating band of the circuit and for power consumption. Modern implementations have managed to reduce power consumption quite well, but this is often achieved by cutting the band. Due to the fact that the detectors in the experiment generate impulses with rise times of single nanoseconds, a circuit with a band of minimum 350 megahertz is needed.

In such a solution, the resistor at the input limits or forces a specific current, the capacitor in the feedback loop charges according to the incoming current, and the amplifier is responsible for maintaining the ground point, i.e. allowing all charge to flow into the capacitor [5]. Equation 1 illustrates the voltage-to-current transfer function of such a circuit.

$$V_{out}(t) = -\frac{1}{RC} \int V_{in}(t)\, dt \qquad (1)$$

In such a solution, resetting (discharging) the capacitor with a MOSFET transistor results in injection of charge into the measurement circuit, and it adds non-linear capacitance only an order of magnitude lower than the integrating capacitor [6].

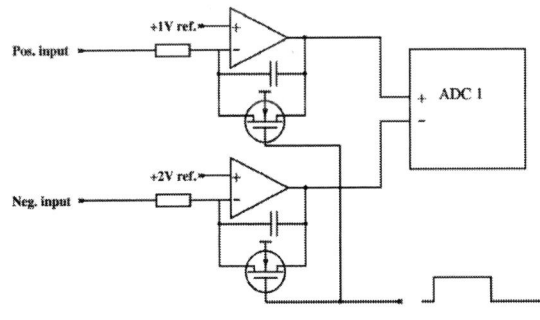

Fig. 1. Present integrator circuit

III. PROPOSED INTEGRATOR ARCHITECTURE

Another way to perform charge integration is to use the OTA as a voltage-current converter and a capacitance at the output. In this solution, the input voltage controls the OTA that generates a current proportional to the input voltage (Eq. 2). The capacitor is charged by this current, allowing one to measure the voltage across it, which corresponds to the input pulse (see Fig. 2).

$$v_{\text{out}}(t) = -\frac{G_m}{C} \int_0^t v_{\text{in}}(\tau)\, d\tau \qquad (2)$$

The discharge of such a circuit could be realized by a controlled current source, which solves the problem of an unstable capacity of the original solution.

This solution is characterized by a lower accuracy in relation to the one presented in Section II, and because the circuit is in open loop, it requires the use of several calibration solutions [7]. However, it allows us to obtain a very wide band and reduce power consumption.

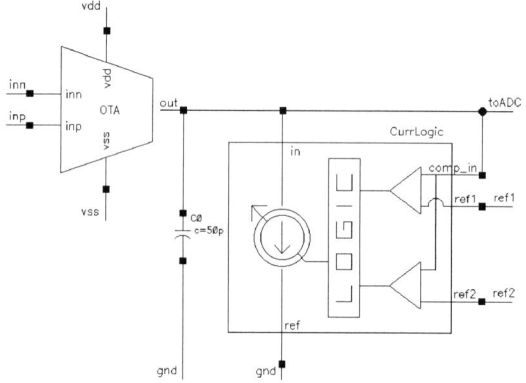

Fig. 2. Proposed integrator circuit

IV. PROPOSED OTA

Several requirements were set for the designed OTA:

- 350 MHz bandwidth - to be able to process nanosecond pulses,
- linearity for input pulses in the range of 3mV-250mV,
- stable output voltage (after integration) for 50 ns,
- amplitude of output signals of the order of tens/hundreds of millivolts,
- relative insensitivity to Monte Carlo scatter.

To meet such requirements, several approaches are needed. High bandwidth will be achieved by use of high currents. Linearity will be achieved when the input differential pair transistors can freely change voltage to current (this requirement is crucial and will be further developed later in the article) [6]. Unfortunately, two other requirements are mutually exclusive. To achieve signal stability at the output (not discharging it), a large RC constant is needed. This constant consists of the load capacitance together with parasitic capacitances seen from the output side and the output resistance of the circuit. To achieve a large RC, we can enlarge C and reduce the current in the circuit by raising the resistance. However, such a procedure reduces the amplitude of the voltages at the output - a low current will not charge a large capacitance to high

voltages. Therefore, a trade-off between these parameters must be established.

The initial phase involves the selection of a suitable OTA architecture. These requirements are met by both the symmetric CMOS OTA architecture with cascodes and the folded cascode [8]. For the needs of this work, the first option was chosen (Fig. 3), because it is slightly easier to implement and optimize.

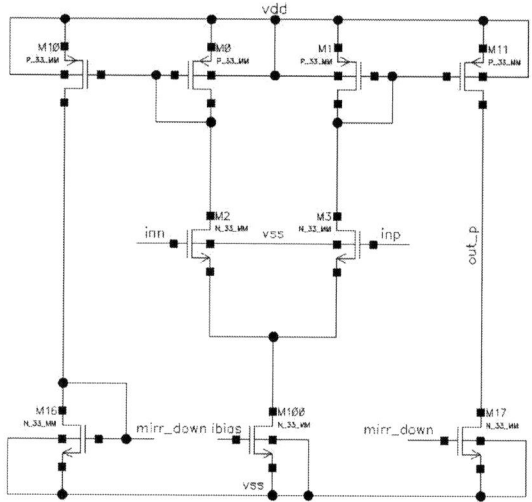

Fig. 3. Scheme of the designed OTA - basic

In the first step, the simplest version of OTA was designed. This streamlined variant was put to work in simulations, but it did not meet all the requirements. The relatively small output resistance resulted in a fast discharge of the voltage on the capacitor (about 10 percent of the maximum value in 50 ns for the largest pulse). The circuit was also not optimized in terms of other parameters, but it was the basis for the target appearance of the OTA. At this point, cascodes were added and the circuit was tuned to the requirements mentioned above.

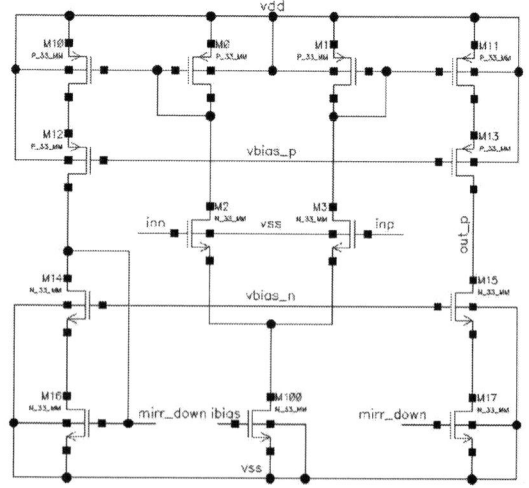

Fig. 4. Scheme of the designed OTA - final

The final transistor sizes and multiplications can be seen in Table II. The majority of the transistors possess uniform dimensions, with variations manifesting solely in their multiplication. This procedure allows us to apply simple principles of good layout (common center geometry, neighborhood, repeatability [6]) only. Thanks to this solution, the scatter caused by the imperfections in the process is reduced.

TABLE II
DIMENSIONS AND MULTIPLICATION OF THE TRANSISTORS

Name	W [μm]	L [μm]	Multiplication
M0/M1	1	0.4	500
M2/M3	1	1.5	100
M10/M11	1	0.4	100
M12/M13	1	0.4	40
M14/M15	1	0.4	40
M16/M17	1	0.4	40
M100	1	0.4	500
M101	1	0.4	50

The circuit was optimized in several steps.

- Linearity/input pulse range – as mentioned earlier, the 1st stage of the circuit is mainly responsible for these features. To correctly integrate the pulse, the current on transistor M2 could not reach 0 mA (as can be seen in Fig. 5. two pulses are within the range and the circuit correctly converts voltage to current; however, for the third - green pulse – the range is too small and the current is cut off around 0 mA). The sizes of the input pair transistors and the current that flows through them were selected so that pulses up to 500 mV were correctly integrated. It was also taken into account that the circuit worked correctly when MC mismatch was considered. This required the use of a current of several milliamps and relatively large transistor dimensions.

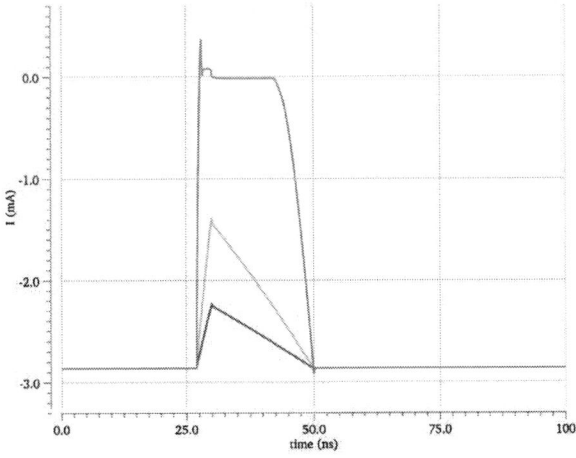

Fig. 5. Current in the branch with transistor M2 (Red - input pulse amplitude 100 mV, Yellow - input pulse amplitude 200 mV, Green - input pulse amplitude 1 V)

- Current bandwidth – in this solution, the OTA converts voltage to current; therefore, for the circuit to function correctly, it must be ensured that the circuit will perform this operation regardless of the input signal frequency. However, the current bandwidth should be relatively high, at the level of minimum 350 megahertz. This measurement was carried out by shorting the OTA output with a 1 ohm resistor to ground. The current was then measured as a function of the input voltage frequency. The bandwidth was considered to be the point at which the current value dropped by 3 dB. The current bandwidth is not the same as the standard bandwidth of the circuit, but simulations have shown that increasing one simultaneously increases the other. The OTA designs achieve a very large bandwidth compared to other solutions [9] [10]. It can be only surpassed by the pure current architectures [11].

- Phase margin – in such an architecture is determined by one non-dominant pole. It lies in the current mirrors (nodes between M0 and M10 and M1 and M11). A detailed explanation of why two nodes create one pole can be found in the book *Analog Design Essentials* [8]. Its value was determined using Equation 3 (where B is the ratio of the size of transistor M11 to M1).

$$f_{nd} = \frac{gm_{M1}}{2\pi(1+B)(C_{GS_{M1}} + C_{DB_{M1}} + C_{DB_{M3}})}, \quad (3)$$

By simplifying the Equation 3, the ensuing relationship is derived:

$$f_{nd} = \frac{gm_{M1}}{(3+B)2\pi C_{GS_{M1}}} \quad (4)$$

Substituting the actual numbers yields the following outcome:

$$f_{nd} = \frac{0.0151}{(3.2) * 2\pi * 814 * 10^{-15}} = 923 \text{ MHz} \quad (5)$$

For the circuit to maintain stability of at least 70 degrees phase margin, the non-dominant pole should be at least 3 bigger than GBW. The GBW of the circuit determined according to the Equation 6 is 4.03 MHz

$$GBW = B\frac{gm_{M2}}{2\pi C_L} = 0.2\frac{6.36mS}{314pF} = 4.03 \text{ MHz} \quad (6)$$

This means that the non-dominant pole is sufficiently distant and the circuit should remain stable above 70 degrees.

- Discharge constant - assuming that the capacitor discharges only through the OTA output resistance, the RC constant can be determined (Eq:7, Eq:8, Eq:9).

$$V(t) = V_0 * e^{-\frac{t}{RC}} \quad (7)$$

$$xV_0 = V_0 * e^{-\frac{t}{RC}} \quad (8)$$

$$\ln(x) = -\frac{t}{RC} \quad (9)$$

The circuit should discharge no more than 5% within 50 ns. Let's assume a capacitance of 50 pF. Substituting these values into the formula results in the resistance seen from the output side to be about $19.5k\Omega$ (Eq:10, Eq:11, Eq:12).

$$R = -\frac{1000}{\ln(x)}\Omega \qquad (10)$$

$$R = -\frac{1000}{\ln(0.95)}\Omega \qquad (11)$$

$$R = 19.496k\Omega \qquad (12)$$

The optimized OTA was simulated using Cadence Virtuoso. When the stimulation was completed, the layout was created as shown in Figure 6. It was created according to the principles of good layout [6] and its dimensions were $350\mu m$ by $125\mu m$.

Fig. 6. Layout of OTA

V. SIMULATION RESULTS

The simulation results and comparison with other circuits are given in Table III. The transfer characteristic can be seen in Figure 7.

Comparison of the schematic (a) and post-extraction (b) simulations shows small differences. This confirms that the layout was done correctly and that parasitic elements have no big impact on these OTA parameters. Based on the comparison to the currently used discrete amplifier, the circuit meets its constraints – it is much smaller and draws less power. However, this comes at the cost of the smaller GBW, which is not a critical parameter in this application.

TABLE III
SUMMARY OF SIMULATION RESULTS AND PERFORMANCE COMPARISON

Parameter [units]	(a)	(b)	(c)	[12]	[13]
CMOS process [nm]	180	180	XFCB	130	180
Supply voltage [V]	±1.65	±1.65	±5	±0.5	1.8
Capacitive load [pF]	50	50	–	120	200
DC gain [dB]	53.31	52.91	63	61.38	72
PM [o]	90.81	90.43	56	60	50
GBW [MHz]	7.978	7.908	410	7.53	86.5
CMRR @DC [dB]	66.5	65.6	90	77	–
PSRR+ @DC [dB]	66.1	65.3	74	63	–
PSRR- @DC [dB]	46.2	45.1	74	75	–
Power [mW]	22.8	22.5	190	0.03	11.9
Area [mm2]	–	0.044	9	0.009	0.070
FoM [MHz·pF/mA]	57.7	58	–	30120	2616.8

(a) Simulation (b) Post-extraction simulation (c) Discrete amplifier

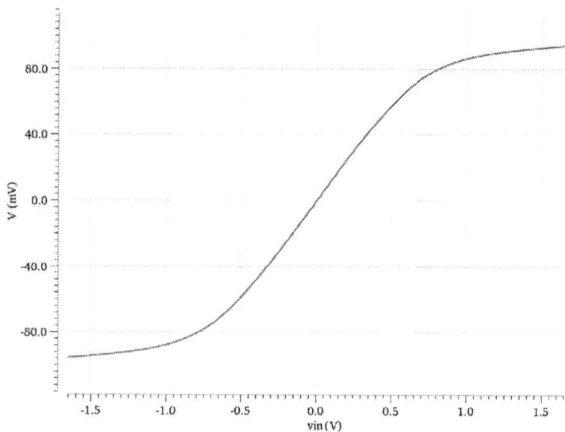

Fig. 7. Transfer curve – Post-extraction simulation – Load 50Ω

Other OTAs, in submicrometer technologies, optimize their parameters differently [12] [13]. However, most of the parameters are of similar order.

The parameters in Table IV are given for pulses in the range of 1 mV to 500 mV with a rise time of 3 ns and a fall time of 21 ns. Transconductance (Gm) and discharge loss were measured for a 500 mV pulse. The response of the OTA to the previously mentioned impulses is visible in Figure 8.

TABLE IV
INTEGRATION PARAMETERS - WORST CASE

Parameter [units]	typical typical	slow slow	fast fast
BW current [MHz]	850.4	536.2	1030
Linearity [%]	3.11	2.22	2.63
Discharge loss [%]	0.41	1.85	4.96
Output base line [mV]	-683	-614	-810
Gm [mS]	2.27	1.49	2.76

typical typical - 25°C, vph no change;
slow slow - 125°C, vph - 10%;
fast fast - 0 °C, vph + 10%

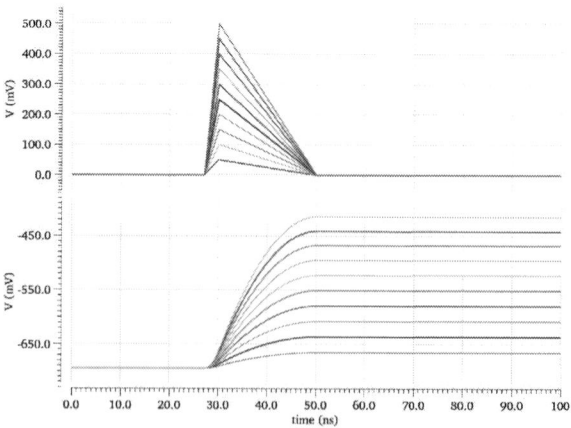

Fig. 8. OTA's response to 5mV - 500mV pulses

The results of the Monte Carlo simulation are presented in Table V, Table VI and Fig. 9.

TABLE V
INTEGRATION PARAMETERS - MONTE CARLO

Parameter [units]	mean. val.	std. dev.
BW current [MHz]	849.9	10.69
Amplitude [mV]	5.03	0.71
Discharge loss [%]	22.26	20.91
Output base line [mV]	-282	1070

Values for 10 mV pulse

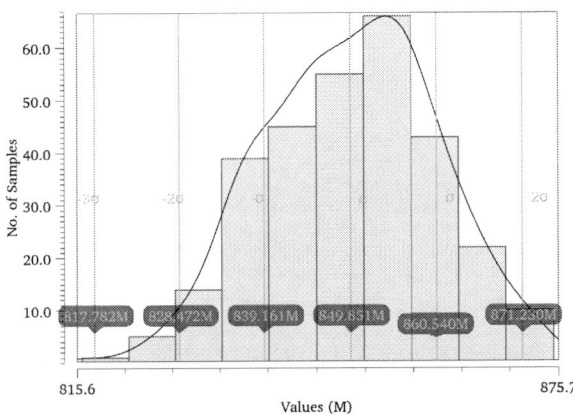

Fig. 9. OTA's current bandwidth - Monte Carlo

Inspection of the Table III allows us to state that the presented OTA has substantial advantages over the discrete ADA4817 circuit. Its size is much smaller, it consumes less power. Also, comparing the current bandwidth in relation to GBW, the designed OTA has an advantage. These advantages are at the cost of the input voltages range reduction (the maximum power supply for this technology is 3.3V), but this problem can be solved by using the appropriate architecture of the entire integrator circuit. When comparing the designed OTA to other works, one must remember that this circuit was optimized for a very specific task. However, it can be asserted that the constructed OTA shows many similarities to other designs. The circuit also performs relatively well in different corners. Table IV presents simulation results in the most extreme cases, i.e. corner "slow slow" with low power and high temperature and "fast fast" with high power and low temperature. Linearity, discharge loss, and output base line differ slightly or are even better than the nominal simulation. Other parameters, such as current bandwidth, are significantly scattered, but remain within acceptable limits for the entire project.

The Monte Carlo simulation results require an appropriate explanation. As can be seen in Table V, the current bandwidth and the pulse amplitude are resistant to scattering. However, discharge loss or output baseline is subject to quite serious

fluctuations. The answer to the question of which element of the circuit causes such fluctuations is given in Table VI.

TABLE VI
INTEGRATION PARAMETERS - MONTE CARLO ONLY I STAGE

Parameter [units]	mean. val.	std. dev.
BW current [MHz]	849.9	10.69
Amplitude [mV]	5.65	0.08
Discharge loss [%]	4.28	2.82
Output base line [mV]	-509	-532

Values for 10 mV pulse

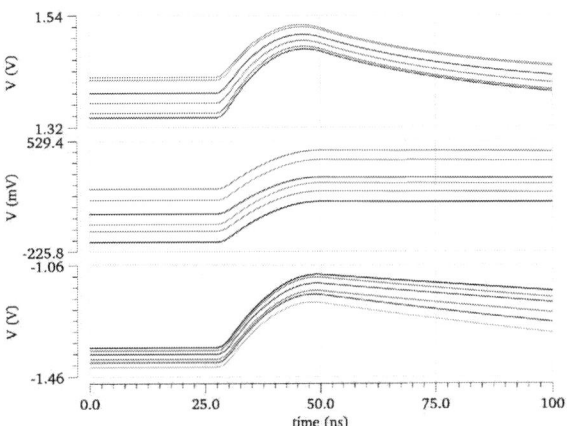

Fig. 10. OTA's response to impulse - Monte Carlo

Table VI results are taken from a simulation that took into account the scatter of transistors from the 1st stage of the OTA that did not directly affect the output. The output baseline is much closer to the middle of the power supply here, and this is crucial for the correct operation of the circuit. This fact is illustrated in Figure 10.

Small amplitude pulses (middle of Figure 10, around 0 V) suffer no problem with discharge; those close to the top or bottom of the power supply line (top and bottom part of Figure 10) discharge in a very short time. This results from the decrease in resistance seen from the output to Vdd or Vss, which significantly reduces the time constant of the circuit. Due to this phenomenon, the discharge loss parameter or the amplitude for larger pulses seems to be very scattered.

This is an expected and inevitable phenomenon in an open-loop OTA architecture [14]. The circuit operates correctly at its optimal operating point, which is approximately half the power supply.

To counteract the mismatch resulting from the Monte Carlo mismatch, an appropriate calibration system was used. Eliminates scatter in the second stage of the OTA by regulating the multiplication of M12 and M14 transistors (Figure 11).

Calibration works according to the principle of successive approximation and, as shown by simulations, allows the base line to be set at $\pm 100mV$ (Figure 12).

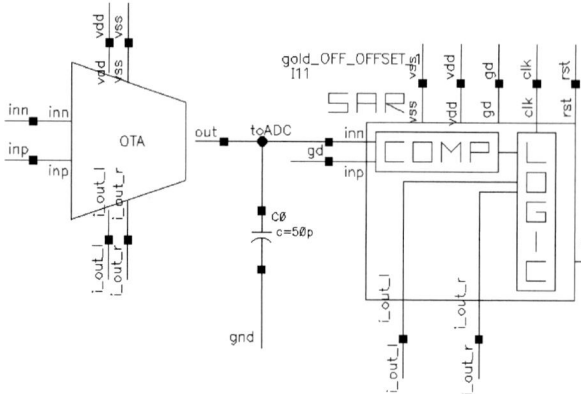

Fig. 11. Mismatch calibration circuit

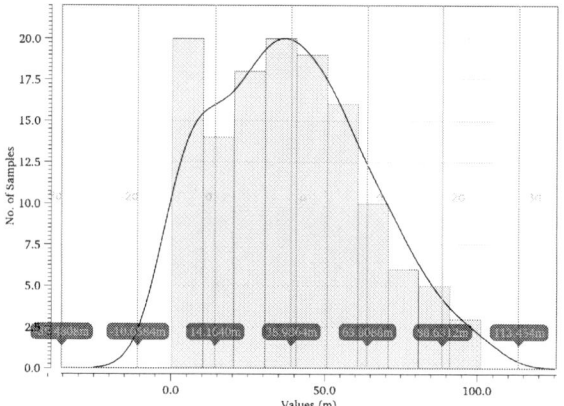

Fig. 12. Absolute value of base line mismatch after calibration - Monte Carlo

VI. CONCLUSION

This paper presents the design and simulations of the CMOS OTA for detector readout electronics integrator in the ALICE FIT detector. The circuit is implemented in CMOS 180 nm technology. To obtain a wide range of input voltages and linear integration, the highest power supply available in this technology and relatively high currents are used. In order to neutralize the negative influence of scattering, appropriate calibration systems were used. After determining the correct operating point, the circuit processes nanosecond pulses efficiently and accurately with a wide voltage range.

REFERENCES

[1] P. Grybos, R. Kleczek, P. Kmon, P. Otfinowski, P. Fajardo, D. Magalhães, and M. Ruat, "Sphird–single photon counting pixel readout asic with pulse pile-up compensation methods," *IEEE Transactions on Circuits and Systems II: Express Briefs*, vol. 70, no. 9, pp. 3248–3252, 2023.

[2] A. Pezzotta, G. Corradi, G. Croci, M. De Matteis, F. Murtas, G. Gorini, and A. Baschirotto, "GEMINI: A triple-GEM detector read-out mixed-signal ASIC in 180nm CMOS," in *2015 IEEE International Symposium on Circuits and Systems (ISCAS)*, pp. 1718–1721, ISSN: 2158-1525. [Online]. Available: https://ieeexplore.ieee.org/document/7168984/?arnumbe r=7168984

[3] J. Kaplon and W. Dabrowski, "Fast CMOS binary front-end for silicon strip detectors at LHC experiments," in *IEEE Symposium Conference Record Nuclear Science 2004.*, vol. 1, pp. 34–38 Vol. 1, ISSN: 1082-3654. [Online]. Available: https://ieeexplore.ieee.org/document/1462063/?arnumbe r=1462063

[4] C. Yang, C. Feng, J. Liu, Y. Teng, S. Liu, Q. An, X. Sun, and P. Yang, "A prototype readout system for the ALPIDE pixel sensor," vol. 66, no. 7, pp. 1088–1094, conference Name: IEEE Transactions on Nuclear Science. [Online]. Available: https://ieeexplore.ieee.org/document/8698804/?arnumbe r=8698804

[5] A. Rivetti, *CMOS Front-End Electronics for Radiation Sensors.* CRC Press, 2015.

[6] B. Razavi, *Fundamentals of microelectronics.* Wiley, 2013.

[7] S. Szczepanski, B. Pankiewicz, and S. Koziel, "Programmable feedforward linearized cmos ota for fully differential continuous-time filter design," *International Journal of Circuit Theory and Applications*, vol. 38, no. 9, pp. 885–899, 2010. [Online]. Available: https://onlinelibrary.wiley.com/doi/abs/10.1002/cta.602

[8] W. M. C. Sansen, *Analog design essentials*, ser. SECS. Springer, 2008, no. 859.

[9] Y. Tang, Y. Zhang, G. K. Fedder, and L. R. Carley, "An ultra-low noise switched capacitor transimpedance amplifier for parallel scanning tunneling microscopy," in *2012 IEEE Sensors.* IEEE, pp. 1–4. [Online]. Available: http://ieeexplore.ieee.org/document/6411083/

[10] G. Ferrari, F. Gozzini, A. Molari, and M. Sampietro, "Transimpedance amplifier for high sensitivity current measurements on nanodevices," *IEEE Journal of Solid-State Circuits*, vol. 44, no. 5, pp. 1609–1616, 2009.

[11] A. Imran, M. Hasan, A. Islam, and S. A. Abbasi, "Optimized design of a 32-nm cnfet-based low-power ultrawideband ccii," *IEEE Transactions on Nanotechnology*, vol. 11, no. 6, pp. 1100–1109, 2012.

[12] M. P. Garde, A. Lopez-Martin, J. M. Algueta-Miguel, J. Beloso-Legarra, and J. Ramirez-Angulo, "Energy-efficient symmetrical cascode ota in a 130 nm cmos process," in *2021 XXXVI Conference on Design of Circuits and Integrated Systems (DCIS)*, 2021, pp. 1–5.

[13] S. Sutula, M. Dei, L. Terés, and F. Serra-Graells, "Variable-mirror amplifier: A new family of process-independent class-ab single-stage otas for low-power sc circuits," *IEEE Transactions on Circuits and Systems I: Regular Papers*, vol. 63, no. 8, pp. 1101–1110, 2016.

[14] P. R. Gray, *Analysis and design of analog integrated circuits*, 5th ed. Wiley, 2010.

Design Considerations for Integrated SiGe BiCMOS Phase-Locked Loops in the Millimeter-Wave Band

Frank Herzel, Arzu Ergintav, Corrado Carta, Gunter Fischer

IHP - Leibniz-Institut für innovative Mikroelektronik
Im Technologiepark 25, 15236 Frankfurt (Oder), Germany
{herzel,ergintav,carta,gfischer}@ihp-microelectronics.com

Abstract—We present design guidelines for analog phase-locked loops (PLL) at millimeter wave (mmWave) frequencies in SiGe BiCMOS technology. Emphasis is placed on a robust functionality with a relatively constant phase noise performance under ionizing radiation in space. The analog tuning range of the voltage-controlled oscillator (VCO) is split into coarse and fine tuning. Using negative feedback in the fine tuning loop of the PLL, the fine tuning control voltage is kept close to the VCO gain maximum for a constant PLL loop bandwidth. Together with self-triggered sub-band switching, a long lifetime of the PLL is expected, since any VCO degradation will be compensated keeping VCO gain and loop bandwidth fairly constant. An integrated SiGe-HBT based phase detector is proposed, which allows mmWave phased-array transceivers to be driven by a low-jitter frequency source in the lower GHz range.

Keywords—Local oscillator, phase-locked loop (PLL), loop bandwidth, radiation hardness, SiGe BiCMOS, beam steering.

I. INTRODUCTION

Flexible local oscillators (LO) realized as phase-locked loops (PLL) are highly desirable in future space applications. An example is a W- to Ka-band frequency converter for ultra-high throughput satellite systems [1]. When realized as a fractional-N PLL, the LO output frequency can be changed in tiny steps [2], which is important for compensating long-term frequency degradation of the crystal oscillator driving the PLL. Silicon-Germanium (SiGe) technology combines high integration levels at low cost with extreme levels of transistor performance. SiGe HBTs showed only modest damage after multi-Mrad TID in total ionizing dose (TID) tests [3]. The frequency degradation of a 30 GHz SiGe VCO under X-ray irradiation was quantified in [4]. At a dose of 1 Mrad(SiO_2) the measured frequency reduction was as small as 60 MHz, which corresponds to 0.2 % of the VCO output frequency. SiGe BiCMOS technology is also widely used in mmWave radar systems. In [5], a low-jitter 76-81 GHz radar PLL with fast settling was presented, where an external opamp was used to generate VCO control voltages between 1.1 V and 9.3 V.

An integration of the VCO and the loop filter together with the PLL core would give a significant cost advantage. However, the low quality factor of the variable capacitors (varactors) in the VCO results in a relatively high phase noise at mmWave frequencies. In order to meet the stringent phase noise requirements in space applications, a large loop bandwidth is mandatory in order to filter out VCO phase noise as much as possible. This conflicts with the requirements to the in-band phase noise, which is low-pass filtered in a PLL.

Theoretically, the VCO phase noise can be reduced by using a large signal power in the VCO core. However, this is limited by the breakdown voltage of SiGe-HBTs, which is lower than that in III/V technologies. A possible way out could be the coherent superposition of several VCO signals in a phased-array transceiver [6],[7],[8]. This would increase the signal power of the downconverted signal in the receiver and improve the bit-error rate.

In order to meet the VCO phase noise and spur requirements at the same time, an integer-N mmWave PLL is preferred. A constant frequency multiplication factor, typically a power of two, would simplify the design and minimize the power consumption. Using several phase-aligned mmWave PLLs would further reduce the phase noise at the output of the receiver frontend. In order to achieve a programmable output frequency, the mmWave PLL (array) should be driven by a fractional-N PLL at a moderate frequency. The jitter performance of a hybrid OFDM system using such an architecture was analyzed in [9]. It was shown that the phase noise of the common low-frequency PLL can be reduced by one global pilot tracking loop, while the phase noise of the mmWave PLL array is much reduced due to noise averaging in conjunction with a large PLL bandwidth. A high frequency at the phase detector (PD) input was found to be essential for a low jitter of the mmWave PLL, since it allows a large loop bandwidth to be used for VCO phase noise reduction. This is because the in-band phase noise and the spurs of the input signal are amplified by $20 \log(N)$ dB, where N is the feedback divider ratio of the PLL. A high input frequency to the PD implies a small N reducing in-band phase noise and spurs. Unfortunately, high-voltage MOSFETs needed for the PD are rather slow in typical BiCMOS technologies compared to their bipolar counterparts. On the other hand, the existence of complementary devices makes the PD design easy by combining a phase-frequency detector (PDF) and a charge pump (CP) into an efficient PD. SiGe-HBTs are much faster, but the lack of a pnp transistor makes the PD design more complicated. Regardless of the PD choice, the VCO fine tuning voltage should be kept roughly constant to keep frequency settling and phase noise performance robust with respect to variations of device parameters with process, voltage and temperature (PVT).

This paper presents a robust 28.3-33 GHz integer-N PLL design in a 130 nm SiGe-BiCMOS technology. It describes and compares two different phase detector versions using MOS-

FETs or SiGe HBTs, respectively. A low VCO gain and a wide tuning range are achieved simultaneously by autonomous sub-band switching of the VCO using analog automated frequency calibration (AFC). The loop bandwidth is equalized over PVT variations by splitting the analog VCO tuning into coarse and fine tuning, with the latter being stabilized using negative voltage feedback within the loop filter.

II. GENERAL DUAL-LOOP PLL ARCHITECTURE

Fig. 1 shows a simplified block diagram of a dual-loop architecture. The VCO frequency is divided by N and then compared with a reference frequency in a phase detector. Here, the term "phase detector" means a circuit which converts the phase difference of the two input signals into a current or a voltage, which is then filtered to generate the two control voltages for the VCO. In integrated CMOS PLLs, the PD is typically composed of a phase-frequency detector (PFD) followed by a charge pump (CP). In BiCMOS technology, high-voltage MOSFETs must be used to generate sufficiently large control voltages. Their speed is typically not higher than a few hundred MHz. Therefore, the charge pump is often replaced with an active filter, which could be external to the chip as in [5] and [10], or fully integrated as in [4]. If VCO and PD are integrated with the PLL core, then the VCO gain usually varies significantly over the analog tuning range of the VCO. As a result, loop bandwidth and phase noise spectrum may strongly vary with PVT variations, if no precautionary measures are taken to keep the fine tuning voltage in a narrow range. The two PD outputs are low-pass filtered in LPF1 and LPF2 to obtain the fine and coarse tuning voltage, respectively. Typically, the fine tuning gain is significantly lower than the coarse tuning gain. This minimizes the phase noise contribution of the fast phase detector in the fine tuning loop. Since the coarse tuning loop is typically extremely slow, frequency settling and phase noise behavior are mainly determined by the fine tuning loop. This offers the opportunity to keep VCO gain, loop bandwidth and phase noise spectrum fairly constant by keeping V_{fine} within a narrow voltage range. Such a soft biasing of V_{fine} was used in [2] and [11] by using a simple resistive voltage divider at the charge pump output. Unfortunately, these biasing resistors

need some additional DC power and increase the sensitivity of the circuit to supply noise. The AFC block senses the strongly filtered coarse tuning voltage. V_{coarse}. Using two Schmitt triggers and a counter, it decides to decrease or to increase the VCO subband number (counted from lower to higher frequencies), if V_{coarse} is close to ground or to the supply voltage, respectively. Since the coarse tuning loop is always overdamped, an oscillating behavior of this analog AFC is excluded, which might otherwise occur in an underdamped single-loop PLL. More details on the AFC circuit are given in [11].

III. PLL ARCHITECTURE WITH CMOS PHASE DETECTOR

A. Circuit design

Fig. 2 shows a CMOS implementation of this idea. Unlike in [2] no resistive voltage divider between VCC and ground is used, which would draw a significant DC current. Rather, the negative feedback for the fine tuning voltage is established by a CMOS opamp with the positive input biased at VCC/2. If the filtered output voltage of CP1 is lower than this value, a high output voltage at the opamp is generated which is fed back to the filter input raising the DC level of V_{FINE}. The bandwidth of the opamp should be small in order to efficiently filter out a possible noise in the reference voltage, V_{REF}. The output resistance of R_F=10 kΩ ensures that the loop dynamics for the PLL phase is not too much influenced by this voltage feedback loop.

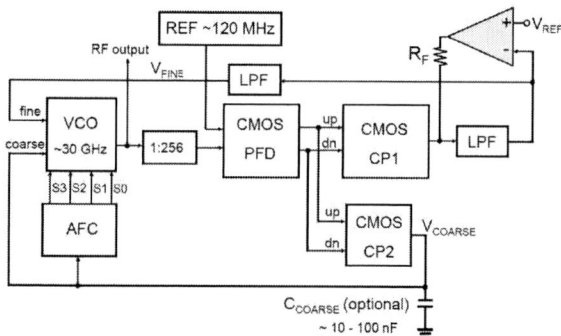

Fig. 2. Block diagram of a dual-loop PLL using a CMOS phase detector.

The circuits were designed in a 130 nm BiCMOS technology [12]. The VCO represents a differential Colpitts oscillator shown in Fig. 3. It employs four digital inputs to generate 16 subbands, where the highest bit is realized by inductor switching. In addition, the analog control input $Vctr$ used in [13] is split into two control inputs for coarse and fine tuning according to Fig. 2. The frequency divider represents a cascade of eight divide-by-two circuits (DTC) to simplify the design and to minimize the DC power consumption. Fig. 4 shows the measured output frequency of the free-running VCO as a function of the coarse tuning voltage for all 16 subbands. The neighboring subbands show a large overlap. This facilitates the AFC design and keeps the VCO coarse tuning voltage away from the rails to avoid overly large variations of the charge pump performance.

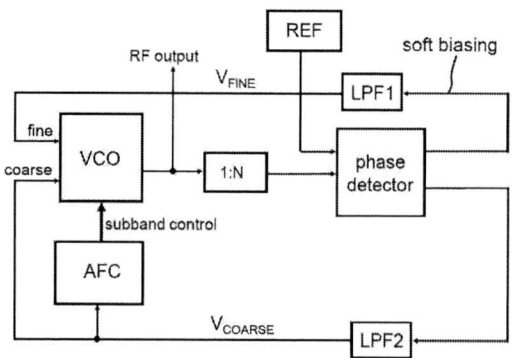

Fig. 1. Block diagram of the general PLL with two parallel analog tuning loops and automated frequency calibration (AFC).

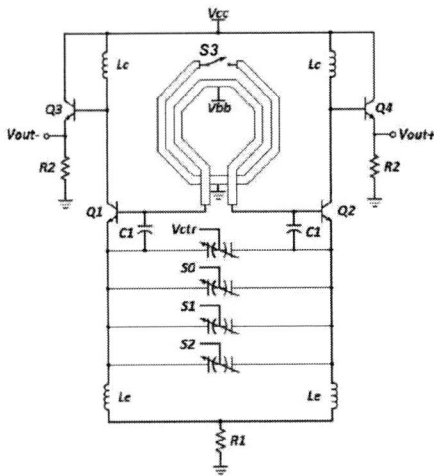

Fig. 3. Schematic of voltage-controlled oscillator taken from [13].

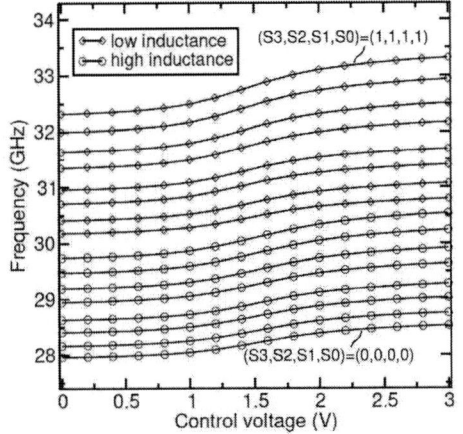

Fig. 4. Measured tuning curves of free-running VCO for the 16 subbands taken from [4].

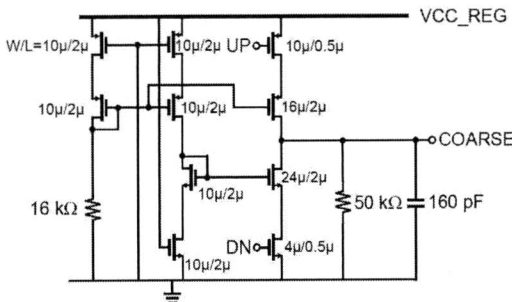

Fig. 5. Design of charge pump CP2 for analog coarse tuning of the VCO.

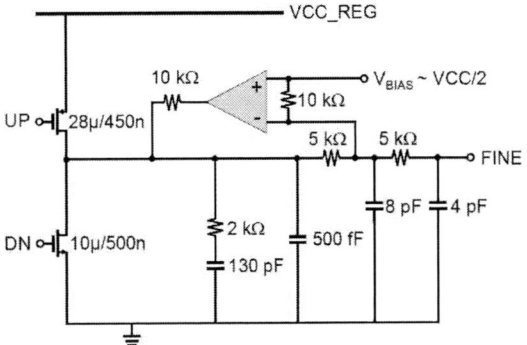

Fig. 6. Design of charge pump CP1 for analog fine tuning of the VCO followed by the two LPFs.

tuning loop. The simple two-transistor architecture results in a large gate-source overdrive voltage, $V_{\mathrm{ov}} = V_{\mathrm{GS}} - V_{\mathrm{th}} = \mathrm{VCC\_REG} - V_{\mathrm{th}}$, which will change only moderately with device aging or under irradiation, provided that the threshold voltage change is significantly smaller than $\mathrm{VCC\_REG} \approx 3\,\mathrm{V}$. Moreover, since the thermal noise of a MOSFET at a given drain current I_D is roughly proportional to I_D/V_{ov}, the noise contribution of this CP is relatively small.

B. Simulation results

We have simulated the whole PLL on transistor level, where the external 10 nF capacitor was reduced to 1 nF for shorter simulation times. Moreover, the actual VCO followed by the 1:256 feedback divider was replaced with an artificial 120 MHz VCO with the same relative tuning range as the 30 GHz VCO. Fig. 7 shows the simulated PLL frequency for different input frequencies. A relatively long slewing phase for charging C_{coarse} is followed by a fast exponential settling with a 1/e time constant around 1 µs, corresponding to a loop bandwidth of about 1 MHz / $(2\pi) \approx 150\,\mathrm{kHz}$. The settling of the coarse tuning voltage is depicted in Fig. 8 for the same input frequencies as in the previous figure. Since V_{coarse} is only slightly biased by the 50 kΩ resistor to ground, the steady-state voltage ranges from 0.4 V to 2 V. By contrast, the fine tuning voltage shown in Fig. 9 varies only slightly. As a result, the loop bandwidth is fairly constant throughout the band. This phenomenon has been experimentally verified in [2] and [11], where resistive biasing at the charge pump output of the 30 GHz fractional-N PLL had been employed.

Fig. 5 shows the design of CP2 including output loads. The task of the coarse tuning loop is the compensation PVT variations, whereas the fine tuning loop defines the small-signal dynamics and the phase noise behavior. The output of the CMOS CP in the coarse tuning loop is loaded with an on-chip capacitor to ground in parallel with an optional large external capacitor C_{COARSE}. The latter reduces the loop bandwidth of the coarse tuning loop to a negligible value and may improve phase noise and spurs, especially, at frequency offsets in the lower kHz range. The 50 kΩ load resistance prevents opposite oscillations of V_{coarse} and V_{fine}. If the PLL is used for beam steering, this resistor should be replaced with an array of binary weighted DC current sources to generate a digitally controlled DC offset current, I_{OS}. By changing this current, the PLL output phase (in rad) is changed by $\Delta\phi = 2\pi N \times (\Delta I_{\mathrm{OS}}/I_{\mathrm{CP2}})$ as described in [9]. If N is very large, the CP current rather than the offset current should be changed to avoid overly small offset currents.

The charge pump CP1 for VCO fine tuning is depicted in Fig. 6 including the two low-pass filters (LPF) in the fine

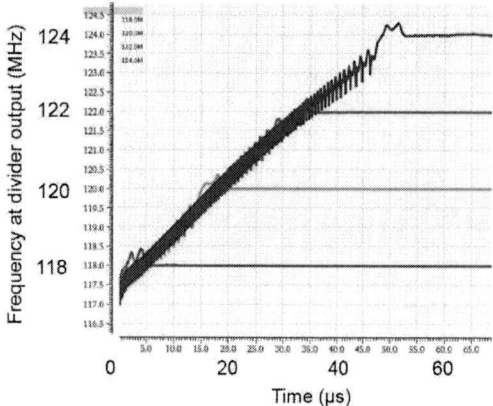

Fig. 7. Simulated settling of the frequency at the 1:256 divider output for different input frequencies.

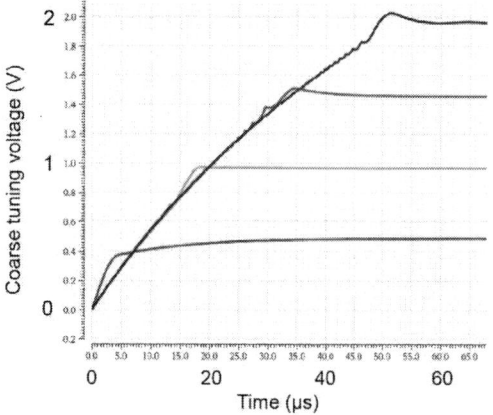

Fig. 8. Simulated settling of the coarse tuning voltage for different frequencies.

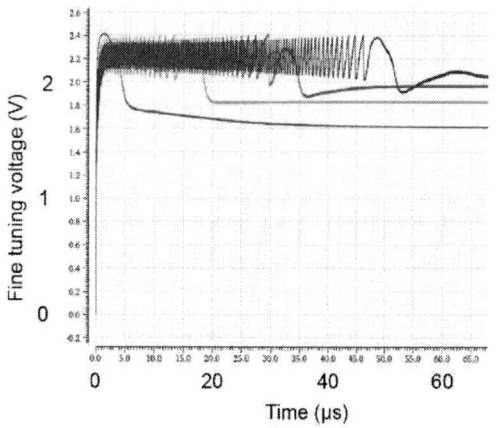

Fig. 9. Simulated settling of the fine tuning voltage for different frequencies.

IV. PLL ARCHITECTURE WITH BIPOLAR PHASE DETECTOR

A. Circuit design

In the previous section, a PD composed of high-voltage MOSFETs has been used in order to generate sufficiently large control voltages for the VCO. The speed of these devices is sufficient for driving a PLL with crystal oscillators, since their

frequency rarely exceeds 200 MHz. In [9] a PLL cascade was proposed for the mmWave frequency range, where a fractional-N PLL (PLL1) drives one or more mmWave PLLs (PLL2) with separate phase control for phased-array transceivers. This architecture has some important advantages over a crystal driven fractional-N mmWave PLL:

- The feedback divider ratio in the mmWave PLL is small reducing the contribution of the PD noise to the phase noise spectrum and allowing a large loop bandwidth for VCO phase noise suppression,
- as a result of the large loop bandwidth of PLL2, the static phase of the mmWave PLL can be switched very fast for supporting beam steering transceivers [9],
- high-frequency noise is reduced by noise averaging, if more than one mmWave PLL is used, and
- the level of fractional spurs is much lower.

The latter advantage results from the fact that a small loop bandwidth of PLL1 can be employed without increasing the phase noise contribution of the mmWave VCO. The phase noise contribution of the low-frequency VCO within PLL1 is less critical due to the high Q-factor of resonators in the lower GHz range. As a result, the PLL phase noise normalized to the output frequency is much lower at frequencies below 10 GHz compared to mmWave frequencies. Using an optoelectronic PLL, extremely low phase noise in the lower GHz can be achieved, see [14] and references therein. As an alternative, a dielectric resonator oscillator (DRO) supported PLL might be used for driving the mmWave PLL array [15].

Our architecture requires a PD working in the lower GHz range, which is not possible for high-voltage MOSFETs in a 130 nm BiCMOS technology. A 30 GHz PLL using an integrated bipolar PD was presented in [4], where a good radiation hardness was demonstrated experimentally. Unlike in that paper, we add a feedback mechanism here to stabilize the VCO fine tuning voltage and, thereby, the loop dynamics. Fig. 10 shows a block diagram of this PLL with an input frequency around 2 GHz. A bipolar PFD followed by an integrated differential bipolar amplifier is employed for a high speed. By using 1:16 frequency dividers, the input frequency to the CMOS PFD is reduced to about 125 MHz, a convenient frequency for the thick-oxide MOSFETs used for CMOS PFD and CP.

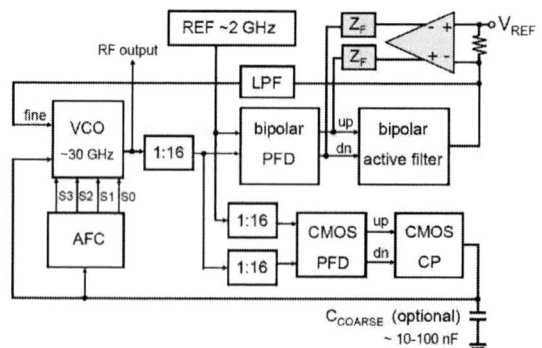

Fig. 10. Block diagram of a dual-loop PLL using a bipolar phase detector.

Fig. 11 shows the schematic of the bipolar PFD. It consists of two edge-triggered resettable D-flipflops designed with SiGe-HBTs for a high speed. Their D inputs are connected to logical ONE. A standard master-slave configuration with an additional reset input was used for the ECL flipflops. The delay element in the reset path eliminates the dead-zone effect. Each D-flipflop consists of two D-latches. Fig. 12 shows the schematic of a resettable D-latch in ECL logic. The latch consists of five differential pairs. The lowest pair receives the differential clock signal and is biased by a current mirror. If the differential reset signal R is LOW, then the current flows through the inner pairs, and the latch behaves like a classical D-latch. If the reset signal R goes HIGH, the current flows through the outer differential pairs and from there through the load resistor R_C connected to the positive output node "out". In other words, the differential output signal is LOW in this case, as desired.

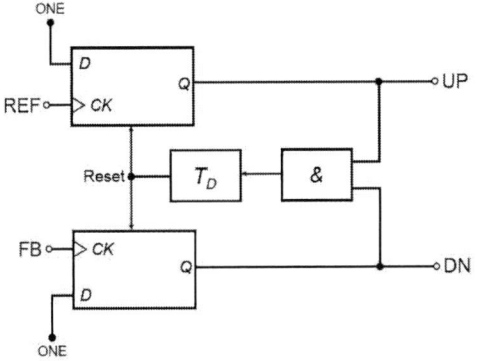

Fig. 11. Schematic of the bipolar phase-frequency detector. All signals are fully differential, but only the positive branches are shown.

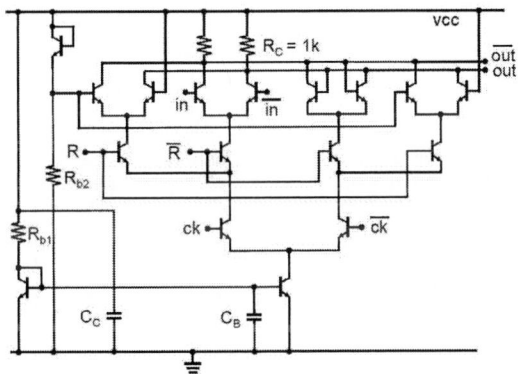

Fig. 12. Schematic of resettable D-latch.

An active loop filter as in [5] was used to avoid MOSFETs in the fine tuning loop. Fig. 13 shows the schematic of the active loop filter. Unlike in [5], our filter is completely integrated and does not require supply voltages higher than 3 V. Both the active filter and the feedback amplifier from Fig. 10 are two-stage differential bipolar amplifiers. Their shot noise contribution to the phase noise spectrum is minimized by using a moderate VCO fine tuning gain of $df_{out}/dV_{fine} \approx 150$ MHz/V at $f_{out} \approx 30$ GHz. The feedback impedance in Fig. 10, Z_F,

represents a low-pass filter consisting of a series resistor $R_F = 8$ kΩ in the feedback path and a capacitor $C_F = 10$ pF connected to chip ground. Another 8 kΩ resistance between the two inputs of the feedback amplifier is used to keep the fine tuning voltage close to V_{REF}. Fig. 14 shows the layout of the PLL chip.

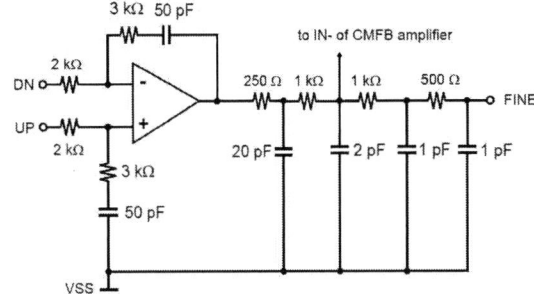

Fig. 13. Schematic of active loop filter according to Fig. 10.

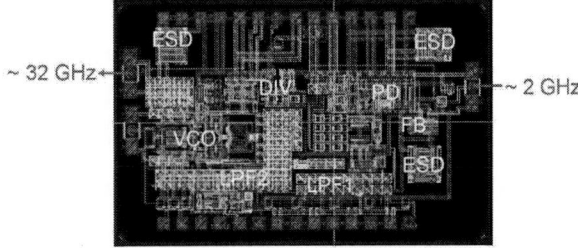

Fig. 14. Layout of the 32 GHz PLL using a 2 GHz input reference and a bipolar phase detector.

B. Simulation results

We have simulated the whole PLL on transistor level. Fig. 15 shows the input and output voltages of the feedback amplifier. Since the amplifier gain is about 15 dB, the distance of the fine tuning voltage V_{IN-} to the reference voltage V_{IN+} remains below 0.2 V. In the steady state, the coarse tuning voltage will slowly move to its final value, and the fine tuning voltage will approach the reference voltage of 1.95 V even more. As a result, fine tuning voltage, fine tuning gain and PLL loop

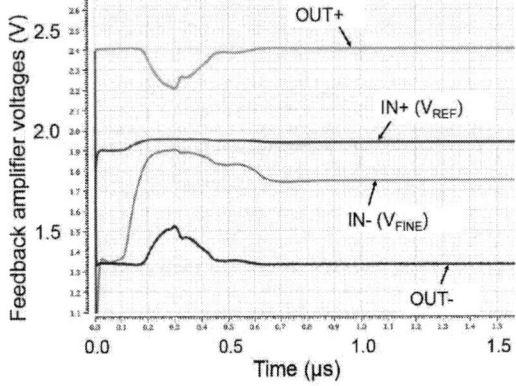

Fig. 15. Simulated input and output voltages of differential feedback amplifier.

bandwidth are fairly independent of the PLL output frequency. This results in a high robustness of the phase noise spectrum with respect to PVT variations.

The coarse tuning voltage is depicted in Fig. 16 together with the fine tuning voltage $V_{\text{FINE}} \approx V_{\text{IN}-}$. By using an external filter capacitor in parallel to the integrated one as indicated in Fig. 2, a low phase noise at very small frequency offsets can be achieved. The simulated frequency settling is shown in Fig. 17. After ramping up, the PLL output frequency error converges exponentially $\propto \exp(-t/\tau_c)$ with a characteristic time of about $\tau_c = 150$ ns. According to [9], this corresponds to a loop bandwidth of $f_L = 1/(2\pi\,\tau_c) \approx 1$ MHz.

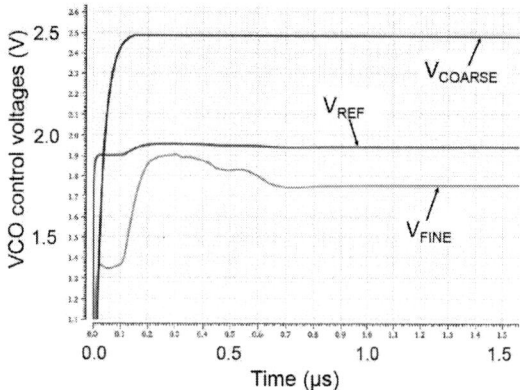

Fig. 16. Simulated VCO tuning voltages during frequency settling.

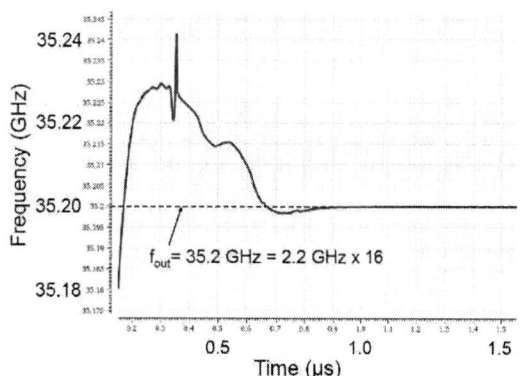

Fig. 17. Simulated PLL output frequency during frequency settling.

V. CONCLUSIONS

We have presented two versions of a 30 GHz PLL design using two parallel analog tuning loops. Unlike in previous designs, operational amplifiers are employed to keep the VCO fine tuning range within a narrow voltage range. This keeps loop bandwidth and phase noise spectrum fairly constant over PVT variations. The first version uses a CMOS PFD/CP phase detector and is limited to moderate frequencies at the PD input. By contrast, the second version uses a bipolar PD and can achieve several GHz at the PD input. The latter is especially suited for mmWave frequency synthesis, where one or several fast integer-N PLLs are driven by a common frequency source

in the lower GHz range. When combined with PLL phase shifting as described in [9], this approach may facilitate the low-noise design of phased array transceivers in the mmWave frequency range, e.g., for 6G wireless systems.

REFERENCES

[1] A. Barigelli, S. Di Nardo, F. Vitulli, E. Limiti, P. Longhi, L. Pace, and F. Deborgies. "W- to Ka-band frequency converter for ultra-high throughput satellite systems," in *Proceedings of the 17th European Microwave Integrated Circuits Conference (EuMIC 2022)*, Milan, Italy, Oct. 2022, pp. 344-347.

[2] F. Herzel, S. A. Osmany, K. Schmalz, W. Winkler, J. C. Scheytt, T. Podrebersek, R. Follmann, and H.-V. Heyer, "An integrated 18 GHz fractional-N PLL in SiGe BiCMOS technology for satellite communications." in *Proc. of 2009 IEEE Radio Frequency Integrated Circuits Symposium (RFIC 2009)*, Boston, USA, June 2009, pp. 329-332.

[3] J. D. Cressler, "Radiation effects in SiGe technology," *IEEE Transactions on Nuclear Science*, vol. 60, pp. 1992-2014, Jun. 2013.

[4] Falk Korndörfer, Frank Herzel, Thomas Mausolf, and Jörg Domke, "Precision measurements of total ionizing dose effects for a 28.3-33 GHz frequency synthesizer in SiGe-BiCMOS technology," 2024 24th European Conference on Radiation and Its Effects on Components and Systems (RADECS), Maspalomas, Spain, Sep. 2024, pp. 1-4.

[5] T. T. Braun, M. van Delden, C. Bredendieck, J. Schoepfel, and N. Pohl, "A low phase noise phase-locked loop with short settling times for automotive radar," in *Proceedings of the 11th European Microwave Integrated Circuits Conference (EuMIC 2021)*, London, UK, Apr. 2022, pp. 205-208.

[6] A. Karakuzulu, W. A. Ahmad, D. Kissinger and A. Malignaggi, "A four-channel bidirectional D-band phased-array transceiver for 200 Gb/s 6G wireless communications in a 130-nm BiCMOS technology." *IEEE Journal of Solid-State Circuits*, vol. 58, no. 5, pp. 1310-1322, May 2023, doi: 10.1109/JSSC.2022.3232948.

[7] T. Maiwald, et al., "A review of integrated systems and components for 6G wireless communication in the D-band," *Proceedings of the IEEE*, vol. 111, no. 3, pp. 220-256, Mar. 2023, doi: 10.1109/JPROC.2023.3240127

[8] L. Steinweg, J. Hebeler, T. Meister, T. Zwick and F. Ellinger, "8.0-pJ/bit BPSK transmitter with LO phase steering and 52-Gbps data rate operating at 246 GHz," *IEEE Transactions on Microwave Theory and Techniques*, doi: 10.1109/TMTT.2023.3239792.

[9] F. Herzel, C. Carta and G. Fischer, "Jitter minimization of phase-locked loops for OFDM-based millimeter-wave communication systems with beam steering," *2024 31st International Conference on Mixed Design of Integrated Circuits and System (MIXDES)*, Gdansk, Poland, 2024, pp. 118-123, doi: 10.23919/MIXDES62605.2024.10613942.

[10] T. Braun. M. Van Delden, C. Bredendiek, J. Schoepfel, S. Hauptmeier, W. Shillue, T. Musch, and N. Pohl, "A phase-locked loop with a jitter of 50 fs for astronomy applications," *International Journal of Microwave and Wireless Technologies*, vol. 15, no. 6. pp. 1012-1020, Jan. 2023, doi:10.1017/S1759078722001386

[11] A. Ergintav, F. Herzel, G. Fischer, D. Kissinger, and C. Carta, "A 30 GHz PLL with automated frequency control option for robust operation in harsh environments," in *Proceedings of the 19th European Microwave Integrated Circuits Conference (EuMIC 2024)*, Paris, France, Sep. 2024, pp. 34-37.

[12] H. Rücker, et al., "A 0.13 μm SiGe BiCMOS technology featuring f_T/f_max of 240/330 GHz and gate delays below 3 ps." *IEEE J. Solid-State Circuits*, vol. 45, no. 9, pp. 1678-1686, Sep. 2010.

[13] M. Kucharski, F. Herzel, H. J. Ng, and D. Kissinger, "A Ka-band BiCMOS LC-VCO with wide tuning range and low phase noise using switched coupled inductors," in *Proceedings of the 11th European Microwave Integrated Circuits Conference (EuMIC 2016)*, London, UK, Oct. 2016, pp. 201-204.

[14] M. Bahmanian and J. C. Scheytt, "Noise processes and nonlinear mechanisms in optoelectronic phase-locked loop using a balanced optical microwave phase detector," *IEEE Transactions on Microwave Theory and Techniques*, vol. 70, no. 10, pp. 4422-4435, Oct. 2022. doi: 10.1109/TMTT.2022.3197621.

[15] T. Mausolf, F. Herzel and G. Fischer, "An integrated circuit to reduce phase noise and spurious tones in radar systems," *2022 IEEE Nordic Circuits and Systems Conference (NorCAS)*, Oslo, Norway, Oct. 2022, pp. 1-5, doi: 10.1109/NorCAS57515.2022.9934454.

Design and Optimization of OTA-C Filters with Shared CMFB and Output Stages: Performance, Power and Area Analysis

Hubert Aleksiuk, Oskar Bogucki, Piotr Halman, Bogdan Pankiewicz
Department of Microelectronic Systems
Gdansk University of Technology
Gdansk, Poland

Abstract—This paper presents an optimized approach to designing OTA-C filters with a focus on reducing the number of components while maintaining or improving performance. By employing shared sub-blocks within the transconductance amplifier architecture, the proposed method minimizes power consumption and circuit area, ensuring efficient integration in modern electronic systems. The paper also discusses the theoretical basis for the design, implementation details, and performance evaluation of the optimized filters. The results demonstrate significant improvements in resource efficiency and scalability compared to conventional designs.

Keywords—OTA, OTA-C filters, optimisation, simulation, design, CMFB, shared sub-blocks, performance enhancement

I. INTRODUCTION

Despite significant advancements in technology, analog filters continue to play a vital role in numerous applications, ranging from communication systems to signal processing. One of the key factors determining the practicality of electronic circuits is power consumption. Reducing the energy usage of a circuit not only minimizes the size of the device but also extends its operational time on a single battery charge.

In recent years, significant advancements have been made in the design of analog filters, particularly in optimizing their performance while minimizing power consumption and component count. For instance, Kumngern et al. presented a fifth-order Butterworth low-pass filter tailored for ECG signal acquisition, utilizing a multiple-input operational transconductance amplifier (MI-OTA) [1]. Their approach demonstrated exceptional efficiency, achieving a power consumption of 34.65 nW while maintaining a wide dynamic range and high reliability through PVT and Monte Carlo analyses [1].

Similarly, Kulej et al. introduced a fully differential multiple-input OTA operating at ultra-low power levels, which enabled the realization of a fifth-order Chebyshev low-pass filter for bio-signal processing [2]. By leveraging bulk-driven MOS transistors in the subthreshold region, they achieved a power consumption of just 60 nW, showcasing the potential of MI-OTA in low-power filter applications [2].

Furthermore, Wyszynski and Schaumann highlighted the advantages of employing multiple-input OTAs in filter design, particularly in reducing the number of components, silicon area, and power dissipation [3]. Their findings emphasized the scalability of MI-OTA-based designs and their impact on enhancing the overall efficiency of OTA-C filters [3].

As shown above, various approaches to optimizing analog filters have lately been explored in the literature. These methods typically focus on reducing power consumption, improving frequency characteristics, or minimizing the number of required components. This paper builds on these concepts by proposing a novel design method for OTA-C filters. The approach involves replacing the integrated OTA amplifier with a version composed of sub-blocks, which allows for the elimination of redundant elements in higher-order OTA-C filters (above the second order). This study demonstrates that such a solution allows for the realization of filters with comparable noise performance and signal distortion, without any degradation in these characteristics, while significantly reducing the number of required MOS transistors and power consumption.

II. DESIGN OF OTA AND CMFB

A. Operational Transconductance Amplifier

To present the idea, the starting point is a simple, folded-cascode OTA, although more sophisticated OTAs are also applicable. Its structure was modified to allow it to be divided into two main sub-circuits: the input stage and the output stage. The input stage is responsible for converting the input voltage into current, while the output stage ensures a high output resistance of the amplifier [4][6]. The structure of the described circuit is shown in Figure 1.

The dimensions of the transistors and their operating points were selected to ensure that, with zero input signal, the

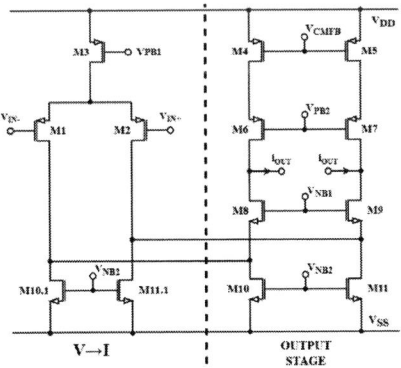

Fig. 1. The designed OTA structure with division into two sub-blocks.

current flowing between the left and right sides of the circuit is equal to zero. This approach isolates the input stage (comprising transistors M1-2, M10.1, M11.1) from the output stage (comprising transistors M4-9, M10.0, M11.0) without affecting the operating points of the transistors. As a result, it becomes possible to connect multiple input stages to a common output stage without affecting the stability of the operating points of the transistors across the entire circuit. An example of such a configuration with two input stages is shown in Fig. 2. This approach makes it possible to eliminate repetitive output stage blocks in designed OTA-C filters.

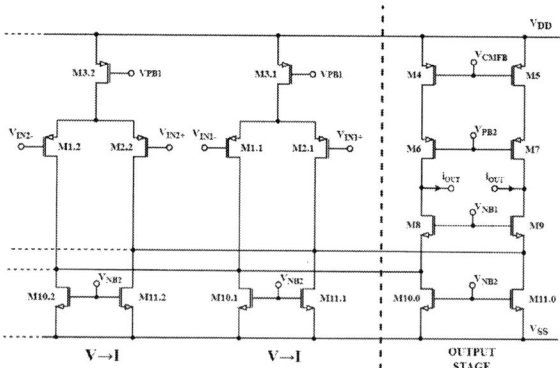

Fig. 2. Example of connecting two input stages to a shared output stage in the OTA-C circuit.

B. Common Mode Feedback Circuit

Since the utilized OTA features a differential output, its proper operation required the implementation of a common-mode voltage stabilization circuit at the output, the schematic of which is shown in Fig. 3. This structure was chosen primarily based on the possibility of sharing the block among the OTA output stages in the filter, which have their outputs connected together.

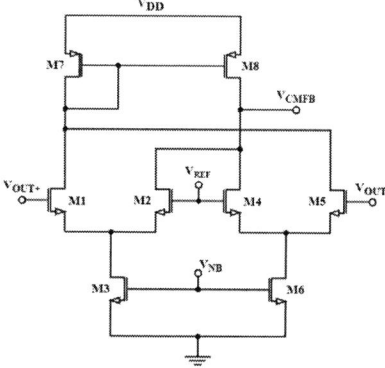

Fig. 3. Schematic of the common mode feedback circuit for the differential output OTA.

III. FILTER DESIGN

A. Design Method

The proposed filter design is based on a RLC ladder shown in Fig. 4 [5]. A 6th order low-pass Butterworth filter was chosen as a representative example due to its wide applicability and the balance of parameters and complexity. The 3-decibel frequency was set to 4.5 MHz as it best fits the OTAs capabilities and is used in real applications, most importantly in analogue audiovisual signals.

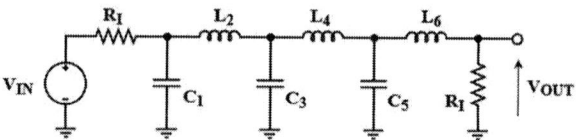

Fig. 4. RLC ladder 6th order filter prototype

Two OTA-C filter designs based on the RLC ladder above were chosen. One was obtained from a direct simulation of the RLC prototype, while the other is based on the signal flow simulation in the aforementioned RLC ladder. Both proposed filters were designed in three versions. The base variants were made from full OTAs, each containing the input stage, output stage, and the CMFB Circuit. Then, in the first approach, we introduced a design with shared CMFB circuits for each filter node, while maintaining OTAs composed of the input and output stages. Lastly, a filter incorporating shared CMFB circuits and output stages was developed, while OTAs from the previous step were reduced to only input stages. The schematics and details of these designs are shown in the paragraphs below.

B. Ladder-based OTA-C filter

The operating principle of this filter is based on simulating the filter's passive components. It directly implements the RLC ladder network shown in Fig. 4. To simulate the inductor, a gyrator topology with a capacitor was employed [5]. Fig. 5 shows the realization of this filter developed using full OTAs that integrate CMFB circuits.

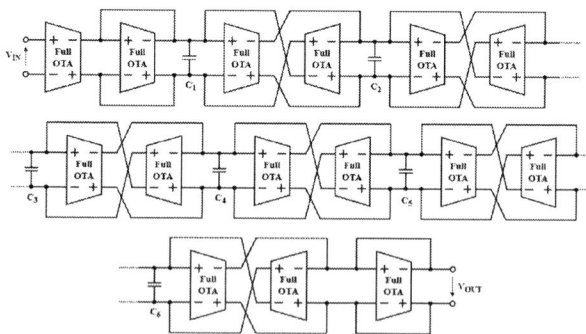

Fig. 5. Ladder based OTA-C filter composed of full OTA circuits.

C. Signal Flow OTA-C filter

The concept of the signal flow OTA-C filter's operation is based on the RLC ladder shown in Fig. 4. However, this filter structure is focused on the simulation of the flow of signals in the previously mentioned ladder and its principle of operation [5]. This approach, similarly to the ladder-based filter, also allows for using solely capacitors as passive components. The schematic diagram showing this type of filter is presented in Fig. 6.

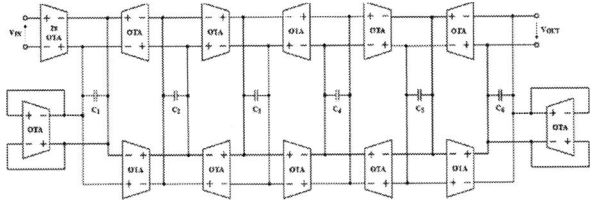

Fig. 6. Signal flow based OTA-C filter composed of full OTA circuits.

D. Optimisation method

The optimization technique applied to the designed filters is based on the division of an OTA into three sub-blocks: input stage, output stage and CMFB. This approach allows for flexible use of these sub-blocks as they can be connected separately.

The first step that was taken builds upon the fact that a CMFB circuit's function is to stabilize the average voltage in the OTA's output node. As its operation is to affect the whole output node, it is possible to use only one such circuit for each of the nodes, where multiple OTAs' outputs meet. Therefore, in this setup the OTAs were cut down only to input and output stages while the CMFB circuit was separated from it. The ladder-based and signal-flow based filters designed with this approach are shown in Fig. 7 and Fig. 8 respectively.

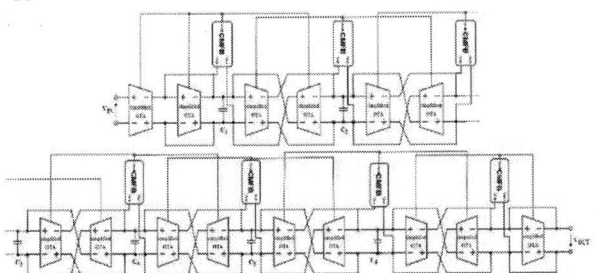

Fig. 7. Ladder based OTA-C filter composed of OTAs with shared CMFB circuits.

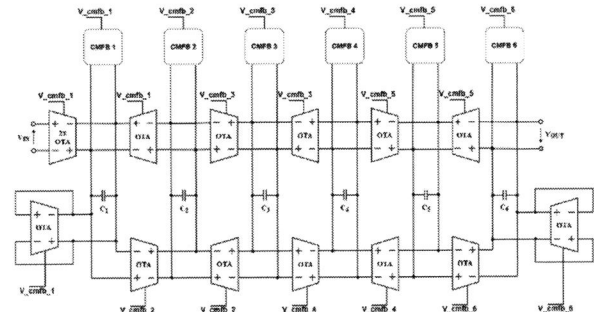

Fig. 8. Signal flow based OTA-C filter composed of OTAs with shared CMFB circuits.

The final filter structure was achieved by extending the method above. The introduced improvements result from the observation that the output stage plays mostly the role in output signal conditioning. It means that multiple input stages can be connected to one output stage, and still work the same as two full OTAs. Therefore, in this approach the OTAs were cut down only to input stages while the output stage and CMFB circuits were separated from it and combined into an output block. Then, the filters were designed with the output blocks shared for every node, in which the OTAs outputs were connected. The ladder-based and signal-flow based filters designed with this approach are shown in Fig. 9 and Fig. 10 respectively.

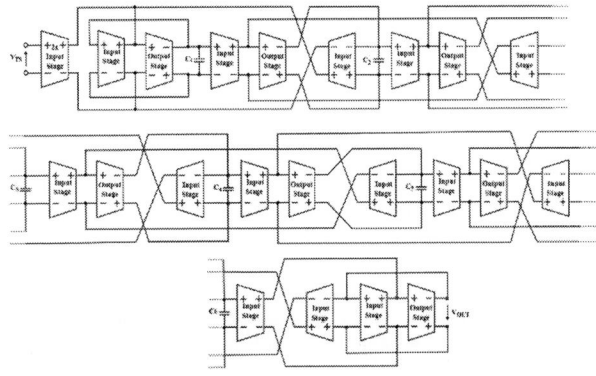

Fig. 9. Ladder based OTA-C filter composed of OTAs with shared output stages and CMFB circuits.

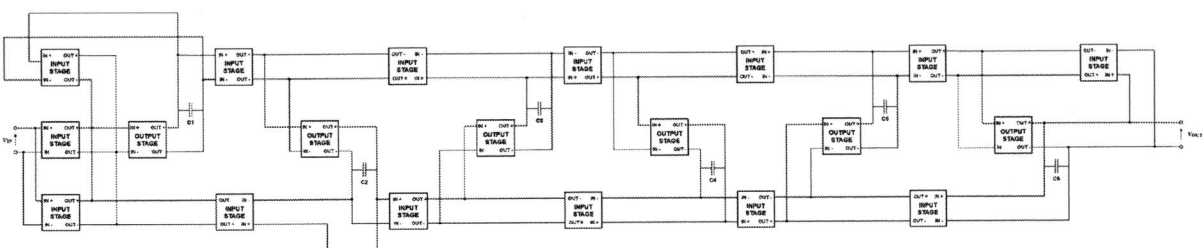

Fig. 10. Signal flow based OTA-C filter composed of OTAs with shared output stages and CMFB circuits.

E. Encountered problems

One of the significant challenges identified during the design process involved the initial transconductance (gm) value of the OTA, which was originally set to 10 μS. While this value was sufficient for the basic operational requirements of the amplifier, it proved inadequate for practical implementation in OTA-C filters targeting a 3-dB cutoff frequency of 4.5 MHz. Specifically, the low gm value resulted in capacitor sizes that were too small for accurate fabrication in integrated circuit technology, posing a significant limitation in achieving the desired performance.

To resolve this issue, the gm of the OTA was increased to 20 μS. This adjustment allowed for larger capacitor values, ensuring that the filters could be implemented with the necessary precision while maintaining stability and performance. However, this change required a reassessment of other design parameters to accommodate the increased transconductance.

Another challenge encountered during the design process was the reduction in effective transconductance observed in configurations utilizing shared output stages in the OTA. Simulations comparing designs with full OTAs and those with multiple input stages sharing a single output stage revealed a noticeable gm decrease of approximately 3.6 μS when three input blocks were connected in parallel. This reduction becomes more pronounced as the number of input blocks connected in parallel increases. This issue primarily stems from the parallel connection of input stages, which introduces a reduction in the output impedance of the input blocks.

The reduced output impedance leads to a portion of the output current from the differential pairs flowing back through the resistance, rather than being fully directed to the output stage. Consequently, the effective transconductance decreases with an increasing number of input stages connected to a shared output. This phenomenon necessitates adjustments in the design process to mitigate its impact on filter performance.

To address this issue, a careful re-calibration of the capacitors used in the OTA-C filters was required. Specifically, the capacitance values had to be decreased to compensate for the reduced transconductance, ensuring that the filters maintained their desired frequency and amplitude characteristics despite the observed transconductance drop. This adjustment highlighted the need for precise parameter tuning when designing filters with shared output stages, to balance power consumption, area optimization, and overall performance.

IV. RESULTS

In order to provide data concerning the filters' performance, area, and power consumption, appropriate simulations were performed. They were based on models provided with the GDPK 45nm technology, the transistors used worked with a power supply of 1.8V, and their dimensions are listed in Table I.

The simulationally measured quantities are presented in Table II. In both filter types – the ladder-based and signal-flow-based – the core filtering parameters remain similar across all three variants. The 3-decibel, -270°, 1dB difference and 1° difference frequencies were maintained, while THD

and noise parameters either stayed the same or improved. Most importantly, however, both transistor area and power consumption saw a meaningful decrease with each optimization step, with only a slight difference between results for both filter types.

For the ladder-based filter, the shared CMFB approach resulted in a 32.4% reduction in power consumption and an 18.9% decrease in transistor area compared to the base variant. Furthermore, incorporating the shared output stage structure reduced power consumption by 42.2% and transistor area by 37.1% compared to the base variant, or by 14.5% and 22.4%, respectively, compared to the shared CMFB variant.

For the signal-flow graph-based filter, a 32.5% reduction in power consumption and a 19.2% decrease in transistor area were achieved by employing the shared CMFB structure. Moreover, introducing the shared output stage approach reduced power consumption by 37.6% and transistor area by 42.4% compared to the base variant. Relative to the shared CMFB variant, these reductions were 14.6% and 22.7%, respectively.

TABLE I.
TRANSISTOR DIMENSIONS IN OTA

Dimensions	Transistors						
	M1-2	M3	M4-5	M6-7	M8-9	M10.0 M10.1	M11.0 M11.1
W [um]	1.8	2	2	2	0.9	0.9	0.9
L [um]	0.9	0.6	0.6	0.6	0.6	0.6	0.6
m[a]	1	2	1	1	1	1	1

[a] transistor multiplier

V. CONCLUSIONS

This paper proposed an optimized design methodology for OTA-C filters, focusing on reducing power consumption and transistor area while maintaining high performance. Two filter types were analyzed: ladder-based and signal-flow-based designs. Both were implemented in three configurations: the base variant with full OTAs, a variant with shared CMFB circuits, and a fully optimized variant with shared output stages and CMFB circuits.

Simulation results demonstrated significant improvements in resource efficiency achieved without worsening the performance. For ladder-based filters, the shared CMFB structure reduced power consumption by 32.4% and transistor area by 18.9%, while the shared output stage further improved these metrics to 42.2% and 37.1%, respectively, compared to the base variant. Similarly, in signal-flow-based filters, the shared CMFB approach achieved reductions of 32.5% and 19.2%, and the shared output stage, compared to the base variant, yielded reductions of 37.6% and 42.4%, respectively.

Despite challenges such as change of OTA's transconductance due to shared output stages, after necessary modifications the proposed designs successfully maintained key filter characteristics, including frequency, noise and distortion characteristics. These findings highlight the potential of shared sub-blocks in OTA-C filters for achieving high integration, low costs, scalability, and energy efficiency in modern electronic systems.

TABLE II.
THE DESIGNED VI-TH ORDER FILTERS' SIMULATED PARAMETERS

Measured parameters	Ladder-based OTA-C filter			Signal-flow-based OTA-C filter		
	Full OTA	Shared CMFB	Shared CMFB and output stage	Full OTA	Shared CMFB	Shared CMFB and output stage
3dB frequency [MHz]	4.504	4.504	4.513	4.512	4.51	4.501
-270° frequency [MHz]	4.489	4.49	4.489	4.492	4.499	4.489
Frequency of 1dB difference from ideal filter [MHz]	142.7	150.6	43.46	295.5	327.5	46.99
Frequency of 1° difference from ideal filter	6.856	6.867	5.139	6.19	10.34	9.654
Input voltage amplitude for 1% THD [mV]	163.28	163.28	179.1	168	168	184
Input referred noise [mV RMS]	5.717	5.717	4.922	5.189	5.131	4.443
Dynamic range [dB]	26.105	26.105	28.209	27.194	27.292	29.333
Power consumption [μW]	1006.74	680.76	582.12	890.5	600.8	513
Transistor area [μm^2]	344.82	279.48	216.84	302.94	244.86	189.18

REFERENCES

[1] Kumngern, M.; Aupithak, N.; Khateb, F.; Kulej, T. 0.5 V Fifth-Order Butterworth Low-Pass Filter Using Multiple-Input OTA for ECG Applications. Sensors 2020, 20, 7343. https://doi.org/10.3390/s20247343

[2] Kulej, T.; Khateb, F.; Kumngern, M. 0.5 V Multiple-Input Fully Differential Operational Transconductance Amplifier and Its Application to a Fifth-Order Chebyshev Low-Pass Filter for Bio-Signal Processing. Sensors 2024, 24, 2150. https://doi.org/10.3390/s24072150

[3] Wyszynski, A., & Schaumann, R. (1992). Using multiple-input transconductors to reduce number of components in OTA-C filter design. Electronics Letters, 28, 217-220.

[4] P. E. Allen and D. R. Holberg, "CMOS Analog Circuit Design," 2nd Edition, Oxford University Press, New York, 2004.

[5] Rolf Schaumann, Haiqiao Xiao, and Van Valkenburg Mac. 2009. Design of Analog Filters 2nd Edition (2nd. ed.). Oxford University Press, Inc., USA.

[6] Behzad Razavi. 2000. Design of Analog CMOS Integrated Circuits (1st. ed.). McGraw-Hill, Inc., USA.

Design of the Charge-Sampling Multiplying PLL in CMOS 40 nm

Jakub Zając, Piotr Kmon

AGH University of Krakow
Krakow, Poland
jakubza@student.agh.edu.pl, kmon@agh.edu.pl

Abstract—This paper presents the design of the Charge Sampling Phase-Locked Loop working on 11 GHz with rms jitter less than 100 fs. The presented PLL works with the 100 MHz reference source and its main advantage is a phase detector PD, working in charge domain, allowing to minimize power consumption while simultaneously keeping the high gain of the phase difference. To alleviate spur transfer from periodic phase detector to PLL high common mode rejection ratio, CMRR operational transconductance amplifier OTA is used with Common Mode CM amplifier, which tune OTA output level improving frequency range. In addition, the differential input class D/F2 Voltage Controlled Oscillator VCO with one inductor turn is used. The proposed circuit is designed in the CMOS 40 nm process, is supplied from two supply sources 1.1 V and 0.6 V and consumes about 5 mW of power.

Keywords—phase locked-loop, subsampling, jitter, phase noise, frequency synthesis, spectral purity, low power applications

I. INTRODUCTION

PLL was invented several dozen years ago as a solution for providing stable clock signal for electronic circuits. The main areas of using PLLs are radio frequency communication, as well as wireline data transmission, clock tree synthesis in digital circuits and ADCs. One of the main parameters of PLL is its phase noise that, especially in contemporary advanced communication protocols, may degrade their parameters or even prevent proper functioning. For example, PAM4's 112Gb/s transmitter requires clock signal with 100 fs jitter, while receiver requires it at the level of 10 fs [1]. Also, the 5G 1024 QAM receiver working with a 39 GHz carrier requires a clock of less than 50 fs of jitter [2]. It is well known that to decrease the phase noise, additional power in oscilator part of PLL needs to be spent. However, considering modern or batteryless working devices, this approach may not be acceptable.

A. Subsampling Architecture Considerations

One of the very popular PLLs is the subsampling-based architecture that main advantage is divider-less operation resulting in cancelation of its noise and also suppression of noise originating in the conventionally used charge pump architecture. Importantly, these are functional problems arising from the sampler circuit, which perturb the oscillator's LC tank by dynamically changing capacitance connected to VCO output nodes in different phase measurements. The

straightforward solution for this problem is to employ buffer with inductive load, however, this is traded for additional power [3]. This downside was addressed by power gated operation in order to minimise ON state period length. Still limitations were set by time needed for common mode setting or resonant steady-state time in the case of the LC tank [4, 5].

B. Charge-Sampling Phase Detection

To overcome the above-mentioned issues, charge integrating phase detector was proposed [6]. Here, the phase difference measurement is realised by converting input voltage into a current in a common source stage (Fig. 1a)) and then integration on a capacitor. The modulation capacitance effect experienced by the PLL tank caused by switching between the ON / OFF state is alleviated by negligible detection time (only 0.45% of the total clock reference period). In Fig. 1b) there are two possible scenarios shown with their corresponding output voltages. Taking into account the synchronisation case (left) the detector output voltage remains equal to the common mode voltage and the VCO is synchronised to the centre of the detection time window, while for lack of synchronisation (right) the output voltage of the charge sampling block is changed, therefore, modifying the VCO frequency.

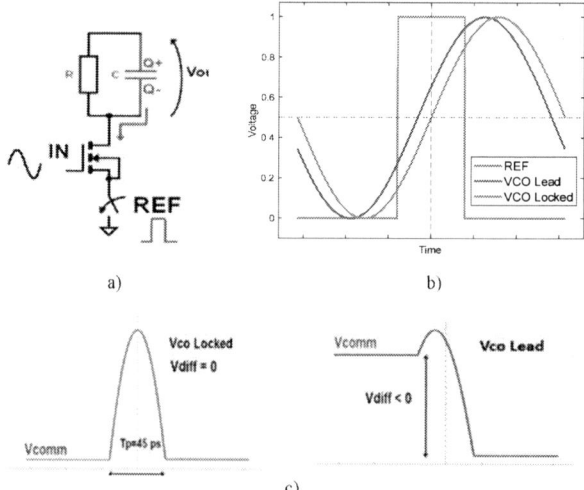

Fig. 1. Phase detection process overview: a) Phase Detector, b) PLL in locked and dislocked state, and c) resulting output voltage.

II. Basic Properties of the CSPLL

A. Gain of the Phase Detector

The charge sampling process is based on integrating current on a capacitance without applying any switches to hold the sampled voltage. The drawback of this solution is the discharging process (see Fig. 2), with time-constant R_SC_S (impedance of common mode stage), by current flowing in parallel connected resistor (needed to avoid extra origin pole in PLL transfer function). On the other hand, lack of dynamic switches solve problems with charge injection, clock feedthrough, and mismatch between switching devices in differential case. The gain of the PD is calculated by averaging over one reference period and is equal to [6]:

$$K_{PD} = \frac{2G_M A_{VCO} R_S}{N\pi} \cdot \sin\left(0.5\omega_{VCO}T_p\right) \qquad (1)$$

where G_M is the transconductance of PD's NMOS transistors and A_{VCO} is VCO amplitude. One can deduce that the maximum available value is achievable for T_p equal to half VCO period, and then it is going to fall because it is a sinusoidal gain characteristic.

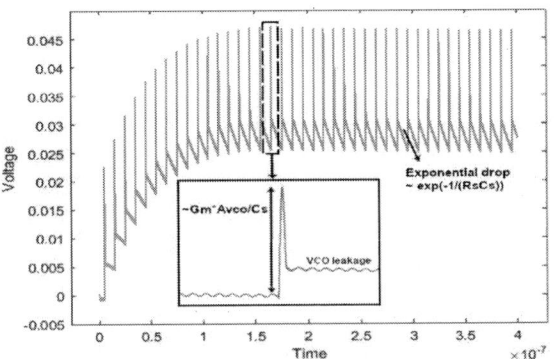

Fig. 2. Phase Detector output behavior.

B. Transfer Function and Phase Margin

Small signal transfer function can be derived from standard 2nd order filter approximation by excluding phase detector pole influence. In terms of natural frequency and damping factor, bandwidth is calculated using the following formula [7]:

$$BW = \sqrt{2\zeta^2 + 1 + \sqrt{((2\zeta^2 + 1)^2 + 1)}} \cdot \frac{\omega_n}{2\pi} \quad (2)$$

where

$$\zeta = \frac{R}{2}\sqrt{K_{pd}K_{VCO}K_{OTA}C} , \ \omega_n = \sqrt{\frac{K_{pd}K_{VCO}K_{OTA}}{C}} \quad (3)$$

where K_{VCO} is the VCO frequency gain and K_{OTA} is the OTA current to voltage gain. To set the bandwidth of the PD to 5 MHz, only 21 pF capacitor is required (as a result of applying symmetric filter version, that reduces its value by half). The second capacitor, needed mainly to suppress spurs coming from resistors voltage fluctuations, must be much lower to prevent stability disturbance and, therefore, is about 300 fF. The phase margin is equal to 64 degrees when the finite output resistance of the PD pole and OTA is taken into account.

C. Reference Spur Power

Capacitances disparity seen from VCO outputs (resulting from PD switching activity) is strongly suppressed by short working time (D_{REF} = 45 ps being 0.45% of reference clock period) and small value of capacitance (C_{MOD}) originating mainly from C_{GS} of PD's transistors. However, this effect results in additional generated spur power given by [6]:

$$S_{REF} = 20 \cdot log_{10}(\sin\left(\pi \cdot D_{REF} \cdot \frac{N}{2\pi} \cdot \frac{C_{MOD}}{C_{TANK}}\right)) \qquad (4)$$

As the oscillator LC tank C_{TANK} is equal to 650 fF, by evaluating (4) we got -83 dBc 100 MHz of offset spur power.

III. Design of the CS-PLL

Fig. 3 shows the complete schematic diagram of Charge Sampling PLL. The proposed PD is followed by a continuously working OTA supported by common-mode loop feedback (CMF). The PLL is also equipped with the symmetric second order filter and LC based voltage controlled oscillator being connected to the slice buffer generating square signal upon its input sinusoidal signal. The reference sinusoidal voltage is transformed by the pulse generator, built from the AND gate and delay stage, to a very short pulse signal (~45ps) that is transmitted to switching PD.

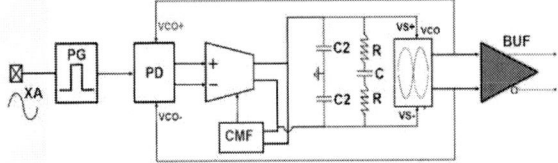

Fig. 3. Charge Sampling PLL block diagram.

A. Phase Detector

The charge sampling block consists of a differential common source amplifier gated by one tail transistor N1 (see Fig. 4). The capacitors and resistors are set in the range 100 kΩ to 300 kΩ and 100 fF to 300 fF, respectively, to tune the circuit gain and establish the correct common mode voltage for the OTA stage (equal to about 820 mV).

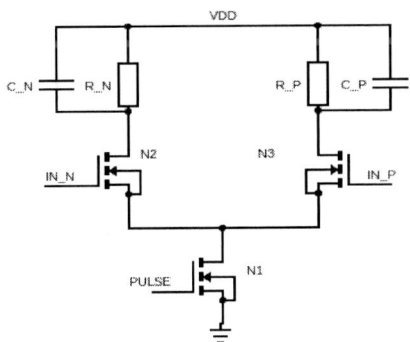

Fig. 4. Charge Sampling Phase Detector schematic idea.

Fig. 5 shows a mask view of the utilized PD that occupies 40.5 x 23.9 μm² of silicon area. Most of its area is occupied by MOM capacitor arrays, below which poly-resistors are located that are connected to NMOS transistors (arranged in the centre).

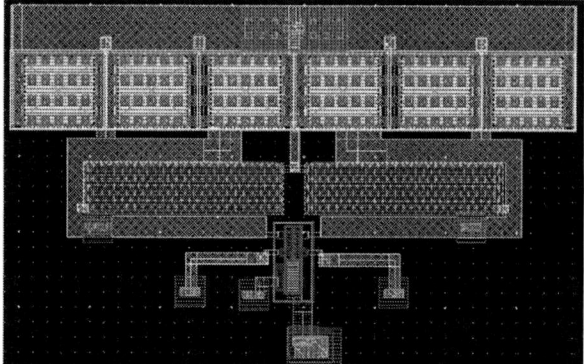

Fig. 5. Mask view of Charge Sampling Phase Detector block.

B. V/I Amplifier

The final block that is connected to the filter (see Fig. 3) is the transconductance amplifier. Here, a differential mode folded cascode configuration is used to provide high output impedance and large voltage swing. For further improvement of the common mode gain, the tail current source is cascaded. The transconductance amplifier is supported by the common mode feedback amplifier CMF and averaging circuit as shown in Fig. 6. That block was designed to control CM level of the OTA amplifier and for reduction of ripples originating from CS detection process. To save power, native NMOS is pertinent, as just one couple is sufficient. Reference voltage is delivered outside of the PLL and a half higher voltage supply is desired (in the typical case 0.55 V). The CMF block was simulated with 200 sweep Monte Carlo analysis resulting in an average OTA CM output voltage equal to practically 0.5 VDD (σ = 30 mV). Corner simulations show for worst case scenario -37.6dB (slow-slow corner) and for the best case -52.47 dB (fast-fast corner) of common-mode gain.

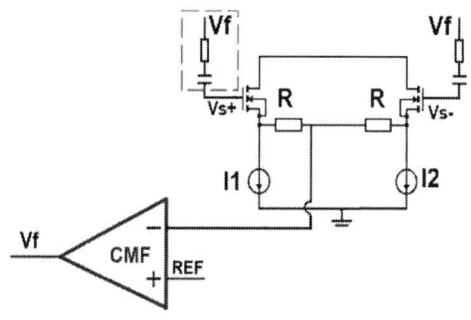

Fig. 6. CMF feedback.

C. VCO

The D/F2 class oscillator, shown in Fig. 7a), is used to achieve low phase noise and high voltage swing. The differential mode frequency gain is 40 MHz/V and is linear in almost all expected voltage ranges, while the common mode

rejection ratio is equal to 20 dB. The C1/C2 (see Fig. 7b)) capacitor pair is chosen to obtain high impedance at the second frequency harmonic (22 GHz), which appears as a common-mode voltage. Such settings prevent up-conversion of flicker noise from cross-coupled NMOS to VCO output [8]. These values are chosen according to formula [9]:

$$\frac{C_{DM}}{C_{CM}} = \frac{3-5k}{1+k} \qquad (5)$$

where C_{DM} is capacitance seen by differential current and is equal $(C_1 + C_2)/2 + (C_3 + C_4)/2$, and C_{CM} is equal to $(C_1 + C_2)/2$, whereas k is magnetic coupling between inductors (here assumed to be zero).

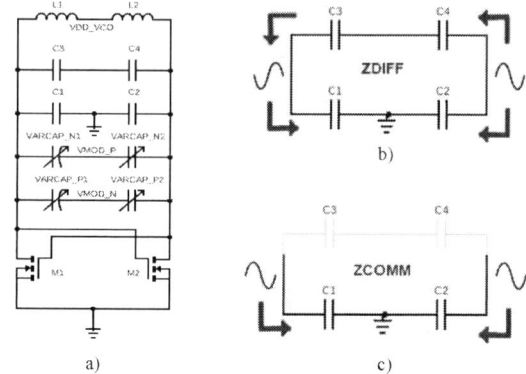

Fig. 7. a) Schematic idea of the VCO D/F₂ Class, b) differential output capacitance, c) common mode output capacitance.

The layout of the VCO takes up the most area occupation among PLLs blocks. Here, both coils, separated from each other to minimise magnetic coupling, are designed as single-turn type to get high Q-factor.

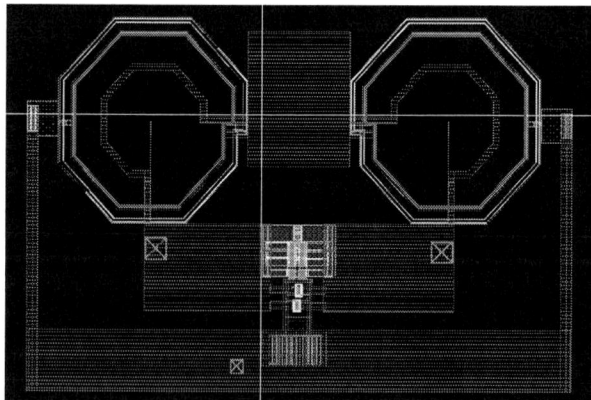

Fig. 8. Mask view of VCO.

IV. CSPLL OPEARTION

To verify the PLL operation, a test bench was prepared that the idea was to provide an additional 0.6 fF capacitance to the PLL node at 1 μs. This was done to see how the PLL reacts to losing its synchronisation. The instantaneous frequency presented in Fig. 9 a) shows stabilisation in nearly 0.7 μs. One

100

can see visible spurs that come from the integration phase in the charge sampling detector. As a consequence of continuously working OTA there a constant phase difference is generated (see Fig. 9 b) between reference and VCO resulted in permanent voltage change as shown in Fig. 9 c).

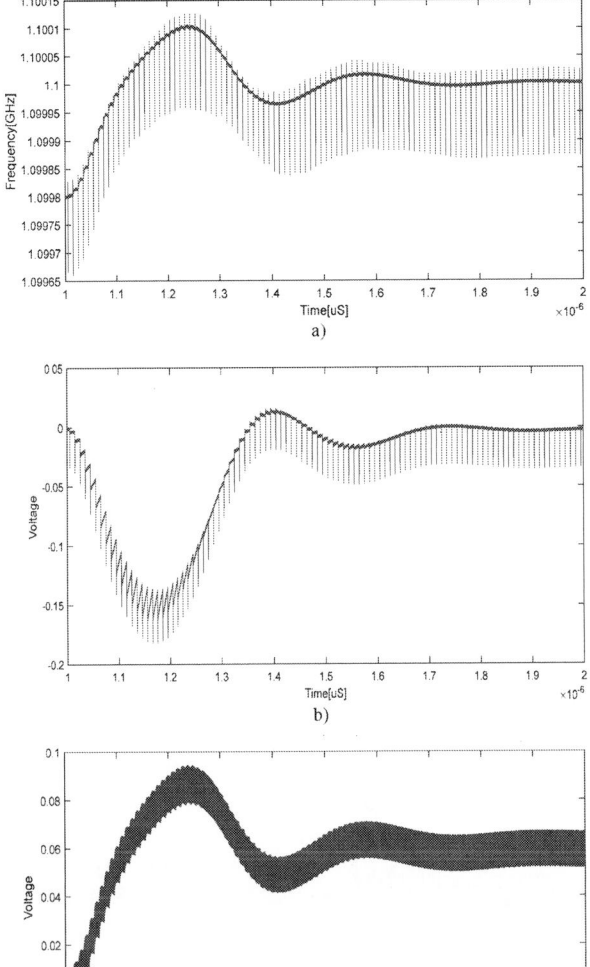

Fig. 9. Locking process in different nodes of PLL: a) PLL's frequency, b) output of the PD, c) output voltage of the OTA.

A. Phase Noise and Frequency Spectrum

Fig 10. shows the phase noise of PLL working at 11 GHz with each block noise contribution. Importantly, almost all low frequency noise comes from the OTA's while in the higher frequency the VCO noise also becomes significant. The overall jitter is equal to 87.6 fs.

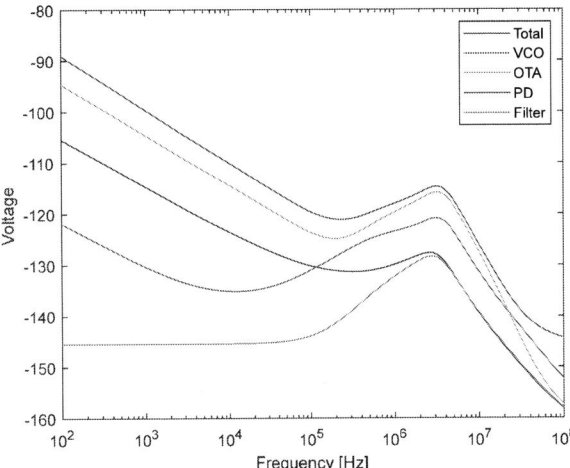

Fig. 10. Simulation results of the phase noise of the PLL blocks in particular.

Due to charge sampling PD, minimising both modulation time and coupling capacitance between switch and VCO spectral impurity peaks drops below -80 dB (compared to 1 W). The carrier allocated at the centre of the Fig. 11 has -1.3 dB with -81.5 dB larger reference spur on the right-hand side, which sum up to -80.2 dBc level.

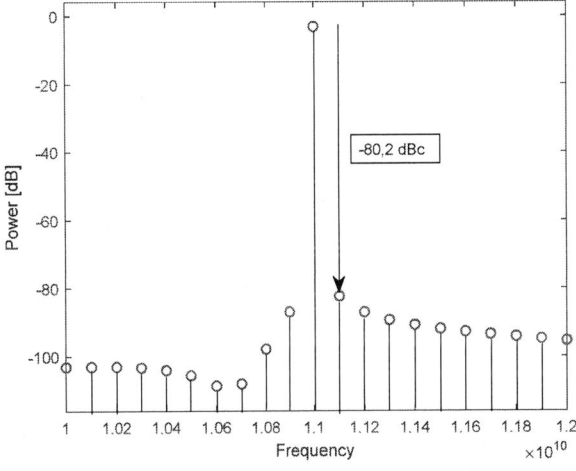

Fig. 11. Frequency spectrum of the generated signal.

B. Power cosumption of the PLLs

Considering the power consumed by the PLL it can be seen that the VCO dissipates about 90% of total power, while the transconductance amplifier with CMF loop consumes 83 μW. Very short ON state time makes the low-power PD block with less than 1% of total (<5 μW) power consumption.

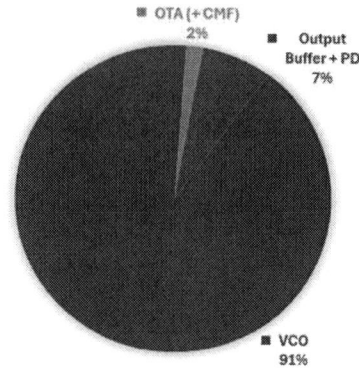

Total Power: 4.17 mW

- OTA (+ CMF) 2%
- Output Buffer + PD 7%
- VCO 91%

Fig. 12. Power consumption of particular blocks of the PLL.

TABLE I.
COMPARISON TABLE WITH STATE-OF-THE-ART. SUBSAMPLING INTEGER PLL

	This work	[6] 2022	[10] 2018	[11] 2020
PLL Architecture	CSPLL	CSPLL	Type-I SSPLL	Type-II SSPLL
F_{PLL} [GHz]	11	11.2	5	13.05
F_{REF} [MHz]	100	100	100	50
F_{REF} [dBc]	-80.2	-77.3	-64.1	-75
P_{DC} [mW]	4.17	5	1.1	6.7
RMS jitter σ [fs]	87.6	48.6	162.2	83
FoM* [dB]	-248.7	-259.2	-255.4	-253
Process node [nm]	40	40	65	65

*FoM = $20\log_{10}(\sigma_{rms}/1s) + 20\log_{10}(P_{DC}/1mW)$

V. CONCLUSIONS

In the following paper, the design of the CSPLL block was shown with particular emphasis on the PD and OTA. Although the simulation results are postlayout simulations, and also not including reference and pulse generator noise, only main PLL parameters (i.e. power consumption, spur power, and the noise

performance) are very promising while compared with other solutions (see Table I). We can see strongly suppressed reference spur power (-80.2 dBc), without using intermediate buffers, as a result of novel PD architecture. Examining jitter value, especially opposing with secondary CSPLL, a number of improvements, including OTA phase noise contribution (see Fig. 10.), can be done.

ACKNOWLEDGMENT

This research is founded by the National Science Center, Poland, project no. 2023/51/B/ST7/01782.

REFERENCES

[1] B. Razavi, "Jitter-Power Trade-Offs in PLLs," IEEE Transactions on Circuits and Systems I, 2021.

[2] T. Siriburanon, R. Staszewski, "Beyond ADPLLs for RF and mm-Wave Frequency Synthesis: Watching out for new techniques: oversampling-reference and charge-sharing locking", IEEE Solid-State Circuits Magazine, 2025.

[3] K. Raczkowski, et al., „A 9.2–12.7 GHz Wideband Fractional-N Subsampling PLL in 28 nm CMOS With 280 fs RMS Jitter", JOURNAL OF SOLID-STATE CIRCUITS, 2015.

[4] Z. Zhang, et al., "A 0.65V 12-to-16GHz Sub-Sampling PLL with 56.4fsrms Integrated Jitter and -256.4dB FoM", IEEE Int. Solid-State Circuits Conf., 2019.

[5] Z. Yang, et al., "A 25.4-to-25.9GHz 10.2mW Isolated Sub-Sampling PLL Achieving -252.9dB Jitter-Power FoM and -63dBc Reference Spur",2019.

[6] G. Jiang, et al., „A Low-Jitter and Low-Spur Charge-Sampling PLL", JOURNAL OF SOLID-STATE CIRCUITS, 2022.

[7] B. Razavi,"Design of CMOS Phase-Locked Loops", Cambridge University Press, 2020.

[8] M. Shahmohammadi, M. Babaie and R. B. Staszewski, „A 1/f Noise Upconversion Reduction Technique for Voltage-Biased RF CMOS Oscillators", IEEE JOURNAL OF SOLID-STATE CIRCUITS 2016.

[9] H. Xu et al. „A Low-Voltage Class-D VCO with Implicit Common-Mode Resonator Implemented in 55 nm CMOS Technology",Electronics, 2023.

[10] A. Sharkia, S. Mirabbasi, and S. Shekhar, "A type-I sub-sampling PLL with a 100×100 μm 2 footprint and −255-dB FOM," IEEE J. Solid-State Circuits, 2018.

[11] Y. Lim, et al., "17.8 A 170 MHz-lock-in-range and −253dB-FoM jitter 12-to-14.5 GHz subsampling PLL with a 150 μW frequency- disturbance-correcting loop using a low-power unevenly spaced edge generator", IEEE Int. Solid-State Circuits Conf.,2020.

Enhancing Test-Driven Development for Reconfigurable Hardware through High-Level Synthesis and Early-Stage Validation

Roman Diachok, Halyna Klym
Specialized Computer Systems Department
Lviv Polytechnic National University
Lviv, Ukraine
rodyachok@gmail.com, halyna.i.klym@lpnu.ua

Abstract—High-level synthesis tools help engineers deal with the challenges of building complex systems that use reconfigurable technologies. Such serves as a precursor to well-established methods in the software industry, such as Test-Driven Development, in the development process of hardware components of an embedded system. However, the assistance offered by the high-level synthesis validation tools could be strengthened and targeted at the early stages of project development. This paper describes a hardware testing framework as a means to quickly evaluate the capabilities of embedded components using a unit testing paradigm, leading to Test-Driven Development implementation on reconfigurable hardware.

Keywords—test, embedded verification, high-level synthesis, FPGA, unit test, infrastructure, test-driven design.

I. INTRODUCTION

High-Level Synthesis (HLS) has recently gained significant popularity, contributing to the acceleration of FPGA-based development and simplifying the verification process [1-3]. HLS expands the capabilities of the FPGA market by allowing even users with basic programming knowledge to evaluate available architectural options quickly. FPGA technology is accessible to software engineers and hardware developers [1,3]. However, HLS does not provide immediate verification of project credibility, which creates particular challenges in the development process.

One of the key stages of verification is testing, which involves generating input vectors for modeling the hardware component and comparing the obtained results with the planned reference model (the so-called "golden model"). Most HLS tools employ similar verification tests to validate synthesized systems [2]. Ensuring the correctness of FPGA-based designs is a crucial task, as errors in verification can lead to severe performance issues or even system failures in critical applications.

Studies indicate that verification poses a significant obstacle in FPGA projects: the effort spent ensuring the correctness of designs that integrate FPGA with other data processing technologies can account for 70–80% of the total development effort [1,2]. As FPGA systems become increasingly complex, the verification process must evolve to address new challenges, such as heterogeneous computing environments and real-time processing constraints. In addition to validating functional correctness, verifying HLS-generated modules imposes additional requirements, including performance validation, power efficiency, and resource utilization assessments.

The verification of FPGA-based systems is particularly relevant in industries such as telecommunications, automotive, aerospace, and consumer electronics, where rapid prototyping and reliable validation are crucial for ensuring high performance and reliability. In these domains, efficient verification methods can significantly reduce development time and improve the accuracy of FPGA implementations. Furthermore, rigorous testing and verification frameworks are essential to meet regulatory compliance and safety standards in mission-critical applications such as autonomous driving, medical devices, and industrial automation.

Various methodologies have been proposed to address these challenges, including Test-Driven Development (TDD), which enables iterative validation of FPGA designs. TDD facilitates the incremental development of complex FPGA-based systems by ensuring that each HLS component undergoes rigorous testing before integration. By incorporating TDD principles, developers can minimize errors early in the design process, reducing debugging efforts and enhancing the overall efficiency of FPGA-based system development [3-5]. Additionally, TDD allows for continuous integration and automated testing, which is crucial for large-scale FPGA-based projects that require frequent updates and modifications.

This paper aims to develop and implement an HLS-based verification framework that leverages the TDD approach to overcome existing challenges in verifying FPGA-based systems. Additionally, this study evaluates the effectiveness of HLS in accelerating FPGA development and improving program verification. The proposed framework is expected to provide a standardized testing methodology that simplifies FPGA verification while enhancing reliability and reducing time-to-market.

II. PRINCIPLES OF TESTING WORKFLOW

HLS technology supports the Specify-Evaluate-Refine (SER) design method, a processor or accelerator design method specific to a particular purpose. Through automation, HLS-based development tools facilitate the transition from complex models that describe the behavior of components to simpler models that describe the behavior of components [6,7].

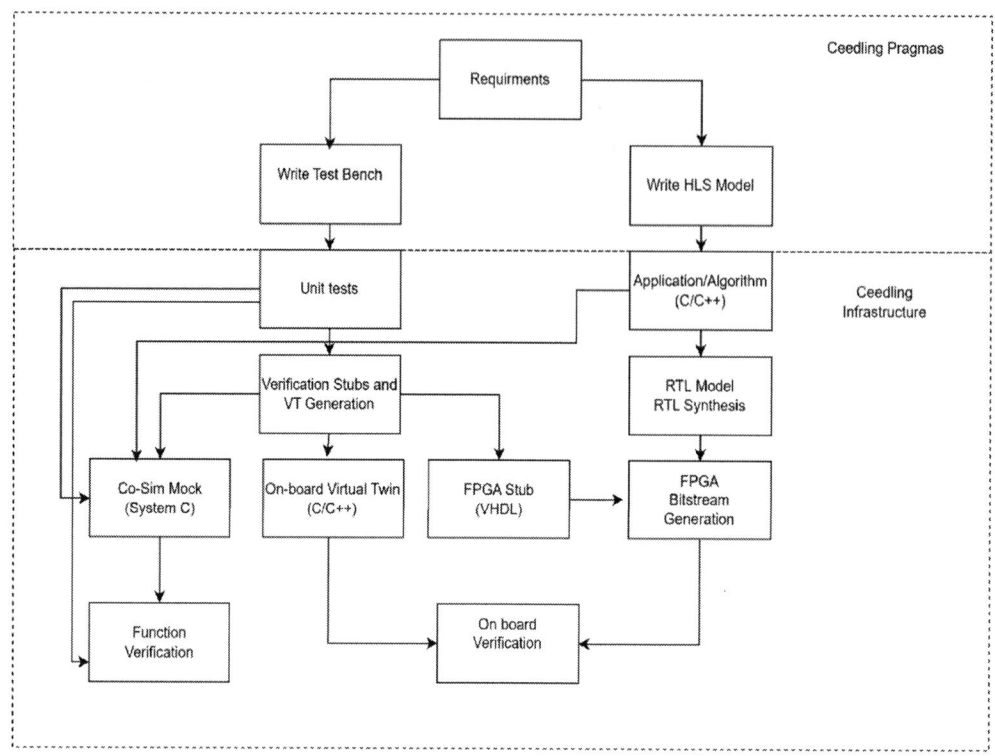

Fig. 1. Ceedling workflow

Project verification occurs at the three points during the design process: 1) processor level, 2) logical level, and 3) peripheral level. However, verifying hardware prototypes is more difficult because developing a testing infrastructure specific to each project requires additional effort. This infrastructure is unlikely to be repeated from one project to another (the problem of having multiple identical projects). The proposed verification methodology in Figure 1. Ceedling, enables temporal and functional analysis of hardware components.

Use TDD during this process and take advantage of increased productivity, quality, and test coverage, all of which have been demonstrated in the software industry. Figure 1 depicts a typical flow of a Ceedling user; it includes manual steps that must be performed and automatic steps that will occur, along with the primary inputs and outputs of each process.

The proposed procedure is high-level and generalizable enough to be re-targeted to specific HLS tools, provided they share the same essential capabilities, design, and generation of FPGA bitstreams.

The framework is employed for the DE0 architecture of Cyclone V FPGA; this results in written and implemented scripts and components that follow the DE0 protocol, the Quartus II toolchain.

The result of the hardware check is a separate report that describes in detail the features of the execution time. To overcome previously known limitations and problems related to

the verification of FPGA-based designs, we propose to extend the software testing framework Ceedling, which facilitates simple implementation of unit testing on hardware accelerators according to two main principles: use of a single test bench, independent of the stage of development, and automated internal path verification. Low cost of the prototype [8].

The hardware infrastructure must be ready during the FPGA/circuit-level design verification. The seedling method is simple and involves adding extra nutrients. To achieve this goal, add additional macros. To enhance the typical design and validation processes, Ceedling creates a testing infrastructure that allows the same set of tests to be repeated throughout the HLS project's life cycle.

The infrastructure necessary for testing includes the following software components.s: 1) a set of unit tests that use the Ceedling extension macros; 2) Test Runner, which sequentially executes each block of tests; 3) co-simulation and virtual twins in FPGA, the FPGA bitstream includes the component to be tested, surrounded by a particular auto-generated stub and test manager from the Ceedling library [9].

The test runner is automatically created using a Ruby script, and its code remains the same regardless of the level of abstraction. To begin testing, the developer invokes the highest test function. The actual implementation of the Design Under Test (DUT) protocol is contingent on the progression of the development process. A functional digital representation of the DUT, an RTL representation, and a physical version of the bit stream on the FPGA are created. To circumvent the DUT's specifics, the program that executes the test must communicate directly with it. The sole exception is functional testing, as both (Test Runner and DUT) are derived from C [10].

A virtual duplicate is a computer-generated code that is automatically triggered during the actual execution of the DUT. The test manager is a hardware component that connects the test tool and the FPGA-based DUT. Test Manager facilitates a standardized protocol and interface for virtual copies of FPGAs and DUTs on the FPGA. A virtual duplicate of the FPGA mimics the Ceedling process or data, which the test manager receives as a transaction. The test manager then interprets these transactions and performs actions based on them. The DUT is equipped with a stub that communicates with the test manager as the primary means of communication with the remainder of the test environment. This mapping concerns the interaction between HW-SW components, addressed using the RMI method [11-14]. In this regard, Ceedling provides a domain solution that combines the test environment and the DUT, despite the latter being implemented in different languages based on the level of abstraction. It was previously discussed that this task was divided into stubs and virtual twins, both created automatically.

Creating these instrumental components is accomplished using the RMI paradigm. RMI was initially intended to communicate in distributed systems with networked components. In this context, RMI permitted components to communicate with each other despite the physical distance between them. The primary advantage of RMI is that it provides a means of differentiating functionality and communication; any communication or functionality must occur between specific components that appear to be directly interacting [15].

III. VALIDATION OF THE PROPOSED METHOD

A. Test Manager

The Test Manager, Device Under Test (DUT), and communication partner are interconnected via AMBA technology, which ensures efficient data transmission. The Test Manager is directly connected to the bus, receiving messages that control the DUT's testing process. It integrates into the development process by delegating data transfer tasks to DDR memory. In this model, test cases provide stimulus data to reserved DDR regions, which the Test Manager reads to configure and trigger the DUT. Subsequently, DUT outputs are written back to DDR, where the Test Manager reads and compares them to predefined reference values. The Test Manager offers additional capabilities not present in Direct Memory Access (DMA), including real-time execution time measurement. This enhances the system's ability to evaluate DUT performance metrics such as latency and throughput. Furthermore, automatic stub generation ensures that virtual duplicates of the DUT mirror RTL/Logic behavior, maintaining consistency across simulation and synthesis. The architecture's modularity allows the DUT's input/output channels to be mapped to memory addresses, with a register enumerating the active channels and a transparent switch directing traffic. A virtual twin serves as a memory-mapped software abstraction of the DUT, communicating solely through memory addresses. This abstraction aligns with RTL and HLS designs, relying on Remote Method Invocation (RMI) to manage data transfer. This approach supports real-time profiling and performance

evaluation, providing valuable insights into the DUT's interaction with the test environment. The test suite includes algorithms from fields such as signal processing, cryptography, and control systems, providing benchmarks for HLS tools and forming the basis of this work's experimental methodology.

B. Design

The proposed architecture is modular, adaptable to FPGA platforms, and utilizes RMI design patterns to simplify component interactions. This modularity facilitates the automatic generation of low-level communication stubs, reducing manual hardware/software integration effort. The primary motivation for this approach is the need for transparent benchmarking of High-Level Synthesis (HLS) tools. Given the diversity of commercial and academic HLS platforms, this design enables objective comparisons by offering a unified, open infrastructure. Free licenses and comprehensive process support make it particularly suitable for educational environments. This design, therefore, serves as both a test infrastructure and research framework, enabling further studies on the efficiency and optimization of FPGA workflows.

Table I compares simulation and execution times for the ten cores in CHStone. Each core was tested differently. The simulation time was determined using the co-simulation function in the Quartus II HLS software. The actual runtime was determined using the Ceedling validation system built into the system. For each core, the average time spent in the simulation and execution processes, and the minimum and maximum difference between the hardware and the simulation, were compared to the core's latency.

TABLE I.
THE DESIGN'S LACK OF ACCURACY IN TERMS OF TIME AND LATENCY

Test Case	Input Value	Latency statistics*		Accuracy-Error	
		Co-Simulation	On-board	Relative	Absolute
Test 1	0.1	22.2	20.9	1.3	4.50%
Test 2	$\pi/16$	390.5	375.3	15.2	5.15%
Test 3	$\pi/8$	938.1	549.2	388.9	41.42%
Test 4	$\pi/5$	1468.3	553.6	914.7	62.51%
Test 5	0.75π	1827.8	729.1	1098.7	60.02%
Test 6	$5\pi/16$	2005.4	723.7	1281.7	63.99%
Test 7	0.33π	2179.6	727.9	1451.7	66.78%
Test 8	$7\pi/16$	2359.2	912.3	1446.9	61.44%
Test 9	0.44π	2543.9	913.2	1630.7	64.09%
Test 10	0.5π	2734.6	918.1	1816.5	66.47%
Test 11	0.56π	2911.3	1078.9	1832.4	62.88%
Test 12	0.61π	3080.8	1076.5	2004.3	64.91%
Test 13	0.67π	3245.3	1081.3	2164	66.61%
Test 14	0.72π	3251.7	1085.6	2166.1	66.48%
Test 15	0.78π	3441.9	1263.9	2178	63,15%
Test 16	0.83π	3621.5	1267.4	2354.1	64.92%

Two strategies are contrasted, and it is observed that the outcomes derived from the joint simulation of Ceedling's on-board delay model and design decisions often exhibit discrepancies due to inherent differences in how Ceedling measures on-board delay. The underlying issue is the inaccurate representation of actual delay behavior in Ceedling, which can lead to inconsistencies in the verification process. However, Ceedling's core functionalities are acknowledged for improving FPGA design verification, particularly in reducing verification time while maintaining method accuracy. Research into High-Level Synthesis (HLS) tools, including Ceedling, indicates that specific optimizations, such as lowering simulation overhead while enhancing the fidelity of time measurements, are critical for improving the overall verification process. Specifically, the back-end adjustments required for platform or technology-specific details (e.g., virtual twins, stubs) are minimal, ensuring that the core framework remains adaptable across different hardware implementations.

Table I illustrates the contrasting simulation and runtime performance across ten cores tested in various conditions. Notably, significant errors are observed in the results, with inaccuracies reaching as high as 66%. These errors highlight a systematic issue in the central portion of the design, which is identified as the primary source of miscalculation. Analyzing Table 1 reveals that the errors become more pronounced as the number of test cases increases, suggesting an amplification effect tied to the accumulation of computational delays and simulation inaccuracies. Further analysis of the Quartus II HLS toolchain indicates the presence of a waitlist mechanism that queues DUT inputs. This design decision leads to the misinterpretation of processing time, as the time spent queuing entries is mistakenly included in the time calculations. As a result, the reported test case processing times are inaccurate, skewing the assessment of DUT performance. Our solution focuses on precise time measurement by calculating the time spent solely on the actual stimulus processing. This refinement enables a more accurate diagnosis of time discrepancies, improving the reliability of time-based analysis in FPGA verification. By eliminating the queuing factor from the calculation, the new methodology enhances the overall accuracy of the simulation, offering a more realistic depiction of DUT behavior in an FPGA environment.

IV. CONCLUSION

This paper describes a framework for verifying the Ceedling protocol. This framework aims to develop and implement an HLS-based design verification environment that utilizes a TDD approach to address the current challenges of verifying systems implemented with FPGA technology. Ceedling is a complete package that uses HLS-based hardware component testing to facilitate and improve the design of embedded systems targeting the FPGA platform. The main objectives of this framework include evaluating the effectiveness of HLS in accelerating FPGA-based development, developing a test environment for modeling hardware components, and implementing TDD principles to ensure the correctness of designs. To the best of our knowledge, this is the first study to create an on-board verification environment that can be used throughout the design cycle, regardless of the level of abstraction associated with system specifications. This facilitates the development of the necessary hardware and software components and saves time and effort. In addition, Ceedling allows you to use the TDD method in system design, enabling you to take advantage of the popular software development method. Ceedling supports functional testing and timing testing; this is a new complexity related to the nature of real-time projects and hardware. As a result, the proposed verification framework includes three primary levels of abstraction regarding the specifics of the embedded system (functional, RTL, and implementation/physical). The main contribution of this framework is creating a configurable and standardized test environment for heterogeneous devices, which allows engineers to reduce the amount of testing and accelerate the path to FPGA implementation. After all, automatic generation makes this proposition accessible and straightforward for software and hardware developers, who do not need extra effort to use it quickly.

REFERENCES

[1] Yang, Hee-Jin; Lee, Jeung-Sub; Lee, Han-Sle. "Implementation of a Window-Masking Method and the Soft-core Processor-based TDD Switching Control SoC FPGA System". The Journal of Korea Institute of Information, Electronics, and Communication Technology, 17.3: 166-175. 2024.

[2] A. Amid et al., "Chipyard: Integrated Design, Simulation, and Implementation Framework for Custom SoCs," in IEEE Micro, vol. 40, no. 4, pp. 10-21, 1 July-Aug. 2020.

[3] G. Rong et al., "LGSVL Simulator: A High Fidelity Simulator for Autonomous Driving," 2020 IEEE 23rd International Conference on Intelligent Transportation Systems (ITSC), Rhodes, Greece, pp. 1-6, 2020.

[4] C. Wen, et al. "Challenges and trends in modern SoC design verification". IEEE Design & Test, 34.5: 7-22, 2017.

[5] T. Havinga, et al. "Improved TDD operation on Software-Defined Radio platforms towards future wireless standards". Computer Communications, 209: 178-187, 2023.

[6] G. Baldini, F. Bonavitacola and J. -M. Chareau, "Wireless Interference Identification With Convolutional Neural Networks Based on the FPGA Implementation of the LTE Cell-Specific Reference Signal (CRS)," in IEEE Transactions on Cognitive Communications and Networking, vol. 10, no. 1, pp. 48-63, Feb. 2024,

[7] T. Fang, G. Yangyang, S. Chaoju, S. Gangle, C. Pengyu, X. Wei, and W. Wenjin "A Multi-Beam XL-MIMO Testbed Based on Hybrid CPU-FPGA Architecture" Electronics 12, no. 2: 380, 2023.

[8] A. Wicaksana, et al. "On-board non-regression test of HLS tools targeting FPGA". In: Proceedings of the 27th International Symposium on Rapid System Prototyping: Shortening the Path from Specification to Prototype. p. 41-47, 2016.

[9] A. Sambas et al., "A New Hyperjerk System With a Half Line Equilibrium: Multistability, Period Doubling Reversals, Antimonotonicity, Electronic Circuit, FPGA Design, and an Application to Image Encryption", in IEEE Access, vol. 12, pp. 9177-9194, 2024,

[10] S. Di Matteo, M. L. Gerfo and S. Saponara, "VLSI Design and FPGA Implementation of an NTT Hardware Accelerator for Homomorphic SEAL-Embedded Library", in IEEE Access, vol. 11, pp. 72498-72508, 2023,

[11] R. Diachok, H. Klym,. "Monitoring Trust Status During Fog Level Data Analysis of the Sensor Network". In: 2022 12th International Conference on Dependable Systems, Services and Technologies (DESSERT). IEEE,. p. 1-6. 2022

[12] L. Aabel, S. Jacobsson, M. Coldrey, F. Olofsson, G. Durisi and C. Fager, "A TDD Distributed MIMO Testbed Using a 1-bit Radio-Over-Fiber Fronthaul Architecture," in IEEE Transactions on Microwave Theory and Techniques, vol., 72, no. 10, pp. 6140-6152, Oct. 2024,

[13] R. Diachok, et al. "Definition of the system of human body position in virtual reality". In: 2022 IEEE 16th International Conference on Advanced Trends in Radioelectronics, Telecommunications and Computer Engineering (TCSET). IEEE,. p. 358-361, 2022.

[14] R. S. Molina, V. Gil-Costa, M. L. Crespo and G. Ramponi, "High-Level Synthesis Hardware Design for FPGA-Based Accelerators: Models, Methodologies, and Frameworks," in IEEE Access, vol. 10, pp. 90429-90455, 2022, .

[15] R.Diachok, H.Klym, V.Lysiak. Development of a unique mathematical computer based on the Altera Cyclone 3 FPGA using the NIOS2 core. Electronics and information technologies/Електроніка та інформаційні технології, 2024, 26..

FSMLock: Sequential Logic Locking Case Study

Jacob LaPietra, Michael Kurdziel
L3Harris Technologies
jacob.lapietra@L3Harris.com, Mike.Kurdziel@L3Harris.com

Marcin Łukowiak
Department of Computer Engineering
Rochester Institute of Technology
mxleec@rit.edu

Abstract—FSMLock is a sequential logic locking technique that has been proposed for protection of intellectual property (IP) of finite state machine (FSM) circuits. While this technique provides security advantages over other sequential logic locking techniques, one major drawback this approach brings is the large amount of memory required for storing data of all states, transitions, and outputs. Finite State Machines with Input Multiplexing (FSMIM) is an optimization methodology and tool that was proposed for efficient mapping of FSMs into memory. This is primarily achieved by reducing the number of effective inputs to the FSM. This paper discusses our work on integrating these two techniques in a practical case study of converting existing state machine into implementation with FSMLock with input multiplexing.

Keywords—logic locking, finite state machine, FPGA

I. INTRODUCTION

There are many security threats to an electronic circuit throughout its product life cycle. An attacker usually intents to either modify an existing design in order to manipulate its operation, or to gain unauthorized knowledge about a circuit's functionality and its implementation details. For example, a Trojan insertion is a type of attack that aims to embed additional malicious components into a design that affects the integrity of the original circuit [1], [2]. This is possible as out-sourcing the physical manufacturing and system integration is a common practice for many companies. Reverse engineering (RE) is another form of attack that aims to gather information by monitoring a circuit's functionality, implementation details, or data leakage through so called side channels [3]–[5]. This type of analysis is legally protected in the United States through the Semiconductor Chip Protection Act as long as no patents or copyrights are infringed [6].

FSMLock is a sequential logic locking technique that has been proposed for the protection of intellectual property (IP) of finite state machine (FSM) circuits through classical encryption [7]–[9]. In FSMLock, the state entry table (SET) is stored as the encrypted binary data with a device specific secret key. Parts of the encrypted state machine are then decrypted and loaded into in-scope memory when needed. Doing so attempts to conceal a circuit's details from third-party malicious players.

The objective of this work was to improve the usability of the original FSMLock tool by applying it to a larger scale existing state machine. This has been achieved by utilizing finite state machines with input multiplexing (FSMIM)

methodology, which was developed for efficient mapping of FSMs into memory based architectures [10].

This paper is structured as follows. Section II gives background into hardware security, FSMLock architecture and methodology, as well as, a brief description of the FSMIM methodology and tools that allow FSMLock to be applied to larger, real circuit scenarios. Section III describes the tool chain that is needed to create a complete FSMLock system. Section IV discusses our simulation and hardware verification setup, and hardware results. Finally, Section V concludes the paper.

II. BACKGROUND

A. Hardware security

The security of IP in the form of hardware circuits commonly refers to risks associated with overproduction, Trojan insertion, or pre- and post-manufacturing RE. There are several techniques that aim to mitigate these threats. These include but are not limited to circuit authentication, logic obfuscation, logic locking, bit stream encryption for field programmable gate arrays (FPGAs), and split manufacturing. Logic obfuscation for example, refers to a transformation of a logic circuit into a concealed yet functionally identical form. Logic locking techniques on the other hand, attempt to protect unauthorized use and duplication of a circuit, by altering it in a way that a secret key is needed for unlocking the device's functionality. Several forms of combinational logic locking have been proposed in the literature [11]–[15]. Slightly different approaches have been recommended for sequential logic locking [16]–[18]. Here, an implicit key in the form of a sequence of input patterns is used to transition through the added locked state space of the original FSMs.

One of the most successful attacks on combinational logic locking, especially early techniques, is the satisfiability (SAT) attack [19]. Here, the secret key needed to unlock the combinational circuit can be identified through conflict-driven pruning of key space, which drastically reduces the time required as compared to a brute force attack. The SAT attack can be also be applied to locked sequential circuits through logic unrolling, alongside model-checking to solve for the key-based initialization sequence [20], [21].

Another attack, based on topological and Boolean functional analysis, is presented in [18]. This paper outlines how, after assuming that locked FSM logic components can be

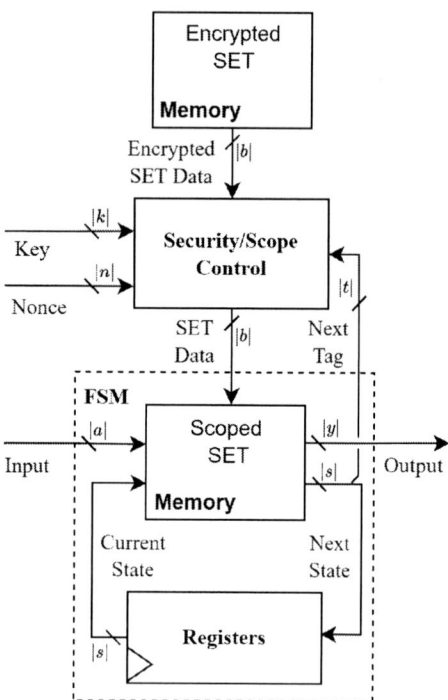

Fig. 1. The FSMLock primitive.

isolated using techniques discussed in [22]–[24], logic locking methodologies such as HARPOON [16] and dynamic state deflection [17] result in identifiable topological patterns in the locked state transiton graph, and retrieval of initialization key sequences.

B. FSMLock

FSMLock is a novel sequential logic locking technique proposed for IP protection of FSMs [7]–[9]. The primary use case is targeting field programmable gate arrays (FPGAs), however, the technique can be extended to other implementation technologies.

Figure 1 outlines high-level architecture of FSMLock. This approach defines an FSM with a state entry table (SET), which stores all of the necessary information that is needed to define all of the inputs, outputs and transitions. This table is then encrypted and stored in either on-chip or off-chip memory - **Encrypted SET**. During runtime, parts of the SET (partitions) are decrypted by the **Security/Scope Control** and brought into **Scoped SET** for the actual operation of the FSM.

C. FSMIM

Finite state machines with input multiplexing (FSMIM) is a technique that reduces memory depth of ROM-based FSMs. The general idea is based on the observation that individual transitions in an FSM are attributed to only a subset of all inputs [10]. This allows the original functionality to be maintained with fewer *effective inputs*, and as a result significantly less memory.

The FSMIM-T is a variation of the main FSMIM approach with transition-based selection of *effective inputs*, while FSMIM-S uses state-based selection. Our work focused on utilizing the FSMIM-T architecture within FSMLock. Detailed discussion on memory requirements for FSMLock is presented in [9]. Figure 2 outlines FSMLock architecture, which allows for input multiplexing. In this modified architecture, Scoped SET and Registers are implemented using a single block of Synchronous RAM.

III. METHODOLOGY

A. Integrated Tool Chain

Before applying the FSMLock methodology to our case study state machine, FSMIM was integrated into a toolchain that produced a set of files ready for an FPGA implementation.

The first step in this process is to define the state transition graph (STG) that will define the inputs, outputs and transitions between all of the states in the FSM. Once the STG is created the SET can be derived from it. The SET formatted into a *.kiss* file [25] is given as an input to the FSMIM tool. The FSMIM optimizes the state machine by reducing the number of effective inputs. The output is a new *.kissim* file that describes the same FSM along side the new definition of the input selection multiplexer. Each state transition also has a new field added to it, the next input selector, which is necessary for the input selection multiplexer to change which inputs are passed to the FSM at any given time. A python script was created to convert the *.kissim* file into a *.cvs* file that is read by the FSMLock tool. This script also generates a VHDL model of the input multiplexer. Once the *.cvs* is created, an encryption key and nonce can be added to it before running the FSMLock tool. The FSMLock tool creates a binary file with the content for the Memory Encrypted SET.

B. Benchmark Results

A set of 152 FSM circuits from the [26] benchmark suite were tested with this tool set in order to determine the general expectations while using FSMIM. The first metric tracked was the percentage of memory that could be saved vs. the number of inputs in the native FSM. In general this value increased as the number of inputs increased, however the benchmark did not contain any FSMs with more than 16 inputs. The second metric that was tracked was the relationship between the width of input multiplexer (effective inputs) and the number of inputs in the native FSMs. This value is highly dependent on each specific design however, in general case, machines with a higher number of inputs tended to yield wider input multiplexers. An additional observation is that not all FSMs were possible to optimize. In order for an FSM to be optimized at least one state must exist where one or more inputs are unused for any transition.

C. Case Study State Machine

The state machine used in this case study has 55 states, 71 transitions, 45 inputs and 63 outputs, and its behavior was originally modeled in Verilog HDL. The state machine

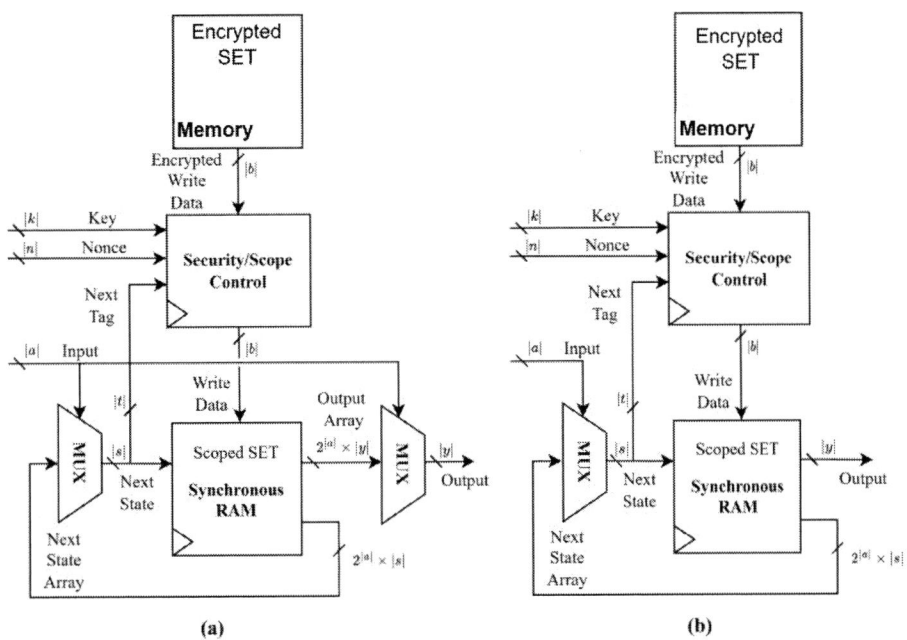

Fig. 2. Model of the FSMLock primitive with FSMIM integrated, utilizing synchronous in-scope random access memory (RAM) while targeting the Mealy (a) or Moore (b) state entry partitioning [8].

is used to control configuration of another hardware system. It has a few states that loop back to the beginning of the state machine if the wrong input is given. These states are used to ensure the system is not configured in an invalid state. There are also some paths that allow the FSM to jump far into the state encodings, which allows for different configurations of the controlled system.

Running this state machine through the FSMIM tool reduced the number of effective inputs to 10, reducing the ROM size of the optimized FSM to 516KB. As this memory requirement was too high for our target hardware platform, (Nexys A7-100T FPGA board), additional optimizations were required.

After careful analysis and manual changes, the effective inputs were reduced down to only four, and the final memory needed for implementing this particular state machine was reduced down to 33KB.

One example of manual optimizations involved vector inputs, which values are being tested for state transitions. In our FSM there were two 8-bit and one 9-bit vector inputs. Each of these was able to be reduced down to 2 bits only (to cover all possible outcomes of the tests), resulting in 27 inputs to the entire state machine down from 45. Table I presents a summary of the memory needed for the original case study state machine and the FSM with manually reduced number of inputs.

TABLE I
CASE STUDY FSM MEMORY REQUIREMENTS

Method	Inputs	Outputs	Memory
FSMIM	45	63	516KB
Manual Optimization + FSMIM	27	63	33KB

D. Security Concerns

In this work we did not investigate any specific vulnerabilities that could be exploited from the manual optimizations of moving some logic from the FSMLock protected state machine to the unprotected part. To combat this potential security risk, we would consider using additional combinational logic locking techniques, especially on the combinational input multiplexer.

IV. RESULTS

A. Simulation

The memory based architecture together with the original Verilog model were simulated in a test bench in order to ensure that functionality of the original FSM was preserved. Figure 3 outlines the architecture of our testing approach. The outputs generated by the original Verilog model of the FSM are used to verify the correctness of the FSMLock outputs. As the FSMLock based architecture will produce delayed outputs whenever fetching and decrypting another portion of the state machine into the Scoped SET, a *Ready* signal was added to ensure proper synchronization for all tests.

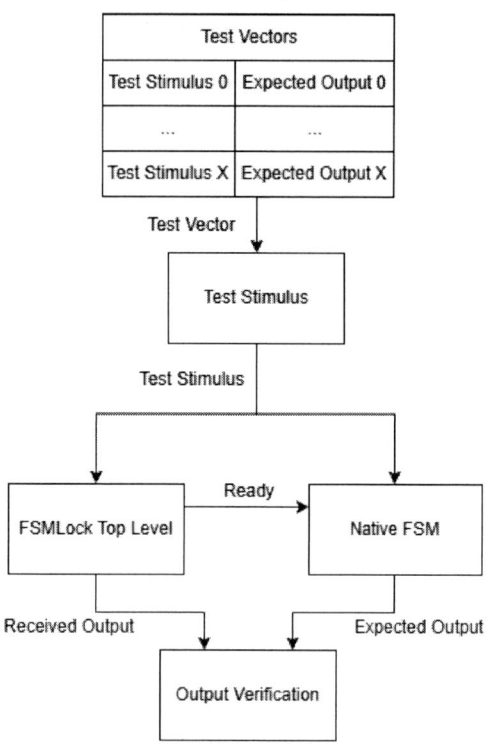

Fig. 3. Simulation setup.

TABLE II
MAIN RESOURCES FOR XC7A100T

Logic slices	LUTs	FFs	BRAMs
15850	63400	126800	135

B. Hardware Setup

The hardware tests were conducted on a Digilent Nexys4 development board with an Artix 7 100T FPGA. Table II shows the total amount of resources available on the target FPGA. The hardware implementation was verified by wrapping the top level design in logic that stimulated and verified the inputs and outputs of the top level FSMLock component as presented in Figure 4. The stimulus and verification vectors for this test were copied from the simulation test bench. The onboard LEDs were utilized to determine status of the entire testing procedure. The following parameters were tracked: all tests passed, tests failed, tests running, system reset status, and a binary encoding of the current test number. The test logic was designed such that if a test failed, the system would stop displaying the current test number. A 100 MHz system clock was used in the hardware tests.

C. Hardware Results

The FSMLock-based case study state machine was implemented and tested in three configurations: with one partition (the entire state machine would fit into Scoped SET memory),

with two partitions, and with four partitions. The synthesis results as well as the number of clock cycles (latency) needed to load a decrypted partition into the Scoped SET memory are presented in Table III. This summary includes resources for the entire state machine (Encrypted SET), Security/Scope Control with the Advanced Encryption Standard (AES) crypto engine, and Scoped SET. As expected, the main difference is in the latency, which is the result of the following: reading the encrypted data from Encrypted SET, decrypting the data using the Advanced Encryption Standard (AES) in the Security/Scope Control, and writing decrypted data into Scoped SET.

TABLE III
FSMLOCK AND MANUAL OPTIMIZATION SYNTHESIS RESOURCE
UTILIZATION REPORT

Partitions	LUTs	FFs	BRAMs	Latency (clock cycles)
1	916	300	73	4354
2	914	359	73	2178
4	911	359	73	1090

In the context of a single partition, the latency refers to the time needed for the entire state machine to be fully operational as it would be without using the FSMLock architecture. For the other two cases, the latency refers to the time needed for the state machine to be operational between selected transitions within the FSM. In the current version of FSMLock, the latency can be controlled only by changing the size of Scoped SET. In the future we will look into two additional, latency mitigation techniques: pre-computing and storing a key stream for a quicker decryption, or preemptive decryption of the next predicted scope and storing it in a ping-pong buffer for instantaneous switching.

For comparison of resources needed for the FSMLock architecture, the synthesis of the original Verilog HDL model of the case study state machine required only 72 LUTs and 55 FFs.

V. CONCLUSION

The FSMLock is a sequential logic locking methodology that provide a unique solution to protecting memory-based finite state machines. In this paper we discussed an integration of the FSMLock with finite state machines with an input multiplexing (FSMIM) technique, and after that results of applying complete methodology to a case study state machine. We analyzed configuration with one partition, where the entire state machine would fit into Scoped SET memory, with two partitions, and with four partitions.

In the future work we will look into architectural improvements to FSMLock to reduce or even circumvent delays when updating in-scope memory content. One possible solution that we consider is based on pre-computing and storing a key stream for quicker decryption, or preemptive decryption of the next predicted scope and storing it in a ping-pong buffer for instantaneous switching.

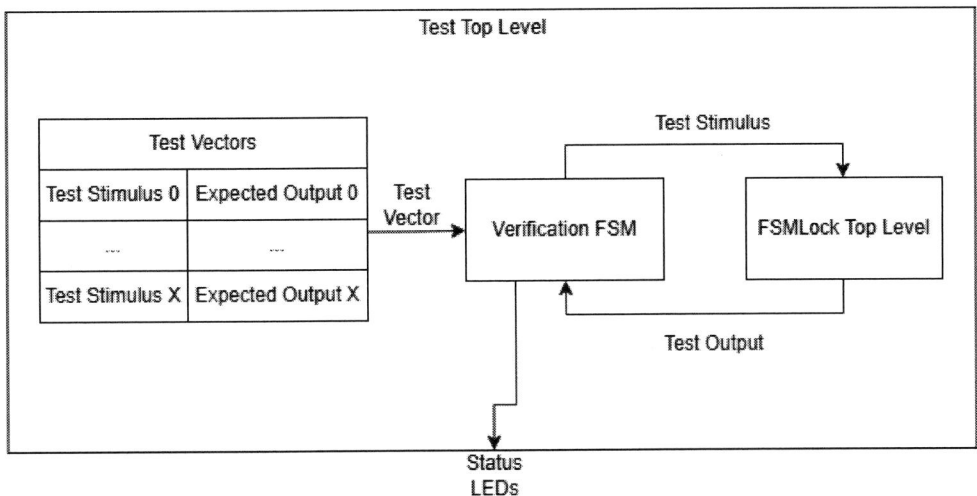

Fig. 4. Hardware testing setup.

REFERENCES

[1] J. Vosatka, *Introduction to Hardware Trojans*. Cham: Springer International Publishing, 2018, pp. 15–51. [Online]. Available: https://doi.org/10.1007/978-3-319-68511-3_2.

[2] S. Dupuis, P.-S. Ba, G. Di Natale, M.-L. Flottes, and B. Rouzeyre, "A novel hardware logic encryption technique for thwarting illegal overproduction and hardware trojans," in *2014 IEEE 20th International On-Line Testing Symposium (IOLTS)*, 2014, pp. 49–54.

[3] L. N. Nguyen, C.-L. Cheng, M. Prvulovic, and A. Zajić, "Creating a backscattering side channel to enable detection of dormant hardware trojans," *IEEE Transactions on Very Large Scale Integration (VLSI) Systems*, vol. 27, no. 7, pp. 1561–1574, 2019.

[4] M. G. Rekoff, "On reverse engineering," *IEEE Transactions on Systems, Man, and Cybernetics*, vol. SMC-15, no. 2, pp. 244–252, 1985.

[5] M. S. Rahman, R. Guo, H. M. Kamali, F. Rahman, F. Farahmandi, and M. Tehranipoor, "ReTrustFSM: Toward RTL Hardware Obfuscation-A Hybrid FSM Approach," in *IEEE Access, vol. 11*, 2023, p. 19741-19761.

[6] "Semiconductor chip protection act of 1984." [Online]. Available: https://www.congress.gov/bill/98th-congress/house-bill/5525?r=1&s=1

[7] M. T. Kurdziel, S. M. Farris, M. Lukowiak, and S. P. Radziszowski, "System and method for obfuscation of sequential logic through encryption - patent application US17/865,519," Jan 2024.

[8] M. Krebs, "FSMLock: Sequential logic locking through encryption," 2023.

[9] M. Krebs, M. Łukowiak, S. Farris, and M. Kurdziel, "FSMLock: Sequential Logic Locking Through Encryption," in *2024 31st International Conference on Mixed Design of Integrated Circuits and System (MIXDES)*, 2024, pp. 98–103.

[10] I. Garcia-Vargas and R. Senhadji-Navarro, "Finite state machines with input multiplexing: A performance study," *IEEE Transactions on Computer-Aided Design of Integrated Circuits and Systems*, vol. 34, no. 5, pp. 867–871, 2015.

[11] J. A. Roy, F. Koushanfar, and I. L. Markov, "EPIC: Ending Piracy of Integrated Circuits," in *2008 Design, Automation and Test in Europe*, 2008, pp. 1069–1074.

[12] M. Yasin, J. J. Rajendran, O. Sinanoglu, and R. Karri, "On Improving the Security of Logic Locking," *IEEE Transactions on Computer-Aided Design of Integrated Circuits and Systems*, vol. 35, no. 9, pp. 1411–1424, 2016.

[13] M. Yasin, B. Mazumdar, J. J. V. Rajendran, and O. Sinanoglu, "SAR-Lock: SAT attack resistant logic locking," in *2016 IEEE International Symposium on Hardware Oriented Security and Trust (HOST)*, 2016, pp. 236–241.

[14] Y. Xie and A. Srivastava, "Anti-SAT: Mitigating SAT Attack on Logic Locking," *IEEE Transactions on Computer-Aided Design of Integrated Circuits and Systems*, vol. 38, no. 2, pp. 199–207, 2019.

[15] H. M. Kamali, K. Z. Azar, H. Homayoun, and A. Sasan, "Full-Lock: Hard Distributions of SAT instances for Obfuscating Circuits using Fully Configurable Logic and Routing Blocks," in *2019 56th ACM/IEEE Design Automation Conference (DAC)*, 2019, pp. 1–6.

[16] R. S. Chakraborty and S. Bhunia, "Security against hardware Trojan through a novel application of design obfuscation," in *2009 IEEE/ACM International Conference on Computer-Aided Design - Digest of Technical Papers*, 2009, pp. 113–116.

[17] J. Dofe and Q. Yu, "Novel Dynamic State-Deflection Method for Gate-Level Design Obfuscation," *IEEE Transactions on Computer-Aided Design of Integrated Circuits and Systems*, vol. 37, no. 2, pp. 273–285, 2018.

[18] M. Fyrbiak, S. Wallat, J. Déchelotte, N. Albartus, S. Böcker, R. Tessier, and C. Paar, "On the Difficulty of FSM-based Hardware Obfuscation," *IACR Trans. Cryptogr. Hardw. Embed. Syst.*, vol. 2018, pp. 293–330, 2018.

[19] P. Subramanyan, S. Ray, and S. Malik, "Evaluating the security of logic encryption algorithms," in *2015 IEEE International Symposium on Hardware Oriented Security and Trust (HOST)*, 2015, pp. 137–143.

[20] Y. Hu, Y. Zhang, K. Yang, D. Chen, P. A. Beerel, and P. Nuzzo, "Fun-SAT: Functional Corruptibility-Guided SAT-Based Attack on Sequential Logic Encryption," in *2021 IEEE International Symposium on Hardware Oriented Security and Trust (HOST)*, 2021, pp. 281–291.

[21] S. Roshanisefat, H. Mardani Kamali, H. Homayoun, and A. Sasan, "RANE: An Open-Source Formal De-obfuscation Attack for Reverse Engineering of Logic Encrypted Circuits," *Great Lakes Symposium on VLSI*, 2021.

[22] Y. Shi, C. W. Ting, B.-H. Gwee, and Y. Ren, "A highly efficient method for extracting FSMs from flattened gate-level netlist," in *Proceedings of 2010 IEEE International Symposium on Circuits and Systems*, 2010, pp. 2610–2613.

[23] T. Meade, S. Zhang, and Y. Jin, "Netlist reverse engineering for high-level functionality reconstruction," in *2016 21st Asia and South Pacific Design Automation Conference (ASP-DAC)*, 2016, pp. 655–660.

[24] T. Meade, Y. Jin, M. Tehranipoor, and S. Zhang, "Gate-level netlist reverse engineering for hardware security: Control logic register identification," in *2016 IEEE International Symposium on Circuits and Systems (ISCAS)*, 2016, pp. 1334–1337.

[25] A. T. Abdel-Hamid, M. H. Zaki, and S. Tahar, "A tool converting finite state machine to VHDL," in *Conference Proceedings of the Canadian Conference on Electrical and Computer Engineering*, June 2004, pp. 1908–1910.

[26] L. Jozwiak, D. Gawlowski, and A. Slusarczyk, "An effective solution of benchmarking problem: FSM benchmark generator and its application to analysis of state assignment methods," in *Euromicro Symposium on Digital System Design, 2004. DSD 2004.*, 2004, pp. 160–167.

Implementation of a PLL Loop Circuit for Frequency Synthesis in 65 nm CMOS Technology

Magdalena Tymińska
Warsaw University of Technology
Institute of Microelectronics and Optoelectronics
Warsaw, Poland

Maciej Kucharski
OmniChip Sp. z o.o.
Warsaw, Poland

Witold Pleskacz
Warsaw University of Technology
Institute of Microelectronics and Optoelectronics
Warsaw, Poland

Abstract—**This paper presents the design and implementation of a phase-locked loop (PLL) circuit in 65 nm CMOS technology, dedicated to frequency synthesis and multiplication in radio frequency (RF) applications. The circuit processes an input signal of 13.56 MHz and generates a multiplied output signal of 867.84 MHz in the ultra high frequency (UHF) band, making it suitable for short-range communication systems such as radio-frequency identification (RFID) and near-field communication (NFC). The circuit was designed to ensure low phase noise, frequency stability, and fast locking time. The results demonstrate the feasibility of the proposed PLL architecture for modern wireless communication systems, highlighting its potential for integration into advanced RF applications.**

Keywords—**VLSI, IC, PLL, phase-locked loop, CMOS, NFC, RFID, RF, HF, UHF, phase noise, stability.**

I. INTRODUCTION

Modern electronic systems require reliable circuits that generate precise clock signals, resistant to interference and ensuring stable data transmission, especially in wireless applications such as radio-frequency identification (RFID) and near-field communication (NFC). Phase-locked loops (PLLs) play a crucial role in meeting these requirements by enabling frequency synthesis and stabilization of radio signals.

In RFID and NFC systems, accurate frequency generation is essential for efficient modulation, demodulation, and synchronization. The 13.56 MHz reference frequency used in these technologies [1] requires multiplication to the UHF band to enable potential operation in both frequency bands.

II. THEORETICAL ANALYSIS

The PLL is inherently a nonlinear system. To simplify the analysis of its behavior, a linearized model in the Laplace domain is commonly used (Fig. 1 [2]).

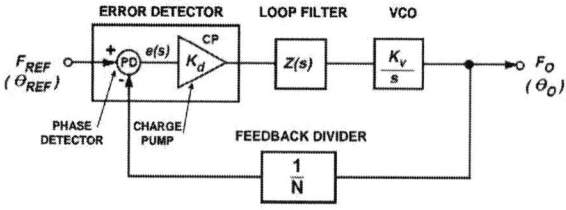

Fig. 1. Basic phase-locked loop linear model [2]

The presented PLL system consists of a phase detector (PD) integrated with a charge pump (CP), forming a phase-frequency detector (PFD) with a gain of K_d:

$$K_d = \frac{I_{CP}}{2\pi} \left[\frac{A}{rad} \right], \tag{1}$$

where:

- I_{CP} - charge pump output current,

a low-pass filter (LPF) with a transfer function $Z(s)$, a voltage-controlled oscillator (VCO) with a gain of K_v/s, and a frequency divider with a division factor N. The PLL in this configuration is called *Integer-N* Charge Pump Phase-Locked Loop (CPPLL) and is of Type II, which provides better bandwidth and stability adjustment compared to Type I, as well as an expanded capture range [3]. According to the theory of negative feedback loops, the transfer function T(s) of the entire PLL is expressed as follows:

$$T(s) = \frac{K_d \cdot Z(s) \cdot K_v}{s + K_d \cdot Z(s) \cdot K_v \cdot \frac{1}{N}}. \tag{2}$$

Incorporating the frequency divider into the loop, which allows the reference frequency F_{REF} to be multiplied N times to reach the output frequency F_O in the UHF band, significantly impacts the operation of the PLL. Specifically, it leads to degradation of key parameters, including stability, bandwidth, and lock time [4]. The reduction in effective gain requires, for example, an increase in the charge pump current to compensate for these effects, which was considered during the design process.

III. DESIGN PROCESS

The circuit was implemented using 65 nm CMOS technology from UMC. The operating conditions were simulated using corner analysis, considering variations in process parameters, supply voltage, and temperature (PVT). Analysis of process variations included cases of fast (F) or slow (S) transistors for NMOS and PMOS, respectively, as well as the typical (T) case. The details are presented in Table I.

A. Phase-Frequency Detector

The typical phase detector based on logic gates, such as XOR or JK flip-flops, is insensitive to frequency differences

TABLE I
PVT PARAMETERS

Parameter	Description
Process corner	TT, FF, FS, SF, SS
Supply voltage	1.2 V±10%
Temperature	-40 °C, 27 °C, 125 °C

and introduces a static phase error [5]. This manifests as narrow pulses at the PLL output, even when the input signals are synchronized.

In this work, the PFD, commonly used in Type II PLL circuits, was implemented. Due to its additional frequency detection capability, the capture range is extended to nearly the entire tuning range of the VCO [4], enabling the PLL to achieve synchronization even with significant frequency differences between the reference input signal and the feedback loop signal.

The PFD, shown in Fig. 2, based on [6], consists of D flip-flops and an AND gate. Unlike an XOR-based phase detector, it generates two output signals – UP and DOWN. The key difference in its operation is that it detects phase differences based only on one edge of the signals, in this case, the rising edge. Depending on which edge is detected first, pulses appear at the corresponding output.

Additionally, inverters were introduced to buffer the outputs of the flip-flops before feeding them into the charge pump. The DOWN output remains low by default, while the UP signal path includes an extra inverter, ensuring that the output is held high in its idle state to correctly drive the charge pump.

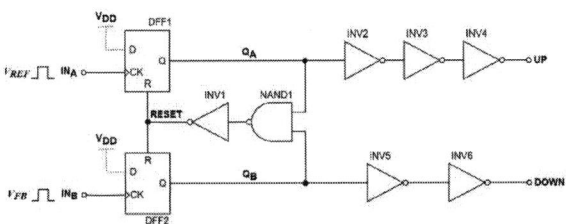

Fig. 2. Implemented Phase-Frequency Detector

B. Charge Pump

To enable the VCO to adjust its frequency according to the instantaneous phase shift, it is necessary to convert the detector's output signals into a stable control signal. This conversion is performed by the charge pump integrated with the PFD. The implemented design utilizes a single-ended charge pump with the switch in source, which ensures fast switching times and keeps the switched transistors in saturation [7]. The circuit presented in Fig. 3 is a modification of the basic version described in [8].

The charge pump consists of three biasing branches and two main branches, which include a current mirror and switches. The I_{BIAS1} and I_{BIAS2} inputs receive I_{b1} and I_{b2} currents

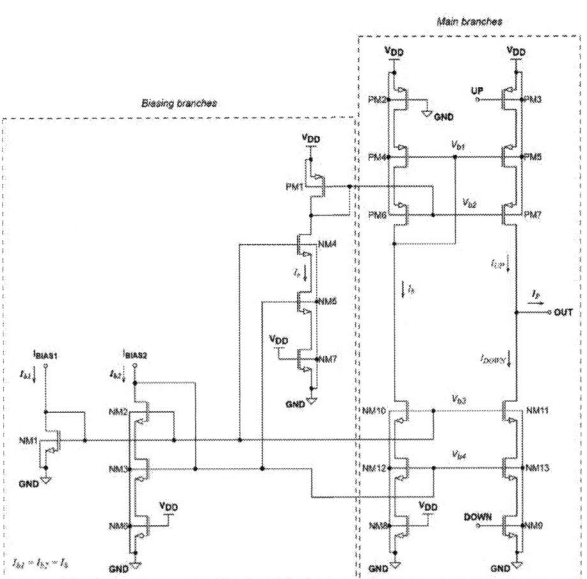

Fig. 3. Implemented Charge Pump

from ideal current sources, each with a magnitude of 8 µA. This value was selected to ensure the appropriate bandwidth for the PLL. The transistors NM1–NM3 generate bias voltages for NM10–NM13, forming a wide-swing cascode current mirror. Due to this configuration the charge pump may operate properly down to an output voltage of 0.2 V [9].

To achieve this, NM1 is approximately 10 times longer than NM3, ensuring that the gate voltage V_{b3} of NM2, NM4, NM10, and NM11 is around $V_{th} + 2V_{DS,sat}$, where V_{th} is the threshold voltage, and $V_{DS,sat}$ is the minimum drain-source voltage required to keep NM2–NM5 and NM10–NM13 in saturation. A similar approach is used to bias transistors PM6–PM7. The cascode configuration in the current sources significantly increases their output resistance, improving current matching between PMOS and NMOS transistors across a wide output voltage range.

This matching is particularly important during reset phases of the PFD, where internal delays, caused by signal propagation through the logic gates, result in narrow pulses on its outputs. During these brief moments, when UP and DOWN signals are simultaneously high, the same current flows through both sources in the charge pump. As a result, the output current I_P becomes effectively zero, preventing any change in the output voltage. The effectiveness of this behavior relies directly on the matching between I_{UP} and I_{DOWN}.

The I-V characteristic of the implemented charge pump for all PVT conditions is shown in Fig. 4 for all PVT conditions. The results indicate that within the 0.64–0.89 V range — identified in Sect. II.D as suitable for the VCO to generate the 867.84 MHz output — the obtained current mismatch does not exceed 10%.

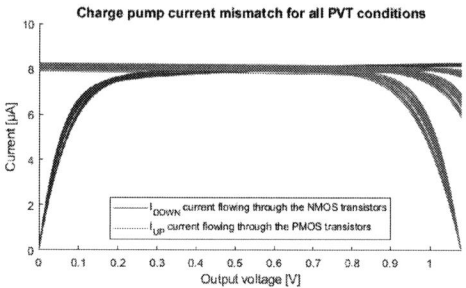

Fig. 4. The UP and DOWN currents of the implemented charge pump as a function of the output voltage for all PVT conditions.

C. Low-Pass Filter

The input voltage of the VCO is subject to significant fluctuations due to real-world disturbances affecting the system's operation. To mitigate these effects and achieve better attenuation at high frequencies, higher-order filters, typically second- or third-order, are used, as shown in Fig. 5 [10].

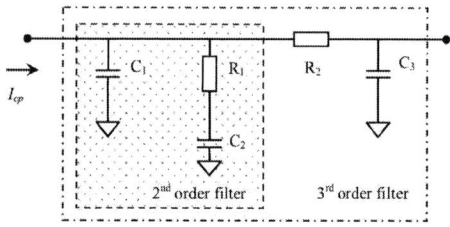

Fig. 5. Second- and third-order low-pass filter [10]

In this design, a third-order low-pass filter was implemented for the PLL loop filter. The component values were selected considering PLL lock time, reference frequency, and filter component sizing, following the steps described in [11].

The PLL bandwidth was set to 500 kHz, approximately 20 times lower than the 13.56 MHz reference frequency. A wider bandwidth enables shorter lock time and smaller filter components but increases susceptibility to noise and potential instability. Conversely, a narrower bandwidth slows down synchronization and complicates the integration of large capacitors in an IC design. The chosen value was deemed optimal, balancing fast synchronization with component integration feasibility.

Using equation (3) [11], which defines the PLL loop cutoff frequency F_C, the charge pump current I_{CP} was estimated at 8 µA, assuming a VCO gain K_v of approximately 4 GHz/V (as described in Sect. II.D), a divider factor N of 64, and a series resistance R_1 of 5 kΩ. The remaining filter component values were determined using a similar approach and are shown in Table II.

$$F_C = \frac{I_{CP} \cdot R_1 \cdot K_v}{N \cdot 2\pi} \tag{3}$$

TABLE II
LPF COMPONENT VALUES

Component	R_1	R_2	C_1	C_2	C_3
Value	5 kΩ	7.5 kΩ	21 pF	210 pF	4.2 pF

D. Voltage-Controlled Oscillator

A ring oscillator was chosen for the frequency generator due to its simple and compact design, eliminating the need for an inductive coil. This architecture enables the generation of the desired output frequencies across a wide tuning range. The design presented in Fig. 6 consists of seven inverting stages, with three inverters implementing the frequency tuning mechanism in a current-starved ring VCO architecture [12]. The remaining stages include Schmitt trigger flip-flops (ST1-ST2) with hysteresis and additional inverters at the outputs, which introduce extra delay, enhance noise immunity, and improve frequency stability.

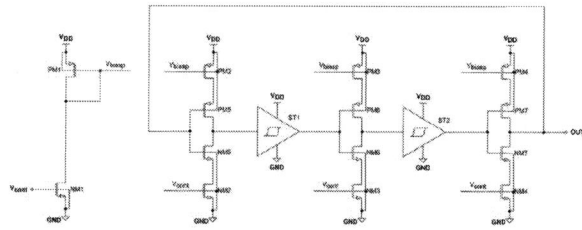

Fig. 6. Implemented Voltage-Controlled Oscillator

Despite their simplicity, ring oscillators are highly sensitive to PVT variations, which was accounted for by performing corner analysis simulations. The tuning characteristic (Fig. 7) was analyzed in the time domain for a control voltage range V_{cont} from 0.45 V to 1.2 V. The results indicate that the target frequency of 867.84 MHz is achievable within the voltage range of approximately 0.64–0.89 V, which limits the range where PLL input signals can synchronize.

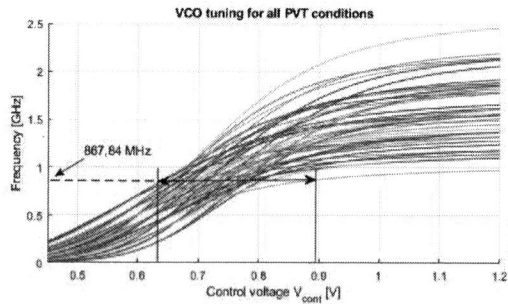

Fig. 7. VCO tuning range for all PVT conditions

Based on these waveforms, the VCO gain was determined, as shown in Fig. 8. The results indicate that for 867.84 MHz, the achieved gain ranges from 0.73 GHz/V to 6.53 GHz/V, with an average value of approximately 4 GHz/V.

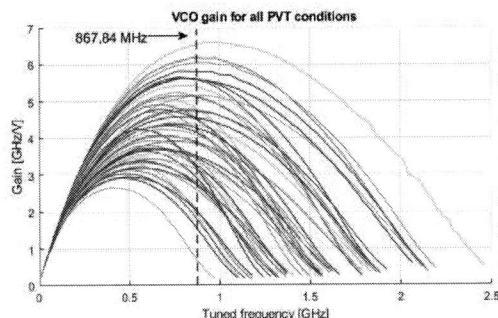

Fig. 8. VCO gain for all PVT conditions

E. Frequency Divider

Generating an output signal in the PLL loop with a frequency higher than the reference is achieved using a frequency divider. Its schematic is shown in Fig. 9.

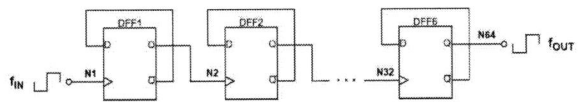

Fig. 9. Implemented divider based on D flip-flops

The circuit consists of D flip-flops (DFF1–DFF6) of Master-Slave architecture and functions similarly to a counter, producing one output pulse for every N input pulses. Each stage divides the frequency of the propagated signal by 2 and passes the result to the clock input of the next stage.

The number of stages was determined based on the PLL design specifications. To achieve the required frequency multiplication, the divider takes the PLL output signal (867.84 MHz) as its input f_{IN}, dividing it until its own output frequency f_{OUT} matches and synchronizes with the PLL input at 13.56 MHz. Since the PLL reference frequency of 13.56 MHz needs to be multiplied $N = 64$ times to achieve 867.84 MHz, which falls precisely within the UHF band, exactly six flip-flops were used.

IV. VERIFICATION OF THE ENTIRE PLL OPERATION

The functionality of the complete PLL system was verified through transient and frequency-domain simulations under various PVT conditions.

A. Loop Stability and Bandwidth

The PLL transfer function was plotted for typical parameters to analyze its stability. The graphs in Fig. 10 were generated using a MATLAB script, where the transfer function of each block was implemented. The open-loop gain and phase response clearly show the zero introduced by the LPF, which ensures stability with a 49° phase margin. The open-loop transfer function confirms the expected bandwidth, which is approximately 370 MHz.

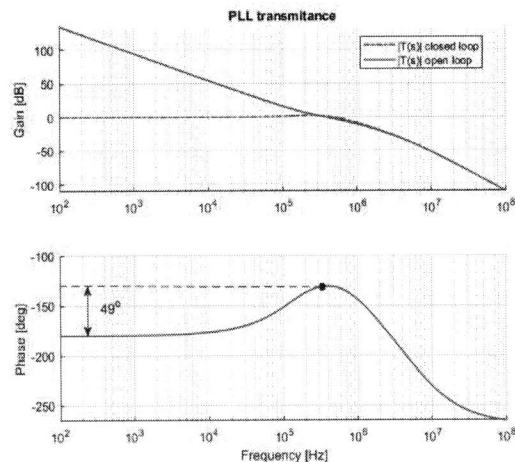

Fig. 10. Open-loop and closed-loop transfer functions of the PLL and phase response

B. Phase Noise

Based on the closed-loop gain, the approximate phase noise level was estimated for each block in the system, as shown in Fig. 11. The analysis is based on the results obtained from periodic steady-state (*pss*) and phase noise (*pnoise*) simulations. The total noise (TOTAL) was calculated as the sum of the contributions from all blocks, including the reference (REF), with its value representing the area under the curve.

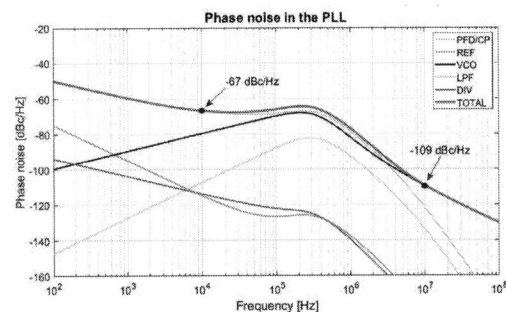

Fig. 11. Phase noise characteristics of the PLL and individual blocks

The total phase noise was measured to be below -60 dBc/Hz at 10 kHz and below -100 dBc/Hz at 10 MHz from the carrier. Within the operating bandwidth, the dominant noise source is the PFD/CP cascade, whose impact decreases outside the loop bandwidth, where the VCO noise becomes dominant. For the loop filter, the primary noise mechanism is thermal noise, which also influences the overall response. The reference clock and frequency divider (DIV) contribute negligibly to the total noise level.

C. Settling Time

A power-up simulation was also conducted to measure the PLL synchronization time. The simulation results are shown in Fig. 12. The supply voltage was applied with a ramp rate of 1 V / 1 μs, while the charge pump current was set to a constant 8 μA.

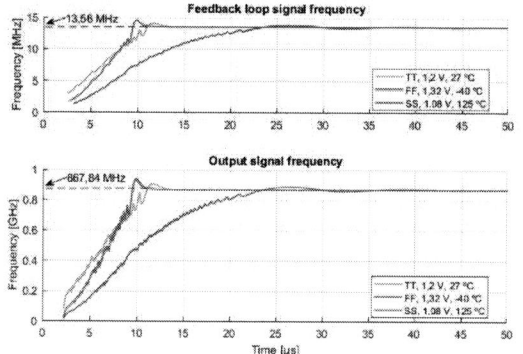

Fig. 12. Time-domain response of the PLL after power-up for three extreme corner cases

The time required for the output frequency to reach a steady state is defined as the settling time. For the TT corner at 1.2 V supply voltage and 27 °C, the settling time was approximately 10 μs. The test was also performed under two extreme operating conditions:

- Fastest case (FF, 1.32 V, -40 °C): ≈ 8 μs
- Slowest case (SS, 1.08 V, 125 °C): ≈ 30 μs

This significant variation in settling time is due to the large spread in VCO gain (K_v). To mitigate this, K_v variations can be compensated by adjusting the charge pump current (I_{CP}), which in turn modifies the PFD gain (K_d). This approach helps stabilize the loop bandwidth and synchronization time.

Based on the above time-domain waveforms, once the output frequency stabilized, the average supply current I_{DD} of the entire phase-locked loop circuit was measured. For the fastest, typical, and slowest corner cases, it amounted to 1.30 mA, 1.01 mA, and 1.01 mA, respectively.

V. CONCLUSION

The simulation results confirmed that the PLL effectively locks to the desired frequency of 867.84 MHz, synchronizing the output signal with the 13.56 MHz reference signal maintaining stable operation across the defined voltage and temperature ranges. The system achieves fast synchronization within 50 μs after activation, a critical requirement for NFC standards. This rapid locking time ensures compliance with the ISO/IEC 14443 standard, which mandates that communication must be established within a few milliseconds after the system wakes up.

An additional feature of the design is the ability to extract a 27.12 MHz signal from the output of the second-to-last stage of the frequency divider, which is twice the PLL reference

frequency (13.56 MHz). This frequency is increasingly used for clocking digital sections in NFC devices.

Future development possibilities include enhancing the ring oscillator's gain stability against PVT variations, which is essential for maintaining the loop bandwidth and preventing instability. One approach is to compensate for gain fluctuations by controlling the charge pump current. This can be achieved by adjusting the reference current or adding extra branches in the charge pump, which could be enabled or disabled as needed (binary control). Another method to mitigate loop parameter variations is trimming resistors and capacitors in the low-pass filter to fine-tune performance.

ACKNOWLEDGMENT

The Authors would like to thank OmniChip Sp. z o.o. for their support, valuable insights, and constructive feedback throughout this work. Appreciation is also extended to the Institute of Microelectronics and Optoelectronics at the Faculty of Electronics and Information Technology, Warsaw University of Technology, for providing access to the necessary software and research resources that significantly contributed to the completion of this project.

REFERENCES

[1] S. Kolev, "Designing a NFC system," *2021 56th International Scientific Conference on Information, Communication and Energy Systems and Technologies (ICEST)*, Sozopol, Bulgaria, 2021, pp. 111-113, doi: 10.1109/ICEST52640.2021.9483482.

[2] "MT-086 TUTORIAL Fundamentals of Phase Locked Loops (PLLs).", Analog Devices. Available: https://www.analog.com/media/en/training-seminars/tutorials/MT-086.pdf

[3] T. H. Lee, *The design of CMOS radio-frequency integrated circuits*. Cambridge, UK; New York: Cambridge University Press, 2004.

[4] Behzad Razavi, *Design of analog CMOS integrated circuits*. New York, NY: Mcgraw-Hill Education, 2017.

[5] D. Banerjee, *PLL Performance, Simulation, and Design*, 6th ed., 2023

[6] S. Soliman, F. Yuan, and K. Raahemifar, "An overview of design techniques for CMOS phase detectors", *2002 IEEE International Symposium on Circuits and Systems. Proceedings (Cat. No.02CH37353)*, vol. 5, pp. V–457V–460, doi: https://doi.org/10.1109/iscas.2002.1010739.

[7] W. Rhee, "Design of high-performance CMOS charge pumps in phase-locked loops", May 1999, doi: https://doi.org/10.1109/iscas.1999.780807.

[8] M. Karbalaei Mohammad Ali and O. Hashemipour, "A simple and high performance charge pump based on the self-cascode transistor", *Analog Integrated Circuits and Signal Processing*, vol. 100, no. 3, pp. 633–638, Jul. 2019, doi: https://doi.org/10.1007/s10470-019-01478-y.

[9] R. Jacob Baker, *CMOS: Circuit Design, Layout, and Simulation*. Hoboken, Nj: John Wiley & Sons; [Piscataway, Nj], 2010.

[10] V. Valenta, G. Baudoin, and M. Villegas, "Phase Noise Analysis of PLL Based Frequency Synthesizers for Multi-Radio Mobile Terminals," *2008 3rd International Conference on Cognitive Radio Oriented Wireless Networks and Communications (CrownCom 2008)*, pp. 1–4, May 2008, doi: https://doi.org/10.1109/crowncom.2008.4562555.

[11] "PLL Loop Filter Design and Fine Tuning", Renesans Electronics, Dec. 05, 2024. https://www.renesas.cn/zh/document/apn/pll-loop-filter-design-and-fine-tuning

[12] S. Suman, K. G. Sharma, and P. K. Ghosh, "Analysis and design of current starved ring VCO," *IEEE Xplore*, Mar. 01, 2016. http://ieeexplore.ieee.org/document/7755299/

Mixed Design of Integrated Circuits and Systems – MIXDES 2025

Optimum Design of a Mostly-Digital Fleischer-Laker Switched-Capacitor Bilinear Bandpass Filter in Standard CMOS Technology

Hugo Serra, João Pedro Oliveira, João Goes
LASI, UNINOVA-CTS, Caparica, Portugal
Department of Electrical and Computer Engineering, NOVA FCT, Caparica, Portugal

Abstract—This paper presents a Fleischer-Laker switched-capacitor (SC) bilinear bandpass filter implemented using an inverter-based amplifier. Due to the generalized scaling used in advanced deep-submicron CMOS technology over the past decades, it is becoming increasingly more difficult to design high-gain high-bandwidth opamps, due to the reduction of the supply voltage and of the intrinsic gain of the transistors. Since the amplifier used is implemented using inverters, it can take advantage of the improved transistor performance in smaller nodes (chip area, transistor transit frequency, and power dissipation), which is mainly exploited by digital circuits. The filter circuit was designed in a 28-nm bulk-CMOS technology, using a supply voltage of 0.9 V and a clock frequency of 100 MHz. Simulation results show that the filter's central frequency is approximately 10 MHz, with a gain of 0 dB, and a quality factor of 10/3. The amplifiers have a typical gain of 42.5 dB, the SC filter has a SNR of 54.4 dB, an IM3 of -63.6 dB, and the circuit's total power dissipation is 2.5 mW.

Keywords—Bandpass filter, bilinear filter, Fleischer-Laker topology, inverter-based amplifier, switched-capacitor circuit

I. INTRODUCTION

The scaling of transistors in advanced deep-submicron CMOS technologies has led to significant improvements in terms of chip area, power dissipation, and transistor transit frequency, which are mostly exploited by digital circuits. In spite of these improvements, there are also negative effects, due to the reduction of the transistors' channel lengths.

When scaling transistor technology, it is desired to maintain the same electric field, which means that the voltages and transistor dimensions should be scaled proportionally. However, in digital circuits, a low current is needed when the transistor is turned OFF, which requires a high V_{th}. This means that the threshold voltage (V_{th}) and the supply voltage can not be scaled in the same proportion. With the scaling performed over the past decades, the maximum output swing of analog circuits has been decreasing, as well as the transistor's intrinsic gain (gm/gds), which has impacted the performance of analog blocks [1], [2]. The process variability also becomes a concern

This work was funded by the European Union, through the European Regional Development Fund, and by national funds, through the Foundation for Science and Technology, under the Lisboa 2030 Program, with references LISBOA2030-FEDER-00816400 and LISBOA2030-FEDER-00917900, and within the scope of the Research Unit CTS - Center for Technology and Systems/UNINOVA/FCT/NOVA, with the reference CTS/00066.

since the variation becomes a larger percentage of the device dimensions in smaller nodes.

This paper proposes a systematic design methodology for the optimization of the gain blocks in filter circuits using inverter-based amplifiers instead of the conventional high-gain high-bandwidth amplifiers. The reason behind this choice is to take advantage performance of digital circuits improving with the scaling-down of CMOS technology.

The paper is organized as follows. Section II describes and analyzes the Fleischer-Laker switched-capacitor (SC) bilinear bandpass filter and the pseudo-differential inverter-based amplifier implemented in this paper. In Section III, the simulation results of inverter-based amplifier and SC filter are presented. Finally, in Section IV, the main conclusions from the work carried out in this paper are drawn.

II. MOSTLY-DIGITAL FLEISCHER-LAKER SC FILTER

A. Bilinear Bandpass Fleischer-Laker SC filter

The bilinear bandpass SC filter circuit, shown in Fig. 1, is based on the Fleischer–Laker architecture [3]. Note that, for simplicity purposes, the equivalent single-ended version of the filter is shown, however, the SC filter was implemented differentially.

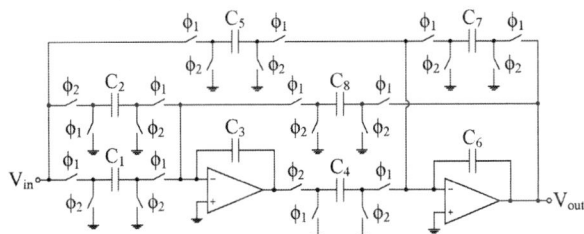

Fig. 1. Bilinear bandpass Fleischer-Laker SC filter.

Considering that the output signal is sampled at the end of clock phase ϕ_2, that the input signal only changes value once per clock period ($1/F_s$), i.e., the input signal is already sampled and not a continuous signal, and that the gain of the amplifiers (A_v) is infinite, the Fleischer-Laker SC filter's transfer function is given by (1).

$$H_{bp}^{\phi_2}(z) = \frac{V_{out}(z)}{V_{in}(z)} = \frac{C_2 C_4 - (C_1 C_4 + C_3 C_5 (z-1)) z}{C_4 C_8 z + C_3 (z-1)(C_6(z-1) + C_7 z)} \quad (1)$$

To obtain a bilinear bandpass frequency response, one of the zeros of the transfer function should be at $z = 1$, while the other zero should be at $z = -1$ (i.e., at $F_s/2$). To place the zeros at $z = \pm1$, the correspondence shown in (2) should be used.

$$Num\left[H_{bp}^{\phi_2}(z)\right] = C_2C_4 - (C_1C_4 + C_3C_5(z-1))\,z = 0$$
$$\Rightarrow z = C_3C_5 - C_1C_4 \pm \sqrt{4\,C_2C_3C_4C_5 + (C_1C_4 - C_3C_5)^2} \quad (2)$$
$$\text{If } \begin{cases} C_1 = C_2 = C_5 \\ C_3 = C_4 = C_6 \end{cases} \Rightarrow z = \pm1 \Rightarrow \text{Bilinear response}$$

With the previously mentioned simplification, the SC filter's transfer function is given by (3).

$$H_{bp}^{\phi_2}(z) = \frac{C_1(z+1)(z-1)}{C_3(z-1)^2 + (C_7(z-1) + C_8)z} \quad (3)$$

Expressions for the filter's gain (K_p), central frequency (f_c), and quality factor (Q_p) can be obtained using the general transfer function of a continuous-time second-order bandpass system (4) [4], where $\omega_0 = 2\pi f_c$.

$$H(s) = K_p \frac{\omega_0/Q_p\, s}{s^2 + \omega_0/Q_p\, s + \omega_0{}^2} \quad (4)$$

Considering that the Fleischer-Laker SC filter described in this work has a bilinear type response, the continuous-time transfer function can be converted, into discrete-time, using the bilinear transform, i.e., by replacing the frequency parameter s, in (4), with $2F_s(z-1)(z+1)^{-1}$. The filter's gain (5), central frequency (6), and quality factor (7) can then be obtained by equating the normalized denominators and numerator of the simplified transfer function (3) with the ones obtained from transfer function (4), after converting it into discrete-time.

$$K_p = \frac{2\,C_1}{C_7} \quad (5)$$

$$\omega_0{}^2 = \frac{4\,C_8F_s{}^2}{4\,C_3 + 2\,C_7 - C_8} \quad (6)$$

$$Q_p = \frac{\sqrt{(4\,C_3 + 2\,C_7 - C_8)\,C_8}}{2\,C_7} \quad (7)$$

All capacitors influence more than one parameter, with the exception of capacitor C_1, which only changes the filter's gain. The relations between the filter parameters and the capacitor values is summarized in (8).

$$\begin{cases} K_p \propto C_1,\, 1/C_7 \\ f_c \propto 1/C_3,\, C_8 \\ Q_p \propto C_3,\, 1/C_7,\, C_8 \end{cases} \quad (8)$$

B. Inverter-based Amplifier

The amplifiers used in the Fleischer-Laker SC filter have been implemented using a three-stage inverter-based architecture [5], [6], which is shown in Fig. 2. The amplifier consists of three paths: a high-gain path consisting of three inverters; a high-speed path with a single inverter; and a common-mode feedback (CMFB) path consisting of two inverters, where

the output common-mode voltage is sensed by resistors R_S and the input common-mode voltage is controlled through resistors R_{FB}. Capacitors C_{Z1} and C_{Z2} are used to stabilize the CMFB loop. The amplifier's phase margin is improved by splitting the poles, through the use of the RC network formed by resistor R_M and capacitor C_M [6]. The high-speed path makes the amplifier more robust to mismatch variations at the cost of reducing the gain. A resistor (R_P) can be connected in parallel with the second inverter of the high-gain path, which also improves the robustness to mismatch variations at the cost of the gain. When both of these approaches are used, to increase robustness to mismatch variations, it is possible to achieve a better solution with a lower impact in the amplifier's gain.

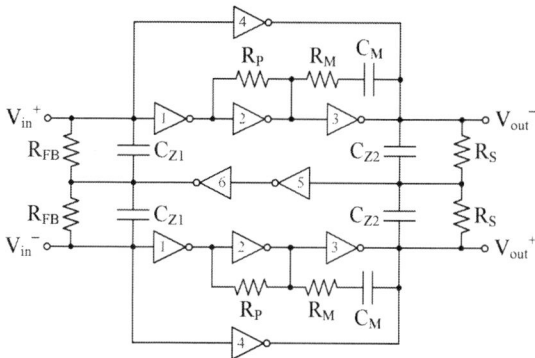

Fig. 2. Three-stage pseudo-differential inverter-based amplifier [5], [6].

III. SIMULATION RESULTS

The bilinear bandpass SC biquad filter was implemented in a standard 28 nm bulk-CMOS technology, and simulated in Spectre, using a nominal supply voltage of 0.9 V and a clock frequency of 100 MHz. The bilinear bandpass SC filter was designed to have a central frequency (f_c) of 10 MHz, a passband between 8.5 MHz and 11.5 MHz, corresponding to a quality factor of 10/3, and maximum attenuation at the lower and higher stopband frequencies (0.1 MHz and 50 MHz, respectively) using the optimization methodology described in [7], considering the amplifier blocks as ideal voltage-controlled voltage sources (VCVSs) with moderate gain (40 dB).

A. Inverter-based Amplifier

The inverter-based amplifier was designed to achieve a settling-time error below 0.1 %. To achieve an error below this value, it is necessary to have a given gain-bandwidth product (GBW), that can be calculated using (9). Since the filter is designed to operate with a clock frequency F_s of 100 MHz, for the given feedback factor, the GBW of the amplifier has to be higher than 220 MHz to ensure a settling error below 0.1 %, in the duration of each clock phase ($1/2F_s$).

$$e^{-GBW\,[\text{rad/s}]\,T_s/2} < 0.1\% \quad \Leftrightarrow$$
$$\Leftrightarrow GBW\,[\text{Hz}] > -\ln(0.1\%)\frac{F_s}{\pi} \quad (9)$$

Another important parameter to consider, when determining the amplifier's GBW, is the equivalent load impedance (worst case) that the filter will impose on the output of each amplifier. In this architecture, this occurs on the second amplifier during phase ϕ_1, and the load impedance is given by (10), where R_{on} represents the switches ON resistance which, for simplicity purposes, are considered to have the same value, and C_L the input (load) capacitance from the following stage.

$$Z_{22}^{\phi_1} = \left(\frac{1}{sC_6}\right) \| \left(\frac{1}{sC_7} + 2R_{on}^{\phi_1}\right) \| \left(\frac{1}{sC_8} + 2R_{on}^{\phi_1}\right) \| \left(\frac{1}{sC_L}\right) \quad (10)$$

The design used in the inverter-based amplifier is shown in Table I, where W_p and W_n represent the width per finger of the PMOS and NMOS devices in the inverters, respectively, L_{inv} their length and F_{inv1} to F_{inv6} the respective number of fingers used in each inverter. Note that the value of $Z_{22}^{\phi_1}$ was estimated using the capacitor values shown in Table II and switches $R_{on} = 100\ \Omega$, i.e., $W_{sw} = 2\ \mu\text{m}$, $F_{sw} = 14$, $L_{sw} = 30\ \text{nm}$, $V_{cm} = 450\ \text{mV}$, where W_{sw} is the transistor's width per finger, F_{sw} is the number of fingers, and V_{cm} is the circuit's common-mode voltage.

TABLE I
DESIGN PARAMETERS USED IN THE INVERTER-BASED AMPLIFIER.

R_{FB} [kΩ]	R_S [kΩ]	R_M [kΩ]	R_P [kΩ]	C_{Z1} [fF]	C_{Z2} [fF]	C_M [fF]
49.2	29.5	1.4	6.0	1200.8	600.4	300.2

W_p [nm]	W_n [nm]	F_{inv1}	$F_{inv2,3}$	F_{inv4}	$F_{inv5,6}$	L_{inv} [nm]
750.0	380.0	40	10	20	1	30.0

The inverter-based amplifier's DC gain as a function of the input common-mode voltage is shown in Fig. 3 and the amplifier's frequency response is shown in Fig. 4, which was obtained using a common-mode voltage of 450 mV and a load capacitance of approximately 4 pF to account for the capacitive load of a chip's padring and measuring probes. Results show a gain of 42.5 dB and a GBW of 989 MHz, which is above the 220 MHz needed to achieve a settling error below 0.1 %.

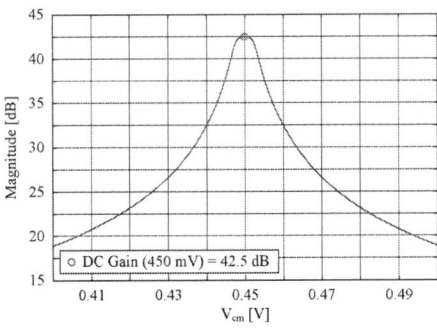

Fig. 3. Gain of the inverter-based amplifier as a function of the input common-mode voltage.

B. Bilinear Bandpass Fleischer-Laker SC filter

The design obtained from the optimization process is shown in Table II and the resulting frequency response of the bilinear bandpass SC filter is shown in Fig. 5, which was obtained from a transient simulation, using an impulse as the input signal, and then by calculating the fast Fourier transform (FFT)

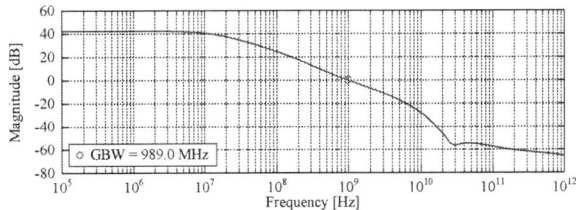

(a) Gain response

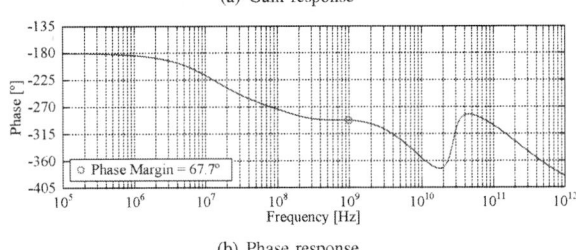

(b) Phase response

Fig. 4. Frequency response of the inverter-based amplifier.

of the sampled output signal at the end of clock phase ϕ_2. Note that, since this filter requires an already sampled input signal to obtain a notch, at $F_s/2$, with a high depth, a SC integrator circuit was used before the bilinear bandpass SC filter to perform the sampling operation. Since it is outside of the scope of this paper, the SC integrator circuit used is not described. Results from the filter's impulse response show a notch depth at $F_s/2$ of -59.3 dB. The SC filter has a total power consumption of 2.5 mW.

TABLE II
DESIGN PARAMETERS USED IN THE
BILINEAR BANDPASS FLEISCHER-LAKER SC FILTER.

C_1 [fF]	C_2 [fF]	C_3 [fF]	C_4 [fF]	C_5 [fF]	C_6 [fF]	C_7 [fF]	C_8 [fF]
200.1	200.1	2003.8	2003.8	200.1	2003.8	362.4	857.1

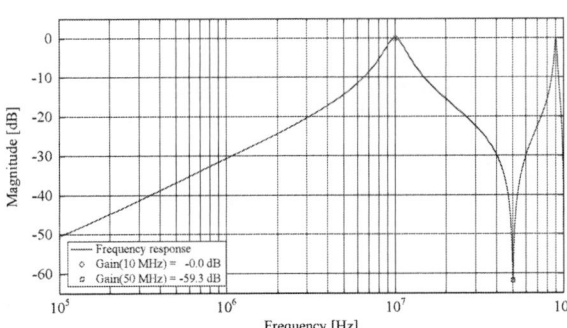

Fig. 5. Frequency response of the bilinear bandpass Fleischer-Laker SC filter.

The simulated output spectrum of the bilinear bandpass SC filter is shown in Fig. 6, for an input signal with two tones ($f_1 = 8.5$ MHz and $f_2 = 11.5$ MHz) each with 100 mV of amplitude, resulting in an 3rd order intermodulation product (IM3) of 63.6 dB. Note that the output spectrum has been normalized with the amplitude of the input signals.

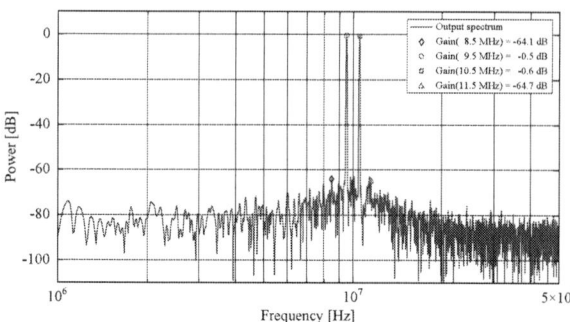

Fig. 6. Normalized output spectrum of the bilinear bandpass Fleischer-Laker SC filter for an input with two tones (f_1 = 9.5 MHz and f_2 = 10.5 MHz) each with an amplitude of 100 mV.

The notch's depth at $F_s/2$ was also validated, by using an input signal with two tones (f_c = 821/8192$\times F_s$ = 10.022 MHz and f_{sh} = 4091/8192$\times F_s$ = 49.939 MHz) each with 100 mV of amplitude. The resulting output spectrum of the bilinear bandpass SC filter is shown in Fig. 7. The depth of the notch is -52.8 dB, at $G_{f_{sh}}$, which is in moderately good agreement with the -59.3 dB expected from the frequency response obtained from the simulation of the impulse response (Fig. 5). For reference proposes, if the input signal is not sampled, the attenuation is only -28.9 dB.

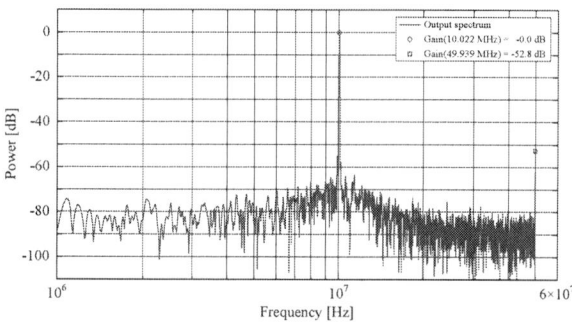

Fig. 7. Normalized output spectrum of the bilinear bandpass Fleischer-Laker SC filter for an input with two tones (f_1 = 10.022 MHz and f_2 = 49.939 MHz) each with an amplitude of 100 mV.

The bilinear bandpass SC filter was also tested under 12 different process, voltage, and temperature (PVT) corners (TT/FF/SS, $V_{DD}\pm5\%\,V_{DD}$, 0°/85°) and under 100 Monte Carlo (MC) cases of mismatch variations, the performance obtained from these two tests is summarized in Table III.

TABLE III
PERFORMANCE OF THE BILINEAR BANDPASS FLEISCHER-LAKER SC FILTER UNDER PVT CORNERS AND MISMATCH VARIATIONS.

		Impulse response performance				Transient performance			
		f_c [MHz]	$G_{f_{sl}}$ [dB]	G_{f_c} [dB]	$G_{f_{sh}}$ [dB]	SNR [dB]	THD [dB]	IM3 [dB]	$G_{f_{sh}}$ [dB]
Nom.		10.00	-50.89	0.03	-59.16	54.38	-80.56	-63.58	-52.75
MC	μ	9.97	-49.70	0.01	-60.81	53.74	-77.14	-63.53	-52.81
MC	σ	0.03	1.42	0.04	1.38	0.08	3.21	2.49	0.13
PVT	μ	9.94	-50.27	-0.02	-60.02	53.91	-72.71	-57.46	-52.41
PVT	σ	0.16	0.65	0.59	4.28	0.97	11.35	12.78	4.64

f_{sl} – lower stopband frequency (0.1 MHz), f_{sh} – higher stopband frequency (49.939 MHz)

Simulation results show that the filter has a central frequency of approximately 10 MHz and a quality factor of 10/3. The pseudo-differential inverter-based amplifier has a typical gain of 42.5 dB and a GBW of 989 MHz for a load of 4 pF. The Fleischer-Laker SC bilinear bandpass filter has a signal-to-noise ratio (SNR) of 54.4 dB, an IM3 of -63.6 dB and a total power dissipation of 2.5 mW. The depth of the notch at $F_s/2$ was also evaluated, by observing the SC filter's output spectrum for an input with two tones, and a gain of -52.8 dB was obtained, which is in moderately good agreement with the -59.3 dB expected from the frequency response obtained from the simulation of the impulse response.

IV. CONCLUSION

This paper presented a Fleischer-Laker SC bilinear bandpass filter implemented using a pseudo-differential inverter-based amplifier instead of a traditional high-gain high-bandwidth amplifier. This type of circuit was chosen since inverter-based amplifiers can take advantage of the improved performance of digital circuits in smaller technology nodes, while traditional high-gain high-bandwidth opamps are affected by the reduction of the intrinsic gain of the transistors and increased process variability, making it more difficult to achieve a high gain with single-stage architectures.

The pseudo-differential inverter-based amplifier was implemented with a high-speed path with a single inverter and with a resistor in parallel with the second inverter of the high-gain path. This was done to improve the robustness of the amplifier to PVT and mismatch variations, which would be otherwise poor. The cost of this approach is a slight reduction in the amplifier's gain.

The filter was designed in a standard 28-nm bulk-CMOS technology, using a supply voltage of 0.9 V and a clock frequency of 100 MHz. Following the proposed design methodology, the biquad is robust against PVT corners and mismatch variations, and the complete filter dissipates only 2.5 mW.

REFERENCES

[1] A. Baschirotto, V. Chironi, G. Cocciolo, S. D'Amico, M. De Matteis, and P. Delizia, "Low power analog design in scaled technologies," in *Proc. Topical Workshop Electron. Particle Phys. (TWEPP)*, Jan. 2009, pp. 103–109.

[2] P. R. Kinget, "Scaling analog circuits into deep nanoscale CMOS: Obstacles and ways to overcome them," in *Proc. IEEE Custom Integrated Circuits Conf. (CICC)*. Sep. 2015, pp. 1–8.

[3] P. Fleischer and K. Laker, "A family of active switched capacitor biquad building blocks," *Bell Syst. Tech. J.*, vol. 58, no. 10, pp. 2235–2269, Dec. 1979.

[4] A. S. Sedra and K. C. Smith, *Microelectronic circuits*, 6th ed. Oxford University Press, 2009.

[5] L. Breems, M. Bolatkale, H. Brekelmans, S. Bajoria, J. Niehof, R. Rutten, B. Oude-Essink, F. Fritschij, J. Singh, and G. Lassche, "A 2.2 GHz Continuous-Time $\Sigma\Delta$ ADC With -102 dBc THD and 25 MHz Bandwidth," *IEEE J. Solid-State Circuits*, vol. 51, no. 12, pp. 2906–2916, Dec. 2016.

[6] M. Neofytou, M. Zhou, M. Bolatkale, Q. Liu, C. Zhang, G. Radulov, P. Baltus, and L. Breems, "A 1.9 mW 250 MHz bandwidth continuous-time $\Sigma\Delta$ modulator for ultra-wideband applications," in *Proc. IEEE Int. Symp. Circuits Syst. (ISCAS)*, May 2018, pp. 1–4.

[7] H. Serra, R. Santos-Tavares, and N. Paulino, "Transistor-level optimization methodology for the complete design of switched-capacitor filter circuits," *Int. J. Circ. Theor. Appl.*, vol. 49, no. 1, pp. 94–113, Jan. 2021.

Mixed Design of Integrated Circuits and Systems – MIXDES 2025

Practical Implementation of Voltage-to-Current and Current-to-Voltage Converter in High Voltage SOI Technology

Mariusz Jankowski

Department of Microelectronics and Computer Science
Lodz University of Technology
Lodz, Poland
mariusz.jankowski@p.lodz.pl

Abstract—**This paper presents an introduction to the analysis of a circuit that can serve as both a voltage-to-current and current to voltage converter. This circuit enables transitions between low-voltage and high voltage ranges of analog signal processing and can produce two counterphase waveforms. The introduction includes the schematic of the circuit, the schematic equipped with overvoltage protection device set and the full layout of the latter schematic version. The consecutive design steps show a degradation in the quality of circuit operation. These changes are presented in relation to the change of selected operation parameters. Conclusion and planned further steps of the analysis are proposed.**

Keywords—**CMOS design; voltage to current converter; current to voltage converter; SOI technology; current mode circuits; high voltage design.**

I. Introduction

High-voltage (HV) signals are used in numerous analog applications. Some of them are related to the automotive industry, where high voltage signals are, among others, to drive antennae of wireless transmission systems like RFID. Numerous automotive applications rely on wireless data transmission, such as localization [1], vehicle authentication [2], or toll collection [3]. Many of such systems use HV waveforms to drive the transmitter antennae [4] for improved data transmission.

The shapes of the utilized waveform may vary in different systems, as presented in [5]. Some waveform shapes may require specific operations, like arithmetic ones. Such operations often are easier to do in current domain, both in case of low voltage (LV) with typical fully planar transistors only, and HV systems [6] that require at least some not fully planar devices. Moreover, differential signals may be used for signal processing or wireless transmission.

The circuit discussed is able to convert LV/HV waveforms into current domain for simples processing and back into voltage mode LV/HV waveforms. Even more interestingly, the discussed implementation of the operation mode discussed in [6] is able to provide two counterphase copies of the developed waveforms e.g. for further differential processing in both voltage and current modes.

This research was supported by National Science Center in Poland, grant number 2013/11/B/ST7/01742 and Lodz University of Technology.

II. Design of the Converter

Basic circuits that share the same mode of operation (Fig. 1) were first introduced as designed in SmartIs 0.8 μm SOI (Silicon on Insulator) process by Atmel [6] (later developed into TeleSmart 0.8 μm by Telefunken). The first differential implementation able to provide two counterphase waveforms was rather referred to as a HV differential amplifier, as the circuit can be used for such applications. Another implementation was attempted with the H35B4 0.35 μm process by AMS and finally the most complete one with XDM 1.0 μm SOI by XFAB (Fig. 2).. This design has reached the furthest stage of advancement with a full layout with post-layout extraction.

As shown in Fig. 1, the operation of the converter relies on technology resistors. This fact may imply limited accuracy of the presented solution, as the values of single resistors can vary in semiconductor technologies due to both process variations and temperature. The presence of resistors in voltage to current and current to voltage converters is discussed [7, 8] and MOS only voltage to current [9, 10, 11] and current to voltage converters are proposed [12, 13].

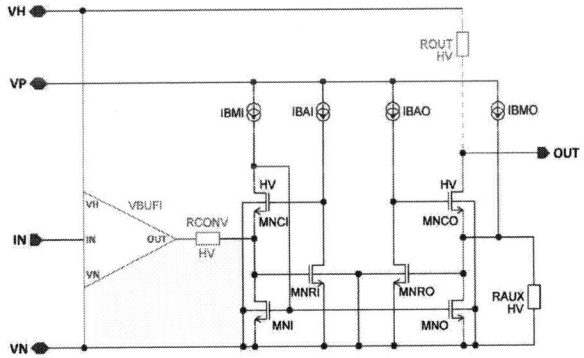

Fig. 1. Operation mode of the circuit discussed.

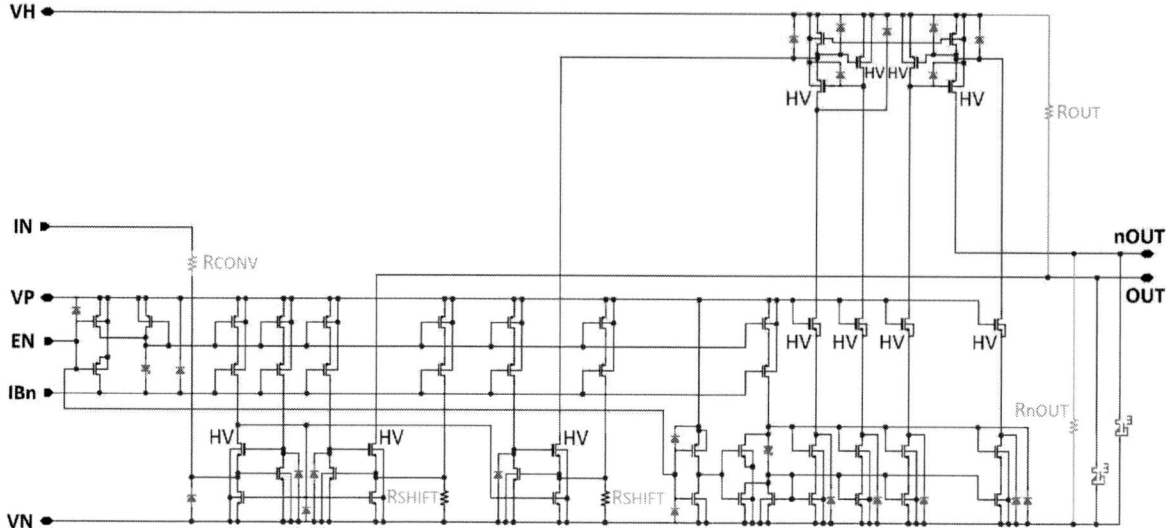

Fig. 2. The discussed implementation of the voltage-to-current / current to voltage converter.

In fact, design of precision resistors can be a challenging task [14], but the presented converted was design as voltage to current and then back to voltage converter [6]. The goal of such an approach is to enable arithmetic operations on HV waveforms by executing these operations in the current domain and then converting the processed current signals into HV waveforms.

Fig. 1 shows that voltage to current conversion commences on two identical RCONV and RAUX resistors that are connected to a regulated cascade mirror for high-quality operation. The RCONV resistor is connected to the source of a cascade transistor on the input side and RAUX on the output side of the mirror, respectively. The RCONV connected between the input and the current mirror produces a current flow proportional to this voltage drop. To produce current proportional to the voltage drop between the input and ground, the RAUX is introduced. It produces a current proportional to the voltage drop between the output side of the mirror and the ground node. This current adds to the output current of the mirror. The voltages at the sources of the input and output side cascade transistors of the utilized precision mirror are nearly identical. Thus, the total output current of the mirror is proportional to the voltage drop between the input and ground nodes of this circuit.

The produced current can be provided to the next signal processing block through the OUT node (Fig. 1) or converted back to the voltage waveform at the optional ROUT resistor (Fig. 1). Thus, the circuit presented in Fig. 1 can operate as a voltage-to-current converter or voltage amplifier. In the case of the latter application, the output node must be buffered against the influence of the following function block. Analogously, the preceding function block should be buffered from the influence of the RCONV resistor, so that the input voltage (the IN node) voltage is unaltered by the current produced on the RCONV resistor. A specialized HV buffer that fits these applications has been introduced [6] and analyzed [15] by the author.

The circuit presented in Fig. 1 can be used as a kind of LV into HV range waveform amplifier. However, such an approach can produce unwanted noise transformation in both frequency and amplitude [5]. Generation of HV waveforms and their subsequent processing can be a better procedure [5].

The circuit presented in Fig. 1 can also be used as a current to voltage converter if the RCONV and RAUX resistors are removed and only the ROUT remains. Also, as it was already shown that both the input and output sides of the discussed circuit can operate in current mode, its general application potential can be described as comprising LV/HV voltage to current converter, current to LV/HV voltage converter, LV/HV to LV/HV voltage amplifier/attenuator, and current voltage amplifier/attenuator. All this functionality is based on operation of a single high-precision current mirror.

Fig. 2 shows the full implementation of the circuit presented in Fig. 1 and is equipped with circuitry and two outputs that provide two counterphase signals. Depending on the presence or absence of specific resistors, it can operate in all the modes or roles. Nevertheless, in this paper, it is discussed as a voltage-to-current and current in voltage converter with two counterphase outputs. Such an operation can be interpreted as an HV differential amplifier.

It can be seen that some of the MOS transistors are HV devices. MOS devices are considered to be HV (or sometimes mid voltage (MV)) if they are not fully planar, owing to which they can operate with higher voltage drops between their terminals. As a side effect, such transistors have different maximum interterminal voltages for specific pairs of their terminals. Moreover, these maximum interterminal voltages are, as a rule, different from those in the case of LV transistors. Thus, cooperation of LV and MV/HV devices can be problematic because of the possibility of exerting overvoltage conditions on LV devices by their HV neighbors. Such safe operating area (SOA) breaches should be avoided, as they can lead to device destruction or lifetime reduction. To avoid such problems, several safety devices have been added to the original schematic. They are marked in red (Fig. 2). These devices are 21 PN and 2 Zener diodes. Both outputs of the converter are additionally protected by single structures consisting of a three-diode-connected HV MOS devices.

However, most of the devices used in the presented converter are LV devices. Among the 48 MOS devices there are only 11 HV ones. The HV devices are cascode transistors of two current sources and additional devices that shield outputs of fully LV NMOS current mirrors from HV conditions on inputs of the following PMOS type current mirrors. Apart from the input RCONV and output ROUT and RnOUT resistors, the whole signal path of the converter operates in current mode and consists of current mirrors. The application of LV devices as main transistors in all these current mirrors ensures the highest possible quality of device matching and thus current copying. HV high-swing conditions are present only on the two HV transistors (one NMOS and one PMOS type) and ROUT and RnOUT resistors.

The presence of safety devices increases the layout area, but in the presented case the measured change is about 10 %, which should not cause any practical problems.

III. MEASUREMENTS

The results presented in this paper are based on the results of simulations made for the converter implemented with XDM 1.0 µm SOI XFAB technology process. All simulations have been performed with the use of Cadence Virtuoso software with the use of SOA check (SOAC) device models. Simulations with these models provide additional information on SOA breaches in the simulation logout. The test setup is presented in Fig. 3. All resistors required for voltage-to-current and current to voltage conversion are considered internal. Thus, apart from the converter itself, the test setup comprises LV and HV supply sources, an external bias current source, and an enable signal. Thus, the properties of the presented converter alone are taken into account.

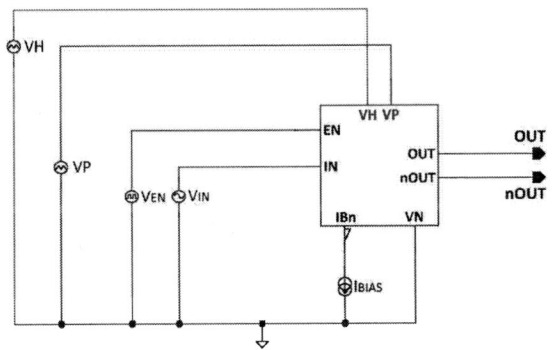

Fig. 3. The test setup for the converter presented in Fig. 2.

All simulations are performed for 5 V LV and 20 V HV supply sources, 10 µA negative bias current, and the enable signal always on. All results of transient (TRAN) simulations are obtained for input sinewave with 7.5 V 0-p amplitude and 10 V offset. The simulated stages of the converter design are:

the original schematic,, the SOAC protected schematic and the final layout.

Fig. 4 presents results of small signal AC gain simulation of all three design stages for both OUT (positive) and nOUT (negative) outputs. It can be observed that results for the original and protected schematics are very similar, while results for the final layout show significant limitation of the converter bandwidth. The bandwidth for the OUT output is about 2.5 higher than for the nOUT, in the case of both schematic versions. The observed deterioration is caused by the OUT signal being processed by a single current mirror and the nOUT signal being processed by two cascaded current mirrors. In case of the final layout the results for OUT and nOUT outputs are much more similar and about an order of magnitude smaller than in case of results for the nOUT obtained for both schematic versions.

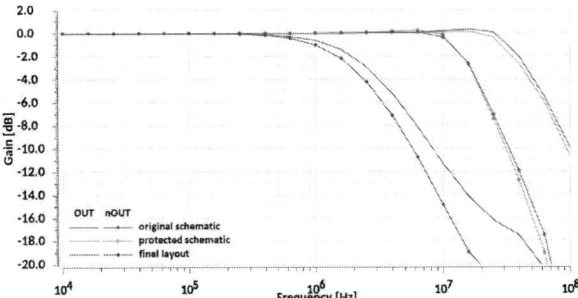

Fig. 4. Small signal AC simulations of gain vs. frequency for all three stages of the converter design.

Fig. 5 shows analogous results obtained with transient simulations for 7.5 V 0-p sinewaves. Both values and relations are very similar as in the case of AC simulation results presented in Fig. 4. This shows that for the selected waveform amplitude the discussed converter operated in linear mode with no significant distortion of the output signals.

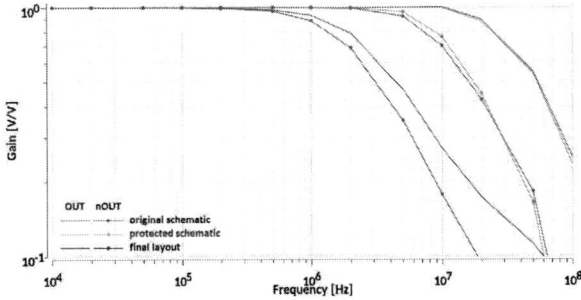

Fig. 5. Large signal transient simulations of gain vs. frequency for all three stages of the converter design.

One of the distortion measures of a sine wave is total harmonic distortion (THD). Fig. 6 shows THD calculations vs. frequency for both outputs of the converter, based on transient simulations. Again, results obtained for both schematic variants are very similar and better than those for the final layout, for frequencies that do not exceed the bandwidths of the relevant converter variants. Furthermore, the results for the OUT node are better than for the nOUT node. The results obtained for the final layout and frequencies higher than its bandwidth are better than in case of both schematic variants as the signal path of the final layout variant works as a low-pass filter.

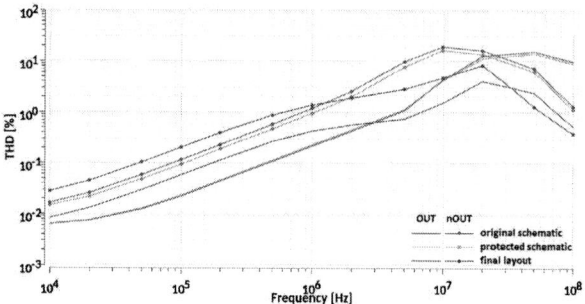

Fig. 6. THD calculations versus frequency, based on transient simulations, for the OUT and nOUT nodes for all three stages of the converter design.

THD can also be applied for the sum of the both output signals. In ideal case this sum should be constant over time (and frequency). In the real case, the amplitude of the summed output signals fluctuates due to the time shift of the OUT and nOUT signals. Moreover, sum of two sine waves should also be a sine wave. Thus, THD and amplitude of such oscillations provide information on the quality of the converter operation.

Fig. 7 shows results of THD vs. frequency calculations for sum of the OUT and nOUT. For low frequencies (up to 100 kHz), the discussed THD for the final layout is higher than for both versions of the schematic. For higher frequencies (above 100 kHz), the final layout yields lower THD values. It may be attributed to the increasing influence of parasitic components that form low-pass filters in the signal path of the final layout variant of the discussed converter. The same effect may be responsible for the slightly lower THD values for the protected schematic than for the original design.

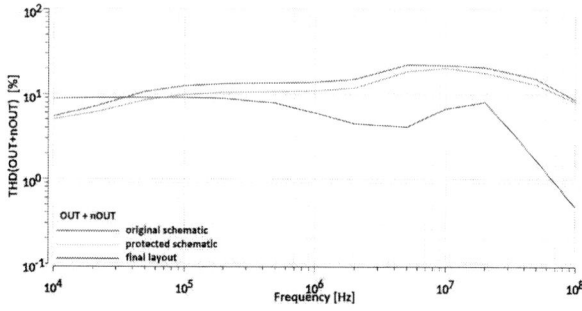

Fig. 7. THD calculations versus frequency, based on transient simulations, for sum of the OUT and nOUT signals for all three stages of the converter design.

Figs. 8 and 9 present sum of the OUT and nOUT signals in time domain for 10 kHz and 200 kHz, respectively. It can be observed that, in fact, the oscillations produced by the final layout for the former frequency is less distorted than those produced by both schematic variants. The result for the latter frequency is the opposite.

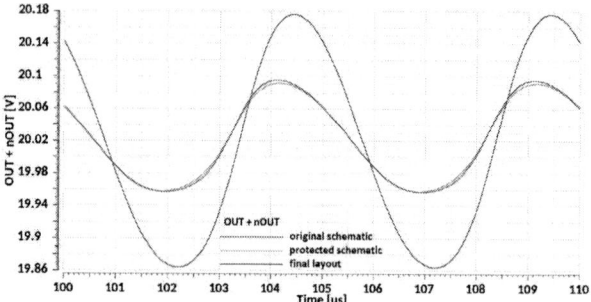

Fig. 8. Sum of the OUT and nOUT signals in time domain for 200 kHz.

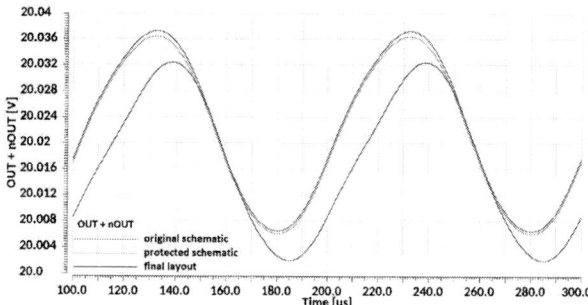

Fig. 9. Sum of the OUT and nOUT signals in time domain for 10 kHz.

The other important property of sum of the OUT and nOUT signals is its amplitude. Fig. 10 shows the amplitude vs. frequency dependence for these oscillations for all three stages of the converter design. The oscillations generated in the final layout are larger than in the case of both schematic versions over the whole simulated frequency range of the final layout.

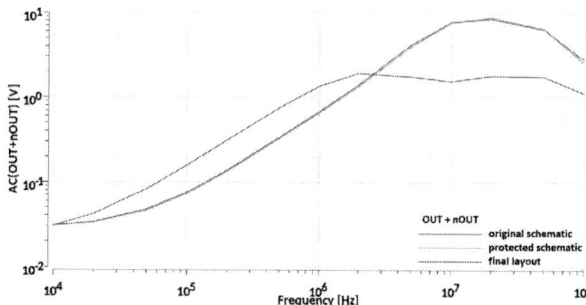

Fig. 10. Amplitude versus frequency dependence for oscillations of the summed OUT and nOUT nodes for all three stages of the design.

This trend reverses for higher frequencies. To eliminate the influence of the changing output gain and show the relative amplitude / intensity of the oscillation, the amplitude vs. frequency dependence for the oscillations has been divided by the relevant output voltage gain (Fig. 11). It can be seen that for frequencies outside the final layout bandwidth, the curves are nearly identical, also for frequencies above the bandwidths of both schematics.

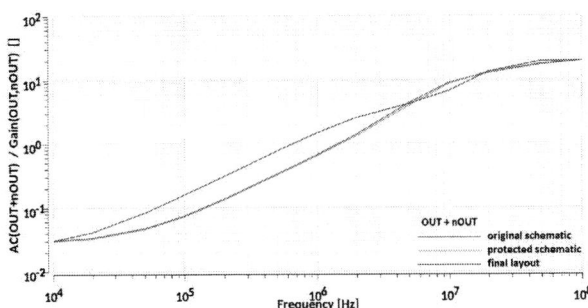

Fig. 11. Amplitude versus frequency dependence for oscillations of the summed OUT and nOUT nodes divided by the output voltage gain.

IV. RESULTS

The presented results show that the influence of the SOA related devices is very limited in comparison to the operation deterioration caused by the laying out the protected schematic. Parameters such as THD or oscillations of the summed output signals are comparable or better in case of the final layout only outside its bandwidth. Thus, such effects cannot be observed during practical applications of the discussed converter.

Very similar operation quality for both versions of the schematic (original and protected) opposed by significantly deteriorated operation of the final layout is consistent with results obtained for the already published analysis results for the HV unity-gain buffer [15] and trapezoidal waveform generator [5], both designed as components of a wider set of HV current-mode-based function blocks. text.

Interestingly, both the mentioned converters, function generators and buffers utilize mixed voltage / current mode of operation, which may cause similar results related to application of similar SOA related solutions.

V. CONCLUSIONS

The operation of consecutive design stages of the presented voltage-to-current and current to voltage converter have been tested and their operation compared. The phase of laying out has been identified as the one that introduces most of the deterioration in the quality of the operation. These results are generally consistent with those obtained during similar analyzes conducted for other HV function blocks designed in the same technology process.

The analysis focused on the implementation of the circuit that can be interpreted as a voltage amplifier with single input and differential output. Applications such as voltage-to-current converter, current to voltage converter or single input differential output current amplifier remain to be analyzed.

Moreover, a comparative analysis with the mentioned implementations in different HV technology processes should be considered to draw more general conclusions. Thus, the presented analysis can be considered as an introductory part to the wider study of application possibilities of the discussed function block.

ACKNOWLEDGMENT

The author expresses his gratitude to late professor Andrzej Napieralski, the Founder of the Department of Microelectronics and Computer Science of the Lodz University of Technology, for the decision to fund the manufacturing of an ASIC consisting, e.g., the buffer presented in this paper, which was essential for the continuation and completion of the whole design process.

REFERENCES

[1] A. Motroni, A. Buffi and P. Nepa, "A Survey on Indoor Vehicle Localization Through RFID Technology," in IEEE Access, vol. 9, pp. 17921-17942, 2021, doi: 10.1109/ACCESS.2021.3052316.

[2] A. Li, J. Li, Y. Zhang, D. Han, T. Li and Y. Zhang, "Secure UHF RFID Authentication With Smart Devices," in IEEE Transactions on Wireless Communications, vol. 22, no. 7, pp. 4520-4533, July 2023, doi: 10.1109/TWC.2022.3226753.

[3] K. Balamurugan, S. Elangovan, R. Mahalakshmi and R. Pavithra, "Automatic check-post and fast track toll system using RFID and GSM module with security system," 2017 International Conference on Advances in Electrical Technology for Green Energy (ICAETGT), Coimbatore, India, 2017, pp. 83-87, doi: 10.1109/ICAETGT.2017.8341461.

[4] M. Jankowski, "Improvements of high-voltage trapezoidal waveform edge-rounding circuit," International Journal of Microelectronics and Computer Science, 2018, Vol. 9, no. 3, pp. 93-100.

[5] M. Jankowski, "Design of a Current-Mode Trapezoidal Waveform Generator in High-Voltage SOI Technology with Modifications Based on Safe Operating Area Limits," in Electronics, vol. 14, no. 3, 2025.

[6] M. Jankowski and A. Napieralski, "Current-mode signal processing implementation in HV SoI integrated systems," Microelectronics Journal, Vol. 45, Issue 7, 2014, pp. 946-959.

[7] V. Srinivasan, R. Chawla and P. Hasler, "Linear current-to-voltage and voltage-to-current converters," 48th Midwest Symposium on Circuits and Systems, 2005., Covington, KY, USA, 2005, pp. 675-678 Vol. 1.

[8] H. Chen, P. Chan. "A Low-Voltage Low-Power Voltage-to-Current Converter with Low Temperature Coefficient Design Awareness," in Sensors, vol. 25, no. 4, 2025.

[9] Chen, R. Y., Lin, S.-F. and Wu, M.-S., A Linear CMOS Voltage-to-Current Converter, Circuits Systems and Signal Processing, vol. 25, no. 4, pp. 497–509, August 1, 2006.

[10] D. Bansal, E. Jolly, "Design and Analysis of Linear Voltage to current converters using CMOS Technology," International Journal of Engineering Research and General Science Volume 3, Issue 3, May-June, 2015, pp. 359-367.

[11] M. Wan, W. Liao, K. Dai, X. Zou, "A Nonlinearity-Compensated All-MOS Voltage-to-Current Converter," in IEEE Transactions on Circuits and Systems II: Express Briefs, vol. 63, no. 2, pp. 156-160, Feb. 2016.

[12] R. Wojtyna, "Upgraded low voltage analog Current-to-Voltage converter with negative feedback," in International Journal of Microelectronics and Computer Science, vol. 8, no. 2, 2017, pp. 80-84.

[13] J. Park, H. Yoon. "Implementation of a Current-to-voltage Converter with a Wide Dynamic Range," in Journal- Korean Physical Society, vol. 56, pp. 863-867, 2010.

[14] H. Aminzadeh, M. Rasekhi, M. Danaie. "Temperature-Insensitive On-Chip Resistors for Linear Voltage-To-Current Conversion in Low-Power Voltage and Current References," inInternational Journal of Numerical Modelling: Electronic Networks, Devices and Fields, vol. 38, 2025.

[15] M. Jankowski. "Influence of Modifications Related to Safe Operating Area Demands on Operation of a Specialized Medium/High-Voltage Unity-Gain Buffer," in Energies, vol. 15, no. 1, 2022.

Mixed Design of Integrated Circuits and Systems – MIXDES 2025

Recording Channel Parameters Influence Analysis on Time-Related X-ray Based Measurements in CMOS 40 nm

Filip Księżyc, Piotr Kmon

AGH University of Krakow
Av. A. Mickiewicza 30, Kraków
fksiezyc@agh.edu.pl, kmon@agh.edu.pl

Abstract—This paper describes the influence analysis of typical detector readout electronics front-end parameters on time-related X-ray based measurements. Here two types of recordings are considered, i.e. the Time over Threshold (ToT) and Time of Arrival (ToA) used for energy and arrival time measurements of incoming particles, respectively. The electronics considered is composed of the core amplifier based on a folded cascode amplifier and the feedback circuit based on the constant current architecture. The recorded channel presented is designed in the CMOS 40nm process. This work provides information on how the precision of time-related measurement depends on the front-end electronics.

Keywords—X-ray detectors, readout electronics, time measurements, ToT, ToA, CSA, folded cascode, constant current feedback

I. INTRODUCTION

X-ray-based imaging cameras are a crucial part of medicine, physics, and industry that allow inspection of the interior of visualised objects without mechanical damage [1, 2]. Contemporary, these are often hybrid pixel detectors (HPDs) that are composed of two pixel-shaped bump-bonded structures, i.e. the detector and the electronic chip. Typically, the purpose of the HPD recording channel is to convert a current pulse, which is a result of charge generated by the impinging into the detector volume photon, into a voltage pulse. The voltage pulse may then be used to extract information regarding the photon energy, its arrival time, or hitting region. The main block of a processing chain is a Charge Sensitive Amplifier (CSA) whose function is to generate an output signal proportional to the input charge collected from the detector; this can be interpreted as current signal integration. Taking into account that the detector signal is a current pulse of amplitude proportional to the energy carried by the photon and is further processed by the transimpedance amplifier, it could be assumed that the CSA output voltage signal is also proportional to the energy of that photon.

Classically, the recording chain is composed of a shaper circuit (preceded by the CSA) to improve the signal-to-noise ratio of the recorded data and analogue-to-digital converter (ADC) for converting analogue signal into digital signal. However, mainly due to the pixel size and its power consumption limits, it is very difficult to use the shaper and ADC in a single pixel. A very attractive alternative is a time-based

measurement that provides many advantages, especially when considering modern processes. The time-based method allows one to evaluate the energy carried by the photon hitting detector by measuring the time that the voltage pulse (related to the energy of the particle) exceeds over the set threshold (Time-over-Threshold, ToT). Also, this approach enables the possibility of measuring the time of hitting the detector by the particle (Time-of-Arrival, ToA) especially important when the particle track needs to be found. A conceptual description of this approach is presented in Fig. 1.

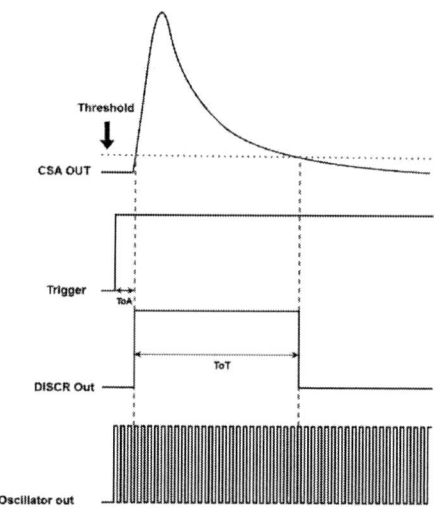

Fig. 1. The conceptual idea of ToT and ToA measurements.

A typical circuit used to measure ToT and ToA, similarly to the classical approach with an ADC, contains a CSA circuit and a discriminator (see Fig. 2). The role of the discriminator is to notify whether the output CSA signal exceeded a preset threshold (this also allows to differentiate among pulses of different energies and to separate them from noise). Whenever the discriminator registers the pulse crossing its threshold, the following actions take place: in the case of ToA the counter is stopped preserving the value proportional to the chip reference clock period, while in the case of ToT the counter is started until the signal amplitude falls below the set threshold.

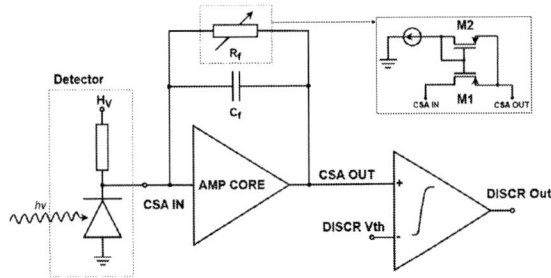

Fig. 2. Schematic diagram of a typical ToT/ToA measurement circuit.

II. STATE OF THE ART

The use of time-based detector methods is widely adopted in both commercial and scientific applications [3, 4]. Table I shows the main parameters comparison of multichannel pixel integrated circuits dedicated to ToT and ToA measurements.

The Timepix chip is developed with the use of 0.25 μm CMOS technology and features a 256 × 256 pixel matrix with each pixel occupying 55 μm × 55 μm. Here, a 100 MHz clock signal is distributed along the columns from the matrix periphery, resulting in a ToA resolution of 10 ns. Timepix allows for measuring ToA by tracking clock ticks from the activation of the discriminator until the shutter closes and also for estimating energy with the use of ToT. The chip employs a frame-based readout, that is, the 14-bit counters of all pixels are sequentially read out after the shutter closes [12].

The Timepix3 chip maintains similar matrix size and pixel dimensions as its predecessor but is fabricated with the use of 130 nm CMOS process. Unlike the previous frame-based approach, it utilizes a data-driven architecture, where data transmission is initiated immediately when a pixel is hit by the photon. Upon activation of the discriminator, a 14-bit timestamp is recorded from an on-chip 40 MHz grey encoded ramp counter, and a 640 MHz ring oscillator begins incrementing a fine time counter until the next 40 MHz clock's rising edge arrives, achieving a fine time resolution of 1.56 ns. The output data packet consists of 48 bits, including ToA and ToT data, along with the pixel's coordinates. Here also the super-pixel group feature was introduced that allowed one to share voltage-controlled ring oscillator between eight pixels [14].

TABLE I.
COMPARATIVE ANALYSIS OF TOT AND TOA EXISTING IC SOLUTIONS

Circuit / Parameter	Timepix [12]	Timepix3 [14]	Timepix4 [4, 11, 13]	TEPIX** [16]	Timespot [8, 9]
ToT resolution [bits]	14 bits	10 bits	11 bits	14 bits	11 bits
ToT precision [keV]	4	< 2	< 1	-	-
ToA resolution [bits]	14	14+4	8 or 16	14	12
ToA precision [ns]	10	1.56	0.2	5	0.01
Front end noise [e-rms]	100	62	68	130	82
Power per pixel [μW]	13.5	12	30.2*	13	40
Pixel size [μm]	55	55	55	50	55
Tech. node [nm]	250	130	65	180	28

* Calculated based on [13], ** simulations only.

The Timepix4 chip, developed in a 65 nm CMOS process, is composed of a four-sided buttable matrix of 448 × 512 pixels, each with a pitch of 55 μm. The chip supports both data-driven and frame-based readout modes. In event-driven mode, individual pixel hits can be time-stamped within a 200 ps bin. Similarly to previous Timepix version, the Timepix4 utilizes single VCO for each super pixel in super pixel group (SPG composed of 4 super-pixels). Additionally, for purpose of clock reproduction across entire pixel matrix (which is necessary to be maintained for the time precision of the circuit), each SPG contains adjustable delay buffers (ADB), that help propagate and synchronize master clock. The ADB's are built of 14 coarse delay elements (controlled by a 4-bit thermometer to enable or skip delay blocks) and 15 finite delay elements (controlled by 4 lower bits to add up to 15 small capacitors for precise tuning) resulting in possible timing adjustment up to 4 ps per step [4, 11, 13].

The TEPIX chip, developed in a 180 nm CMOS process, features a 64 × 2 pixel array. It operates in a frame-based mode and incorporates a charge integration front-end that continuously integrates signals while allowing event-driven data recording. The time of arrival (ToA) is recorded using a 14-bit counter, while a Wilkinson-type 14-bit ADC measures the signal amplitude to perform the energy measurement [16].

The TimeSPOT sensor, developed using 3D trench-type silicon pixel technology, features a 55 μm × 55 μm pixel matrix with a 150 μm thick active volume. Readout channels, designed in a 28 nm process, utilize Si-Ge bipolar transistors with 85 GHz transition frequency as active elements in the core of the fast transimpedance amplifier (TIA). The TIA is a block with two amplification stages to boost signal amplitude while keeping low noise. The arrival time as well as ToT is recorded using Time-to-Digital converter (TDC) based on a Vernier architecture, with two identical Digital Controlled Oscillators working in slightly different frequency, which enabled the possibility to acquire a time resolution close to 10 ps. This result was possible to achieve thanks to software-based amplitude correction algorithms, however, using only simple leading-edge discrimination techniques it was feasible to achieve a time resolution of around 25 ps [8, 9].

III. ANALYSIS OF THE RECORDING CHANNEL

In this paper, the particular blocks of the recording channel (see Fig. 2) are analysed in terms of their influence on the ToT and ToA measurement.

Based on our previous works [5 - 7] we decided to use a folded cascode based amplifier as a core of the CSA. This architecture provides reasonable voltage gain, acceptable noise performance, and good parameter controllability, which is especially important while considering process, temperature, and voltage (PVT) variations mitigation in case of multichannel architecture. Whenever a high gain of the CSA core amplifier is assumed, its output voltage may be given in the following way:

$$V_{out} = \frac{Q_{IN}}{C_F} \cdot \frac{\tau_f}{\tau_r - \tau_f} \cdot \left(e^{-t/\tau_R} - e^{-t/\tau_F} \right) \quad (1)$$

where τ_F and τ_R refers to the fall and rise time of the CSA output voltage signal respectively and these are given as follows:

$$\tau_F = R_F C_F \quad (2)$$

$$\tau_R \approx \frac{C_T C_L}{g_m C_F} \quad (3)$$

C_F and R_F are the feedback capacitance of the amplifier and the effective resistance, respectively, C_T and C_L are the effective input and output capacitances of the CSA, g_m is the transconductance of the input CSA transistor while Q_{IN} is the charge deposited in the detector.

Time-based measurement methods should by definition be more resistant to voltage noise. However, due to the comparator threshold-based recordings (used to start/stop counters), voltage variations induced by noise may cause that the threshold is crossed in different time compared to ideal, noiseless signal (see Fig. 3). Additionally, signal at the CSA output is affected by slope variation induced by change of signal amplitude due to different input charge. All of these lead to uncertainty in measured time and are defined as time jitter s_{rms}. The time jitter can be expressed in a following way:

$$s_{rms} = ENC k_q \frac{1}{\frac{dV_{out}(t)}{dt}} \qquad (4)$$

where ENC is total equivalent noise charge of the recording channel, k_q is a charge gain of the CSA, and $dV_{out}(t)/dt$ is CSA's output signal rising slope.

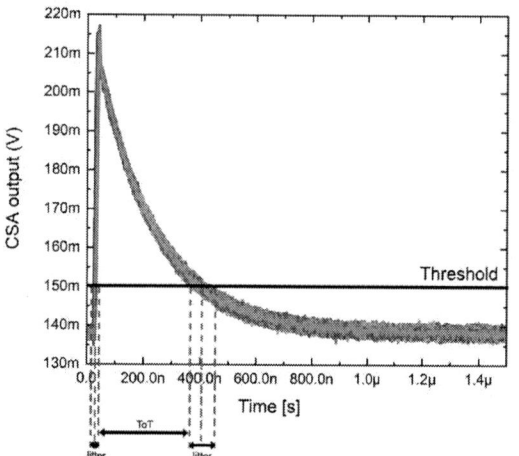

Fig. 3. Simulated CSA output signal for input charge of 10 ke⁻.

We have investigated the influence of the current biasing the CSA on its main parameters (see Table II). It can be seen that the larger the current, the lower the noise of the amplifier although the amplifier noise is not as effectively decreased as the transconductance of the input transistors increases with its current (input transistor dimensioning is set at W / L = 11.3 μm/0.15 μm). This effect is mainly due to the fact that the bandwidth and input capacitance of the CSA increase while increasing the input transistor's current. Both of these effects lead to higher noise of the amplifier.

When it comes to the time-related parameters of the amplifier (see Table III) and its dependence on the amplifier current, it can be seen that the rising slope actually does not increase with the current and is mainly caused by the input capacitance dependence on the transistor current (see Eq. 3). However the jitter may be lowered with the amplifier's current as a result of lower ENC (see Eq. 4).

TABLE II.
MAIN CSA PARAMETERS FOR DIFFERENT OPERATING XONDITIONS

Biasing Current [μA]	gm [μS]	Gain [V/V]	GBW [GHz]	3 dB cut-off [MHz]	Input transistor capacitance [fF]	Noise RMS [V]
3	66	241	1.9	7.9	8.3	1.7 m
6	132	256	2.6	10.0	9.6	1.4 m
12	226	229	2.8	12.0	10.7	1.1 m
20	337	160	2.3	14.5	11.6	1.0 m

TABLE III.
FRONT-END PARAMETERS MEASURED FOR 10 KE⁻ WITH THE USE OF CONSTANT CURRENT FEEDBACK

Biasing Current [μA]	ENC [e-]	Rising Time [ns] (10%-90%)	Falling Time [ns] (90%-10%)	ToT [ns]	ToA [ns]
3	110.5	9.0	91.1	92.5	30.236
6	74.0	9.3	95.1	96.1	29.861
12	73.9	9.4	95.4	98.8	29.751
20	66.6	9.2	94.9	98.1	29.910

Another effect that needs to be taken into account is the amplifier's biasing current influence on the ToA and ToT measurement results. Fig. 4 shows how the ToA measurement may be influenced by both the current sourcing the amplifier and the charge deposited in the detector. It can be seen that whenever very precise ToA measurements are required (in the picosecond range), the particular recording channel parameterisation or calibration must be performed to minimise time walk effects [10, 17].

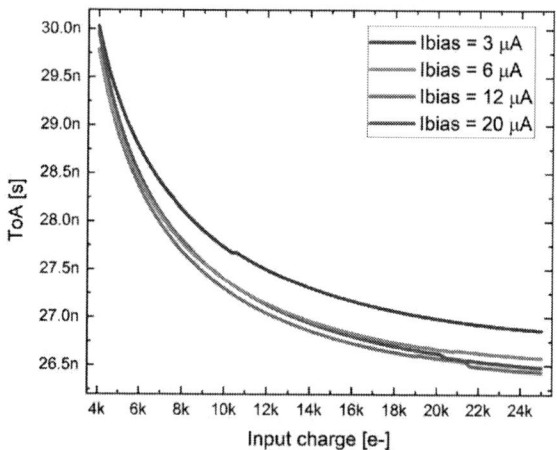

Fig. 4. Simulated value of ToA as a function of input charge with constant discrimination threshold equal to half of CSA output amplitude for input charge of 4 ke⁻.

This effect is less prominent in the ToT measurements (see Fig. 5) as ToT requires nanosecond resolution rather than picoseconds. Change of threshold value to much lower (tenth of 4 ke⁻ compared to half of this amplitude) results in ToT dynamic range lowering (see Fig. 5 considering Ibias change and two different thresholds) from around 450 ns to around 350 ns. Also non-linearities at lower input charge become more visible.

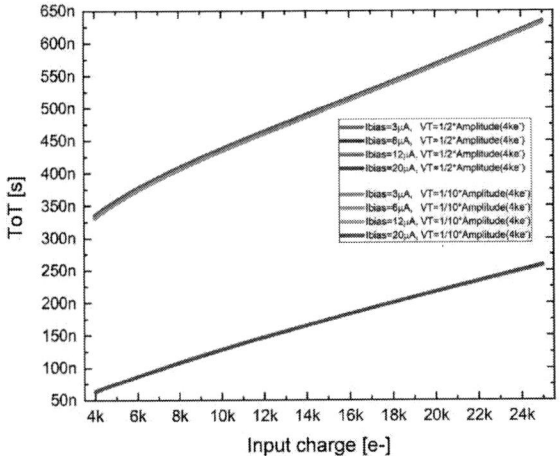

Fig. 5. Simulated value of ToT versus input charge with constant discrimination threshold equal to half of CSA output amplitude for input charge of 4 ke⁻, and for treshold equal to tenth of CSA output amplitude for same input charge.

The jitter analysis of the readout channel was also performed and this was done for input charge equal to 10 ke⁻. It can be seen (see Fig. 6) that for a given channel configuration, the ToA jitter is limited to about 75 ps$_{RMS}$ even for a 20 μA of the amplifier's current. The time jitter of the ToT is much higher (see Fig. 7) and this is caused mainly by the falling slope of the CSA's output pulse (the feedback capacitance C_F is discharged by the M1 based current source – see Fig. 2 and Fig. 3).

Taking into account the particular contribution of time jitter, it should be noted that the dominant uncertainty of the time measurement comes from the CSA (for an I_{bias} = 10 μA it is 75 ps$_{RMS}$) while the remaining part of the time jitter (5 ps$_{RMS}$) is contributed by the discriminator, proving that the most important block from the precision point of view of time measurement is a front-end amplifier.

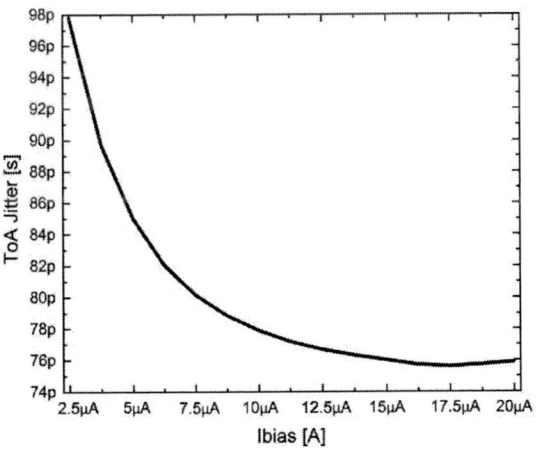

Fig. 6. Simulated value of ToA jitter verus current sourcing the CSA with constant discrimination threshold equal to half of CSA output amplitude for input charge of 4 ke⁻.

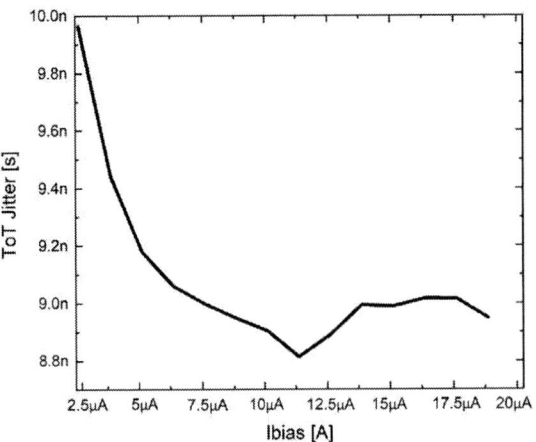

Fig. 7. Simulated value of ToT jitter versus current sourcing the CSA with constant discrimination threshold equal to half of CSA output amplitude for input charge of 4 ke⁻.

IV. CONCLUSIONS

The presented work shows the state of the art of integrated circuits dedicated to time-based measurements adopted in X-ray imaging. We also showed an analysis of a typical recording channel in terms of its influence on time-related accuracy measurements while considering ToT and ToA recordings. These shown that the front-end amplifier is the most important component of the recording channel in terms of time jitter. The time jitter of the CSA is especially prominent in the ToA measurements and this may be lowered by the current sourcing the amplifier; however, other aspects like input and output capacitances need to be taken into account as well.

ACKNOWLEDGMENT

This research is founded by the National Science Center, Poland, project no. 2023/51/B/ST7/01782.

REFERENCES

[1] C. Hansson, K. Iniewski, "X-ray Photon Processing Detectors: Space, Industrial, and Medical applications", Springer, 2023.

[2] K. Iniewski, et al, "Emerging Radiation Detection", Springer, 2024.

[3] R. Ballabriga, M. Campbell and X. Llopart, Asic developments for radiation imaging applications: The medipix and timepix family, Nucl. Instrum. Meth. A 878 (2018) 10.

[4] R. Ballabriga, et al., „The Timepix4 analog front-end design: Lessons learnt on fundamental limits to noise and time resolution in highly segmented hybrid pixel detectors". W:Nuclear Instruments and Methods in Physics Research Section A: Accelerators, Spectrometers, Detectors and Associated Equipment 1045 (2023), s. 167489. DOI: https://doi.org/10.1016/j.nima.2022.167489.

[5] R. Kleczek, P. Kmon, P. Maj, R. Szczygiel, M. Zoladz, P. Grybos, "Single Photon Counting Readout IC with 44 e- rms ENC and 5.5 e- rms Offset Spread with Charge Sensitive Amplifier Active Feedback Discharge," IEEE Transactions on Circuits and Systems I, 2023.

[6] P. Grybos, R. Kleczek, P. Kmon, P. Otfinowski, P. Fajardo, D. Magalhaes, M. Ruat, "SPHIRD–Single Photon Counting Pixel Readout ASIC With Pulse Pile-Up Compensation Methods," IEEE Transactions on Circuits and Systems II, 2023.

[7] L.A. Kadlubowski, P. Kmon, „Recording channel design for time-based measurements in 28nm CMOS", Journal of Instrumentation, 2023.

[8] S. Cadeddu, , et al., „Timespot1: a 28 nm CMOS Pixel Read-Out ASIC for 4D Tracking at High Rates", Journal of Instrumentation, 2023.

[9] G. M. Cossu, A. Lai. „Front-end Electronics for Timing with pico-second precision using 3D Trench Silicon Sensors", Journal of Instrumentation 2023.

[10] B. S. Fong, M. Davies, and P. Deschamps. „Timing resolution and time walk in SLiK SPAD: measurement and optimization", Society of Photo-Optical Instrumentation Engineers (SPIE) 2017.

[11] K. Heijhoff, , et al., „Timing performance of the Timepix4 front-end", Journal of Instrumentation, 2022.

[12] X. Llopart, , et al., „Timepix, a 65k programmable pixel readout chip for arrival time, energy and/or photon counting measurements", Nuclear Instruments and Methods in Physics Research Section A, 2007.

[13] X. Llopart, et al., „Timepix4, a large area pixel detector readout chip which can be tiled on 4 sides providing sub-200 ps timestamp binning", Journal of Instrumentation, 2022.

[14] T. Poikela, et al., „Timepix3: a 65K channel hybrid pixel readout chip with simultaneous ToA/ToT and sparse readout", Journal of Instrumentation, 2014.

[15] A. Rivetti, "CMOS: Front-End Electronics for Radiation Sensors. Devices, Circuits, and Systems", CRC Press, 2018.

[16] T. Wei, D. Zhi, X. Wang. „An Energy and Time Measurement ASIC for Large Pixel Semiconductor Detectors for Spectroscopic and Imaging Applications", IEEE Nuclear Science Symposium and Medical Imaging Conference, 2022.

[17] M. Nakhostin, et al., "Time walk correction of CdTe detectors using depth sensing technique", Nuclear Instruments and Methods in Physics Research Section A, 2010.

SHA-256 Hash Generator in Verilog HDL

Bartosz Rulka[1,2], Paweł Pieńczuk[1,2], Witold Pleskacz[2]

[1] Łukasiewicz Research Network – Institute of Microelectronics and Photonics, Warsaw, Poland

[2] Institute of Microelectronics and Optoelectronics, Warsaw University of Technology, Warsaw, Poland

e-mail: bartosz.rulka@imif.lukasiewicz.gov.pl

Abstract—**An implementation of SHA-256 hash generator is presented. A block has been described in Verilog HDL. A generator code is written with basic logical and arithmetic operations to create a easily-synthesizable block. A design process a 512-bit block in 67 cycles. The generator is tested in simulations with the test vectors and reference digests published by NIST. The simulation testbench has been designed in SystemVerilog. The design passed all test cases used.**

Keywords—**hash function, SHA-256, HDL, Verilog.**

I. INTRODUCTION

Cryptographic hash functions convert the input data with variable length to short output with fixed length [1]. Ideally, they are deterministic, one-directional functions made from fast and basic operators (like AND, OR, +, bit-shift, etc.). They are used for data integrity verification, authentication, and other topics related to the information-security. .

SHA-256 is one of the SHA-2 hash function family participant published by U.S. National Institute of Standards and Technology (NIST) [2]. The algorithm takes an input data with maximum length of $L_{MAX} = 2^{64} - 1$ bits and gives a 256-bit output. SHA-256 is used for transmission security, data authentication, or digital signature [3].

Hardware implementations can have significant impact on the system performance. First of all, hardware implementation can be faster than software implementations [4]. Secondly, they can limit the number of operations performed by the processor core, hence enable the processor to make other tasks. Hardware implementations of the digital circuits can be described by Hardware Description Languages (HDLs) like VHDL or Verilog. This code blocks can be further synthesized on a selected platform: Field Programmable Gate Array (FPGA) or Application-Specific Integrated Circuit (ASIC).

II. SHA-256 ALGORITHM

The process of the SHA-256 algorithm is divided into several stages, each of which is crucial to receive the final result - a 256-bit hash value. Each step describes how the input data are computed and what operations are performed.

A. Preprocessing

The first step in the SHA-256 algorithm is to prepare the input message for computation. This process includes two main stages:

- **Padding** - The input message is padded to achieve a length of a multiple of 512 bits. Firstly, a bit of 1 is added to the message, then enough bits of 0 are added to make the message length 448 bits modulo 512. Finally, a 64-bit value is added to represent the length of the message before padding.
- **Parsing** - The message is divided into 512-bit blocks. If the last block's length is greater than 447 bits and less than 512 bits, then an additional block is added. This extra block consists of 448 zeros and a 64-bit representation of the message length.

B. Initialization of initial values

At the beginning of the algorithm eight values $H_0^{(0)}$, ..., $H_7^{(0)}$ are initialized as the first 32 bits of the square roots of the decimal expansion of the first eight prime numbers.

C. Computing the hash value

The hash value is calculated iteratively for each i 512-bit block (M^i). For each M block process includes the following steps:

- **Preparing a message schedule** - For each block the so-called *message scheduler*, which consists of 64 32-bit words, is being prepared. The first 16 words are the words of the message block M^i, and the remaining 48 are generated based on the formulas:

for $0 \leq t \leq 15$:

$$W_t^{(i)} = M_t^{(i)} \tag{1}$$

for $16 \leq t \leq 63$:

$$W_t^{(i)} = \sigma_1(W_{t-2})^{\{256\}} + W_{t-7} + \sigma_0(W_{t-15})^{\{256\}} + W_{t-16} \tag{2}$$

where:

$$\sigma_0(x)^{\{256\}} = ROTR^7(x) \oplus ROTR^{18}(x) \oplus SHR^3(x), \tag{3}$$

$$\sigma_1(x)^{\{256\}} = ROTR^{17}(x) \oplus ROTR^{19}(x) \oplus SHR^{10}(x). \tag{4}$$

- **Main compression loop** - The main compression loop uses 8 working variables a, b, c, d, e, f, g, h, which are initialized with initial values $H_{0-7}^{(i)}$. For each word W_t of the schedule the temporary values T_1 and T_2 are calculated and the values of the working variables are updated:

$$T_1 = \Sigma_1(e)^{\{256\}} + Ch(e, f, g) + K_t^{\{256\}} + W_t \tag{5}$$

$$T_2 = \Sigma_0(a)^{\{256\}} + Maj(a, b, c) \tag{6}$$

where:

$$\Sigma_0(x)^{\{256\}} = ROTR^2(x) \oplus ROTR^{13}(x) \oplus ROTR^{22}(x), \tag{7}$$

$$\Sigma_1(x)^{\{256\}} = ROTR^6(x) \oplus ROTR^{11}(x) \oplus ROTR^{25}(x), \tag{8}$$

$$Ch(x,y,z) = (x \wedge y) \oplus (\neg x \wedge z), \tag{9}$$

$$Maj(x,y,z) = (x \wedge y) \oplus (x \wedge z) \oplus (y \wedge z). \tag{10}$$

The update of working variables follows the formulas:

$$a = T_1 + T_2 \tag{11}$$

$$b = a \tag{12}$$

$$c = b \tag{13}$$

$$d = c \tag{14}$$

$$e = d + T_1 \tag{15}$$

$$f = e \tag{16}$$

$$g = f \tag{17}$$

$$h = g \tag{18}$$

After all loop iterations are completed, the working values a - h from the compression loops of the last iteration are summed with the initial values of the 32-bit variables $H_{0-7}^{(i-1)}$. All received values are then sequentially written to new initial variables needed to process the next block of data, or to create an output hash.

$$H_0^{(i)} = a + H_0^{(i-1)} \tag{19}$$

$$H_1^{(i)} = b + H_1^{(i-1)} \tag{20}$$

$$H_2^{(i)} = c + H_2^{(i-1)} \tag{21}$$

$$H_3^{(i)} = d + H_3^{(i-1)} \tag{22}$$

$$H_4^{(i)} = e + H_4^{(i-1)} \tag{23}$$

$$H_5^{(i)} = f + H_5^{(i-1)} \tag{24}$$

$$H_6^{(i)} = g + H_6^{(i-1)} \tag{25}$$

$$H_7^{(i)} = h + H_7^{(i-1)} \tag{26}$$

D. Final hash

After processing all the blocks, the output of the algorithm is a 256-bit hash value, which is a concatenation of the eight 32-bit working values H_i. The final result is as follows:

$$\text{HASH} = H_0^{(N)} \parallel H_1^{(N)} \parallel H_2^{(N)} \parallel H_3^{(N)} \parallel H_4^{(N)} \parallel H_5^{(N)} \parallel H_6^{(N)} \parallel H_7^{(N)} \tag{27}$$

III. SOLUTION PROPOSAL

A. Top level

The SHA-256 generator block has been designed in Verilog HDL using the SHA-256 hash function from the SHA-2 hash functions family. The generator returns the output 32-bytes length *hash*, which is based on input data and input data length.

Figure 1 shows a simplified diagram of the SHA256 module. The generator is created with 21 interconnected submodules that execute appropriate logical and arithmetic operations that control and operate the course of subsequent rounds across the algorithm. This is the highest level of the hierarchy that contains the main processing block. The built-in controller controls the block submodules in compliance with the *FIPS 180-4* standard. The entire block is resetable by asynchoronous *RESET* input.

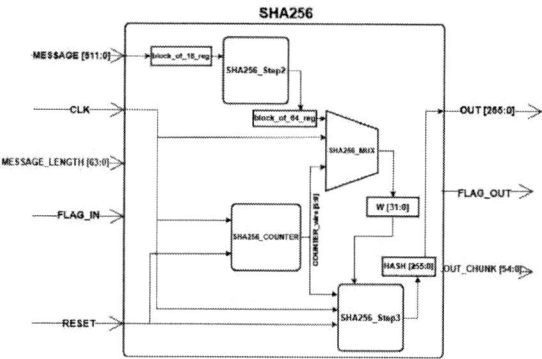

Fig. 1. Diagram of connecting modules at the top level of the hierarchy.

B. Preprocessing block

The generator starts a processing at the rising edge of *CLK* when the *FLAG_IN* flag is raised. It accepts the 512-bit input data *MESSAGE*, and 64-bit input data length *MESSAGE_LENGTH*. Immediately, the number of 512-bit data blocks is computed by division of *MESSAGE_LENGTH* by $2^9 = 512$ (realized by the 9-bit shift). This *OUT_CHUNK* value is needed by the controller to decide when to stop the hashing procedure.

There are already several publicly available SHA-256 implementations [5] [6].In this work we decided to implement the padding step in the hardware in order to limit the number of operations performed by the software. The preparation of data and padding is done within three clock cycles, but the padding itself is done in one clock cycle. However, these steps are rather complex since these require e.g. bit rotation by variable bit number.

Our preliminary synthesis and static timing analysis (STA) tests with several CMOS standard libraries shown that the critical path is mostly hidden in padding operation. This issue needs further investigations, but we see two potential approaches to deal with it:

- pipelining the operation in preprocessing;
- move the padding operations to the software domain.

C. Main loop

Main SHA256 loop block strictly follows the *FIPS 180-4* [2] standard. Since the SHA-256 algorithm is designed with simple arithmetic and logical operations (AND, OR, ROTR, etc.), the implementation in any HDL is rather straightforward.

After the preprocessing, the following operations are performed

- $H_{0:7}^{(0)}$ values initialization;
- $W_t^{(i)}$ message schedule preparation;
- main compression loop;
- intermediate hash $H_{0:7}^{(i)}$ update;

After processing a 512-bit block, the intermediate hash values are used in the next round (with next 512-bit data block). When the entire message is processed, the final hash values are concatenated to the final 256-bit output hash value. Simultaneously, the output flag *FLAG_OUT* is raised, indicating the end of the calculations. Now the outer system can collect the output hash.

IV. SIMULATION RESULTS

The generator tests start with reading all test vectors from the text files

- *SHA256ShortByteMsg.rsp*,
- *SHA256ShortBitMsg.rsp*,
- *SHA256LongByteMsg.rsp*,
- *SHA256LongBitMsg.rsp* [7].

provided by *NIST*. These files contain message (*Msg*), length (*Len*) and reference message digest (*MD*). The example is presented in Figure 2. Then, these vectors are transmitted to the *MESSAGE* input, and the length of their binary notation to the *MESSAGE_LENGTH* input. After the calculations, the testbench save the generator's output hash value. After that, the calculated hash is compared to the reference hash *MD (Message Digest)*. If the hashes are the same, the *SUCCESS* message is written in the log file. Otherwise, the *ERROR* message is written in the log file.

```
Len = 80
Msg = 74cb9381d89f5aa73368
MD = 73d6fad1caaa75b43b21733561fd3958bdc555194a037c2addec19dc2d7a52bd
```

Fig. 2. Example of a test vector and its length in a test file *SHA256ShortByteMsg.rsp*

The entire hashing process is presented in the waveforms in Figures 3, 4, and 5. Fig. 3 presents the input data collection (input data and data length set, raising of the *FLAG_IN* by the test system). Fig. 4 depicts the start of the main compression loop. Here, the initialization of the working variables can be seen, as well as the first iterations of the algorithms. Fig. 5 shows outputs at the end of the message processing. The output hash value *OUT_MESSAGE* is presented together with *FLAG_OUT* flag, indicating the end of the calculations.

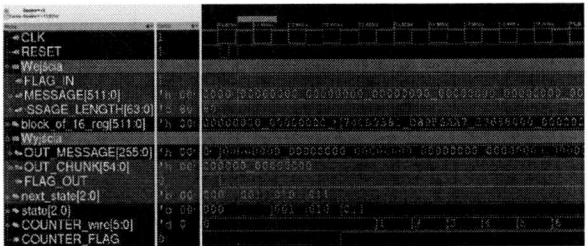

Fig. 3. Signal waveforms after receiving the input data

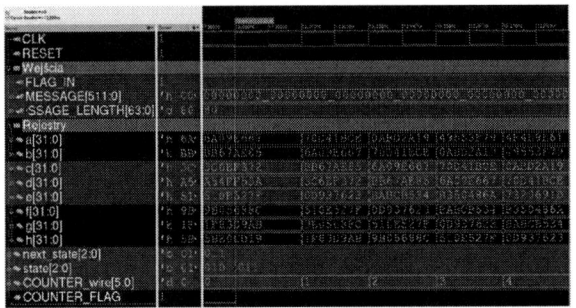

Fig. 4. Starting the hash loop

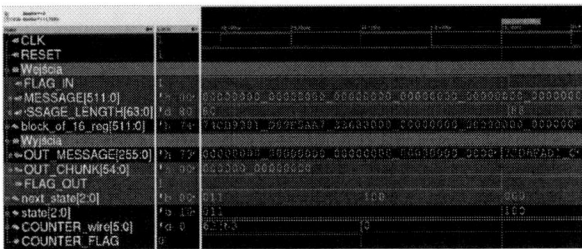

Fig. 5. Signal waveforms after processing the message and setting the hash value at the output

All of the test cases results are written in text log file. This log file contains all the test cases results and overall statistics, namely:

- number of errors;
- number of processed messages;

The end statistics are presented in Figure 6.

```
ERRORS IN ALL:              0
NUMBER OF MESSAGES:         1155
TEST COMPLETED SUCCESSFULLY!
```

Fig. 6. Output from simulation tests of all test vectors through the system

V. CONCLUSION

In this work, a design of the SHA-256 hash generator hardware block is presented. The code is written in Verilog HDL. The implementation process the 512-bit block within 70 cycles. The block has been tested in simulations with NIST test vectors and reference digests. A design has passed all test cases correctly. However, there are couple of issues (like preprocessing) that should be further investigated.

In future, the authors have planned to add the standard interface (such as Wishbone or AXI4-Lite) to improve adaptability to microprocessor architecture. In addition, it is planned to verify a design in a FPGA platform to perform experimental tests.

ACKNOWLEDGEMENTS

This work has been supported by the Łukasiewicz-IMiF subvention funds (20.31.120031) for 2024 and 2025 as "SHA256 Hash Generator in Verilog HDL" project.

REFERENCES

[1] J. Katz and Y. Lindell, *Introduction to Modern Cryptography, Third edition.* CRC Press, 2021.

[2] NIST, "FIPS 180-4: Secure Hash Standard (SHS)," NIST, Tech. Rep., 2015.

[3] W. Stallings, *Cryptography and Network Security, Principles and Practice, 7th edition.* Pearson, 2017.

[4] U. Banerjee, "Energy-Efficient Cryptographic Acceleration using Hardware-Software Co-Design with RISC-V," in *2023 IEEE International Symposium on Smart Electronic Systems (iSES)*, 2023, pp. 197–198.

[5] J. Eklund, "Sha-256 core," https://github.com/secworks/sha256, 2014.

[6] T. Chitipolu, "Sha-256 verilog hdl," https://github.com/tharunchitipolu/SHA-256-Verilog-HDL, 2021.

[7] https://csrc.nist.gov/projects/cryptographic-algorithm-validation-program/secure-hashing.

SYNAPSE - A New Approach to Semi-automated Design of Ultra-low-power Application-specific Embedded Processors

Xuan Ji, Tom J Kaźmierski , Basel Halak

School of Electronics & Computer Science

University of Southampton

Southampton, United Kingdom

Emails: xj2n21@soton.ac.uk, tjk@ecs.soton.ac.uk, basel.halak@soton.ac.uk

Abstract—The main contribution of this paper is a new approach to semi-automated synthesis of Application Specific Embedded Processors (ASEPs). Designers of ASEPs do not have the comfort of standardized software support because the instruction sets are customised. Therefore, ASEP designs are frequently performed manually. However, in recent years a growing interest in ASEPs has been observed. This can be explained by the following two factors. Firstly, in contrast to general-purpose processors, ASEPs are particularly resilient to cybersecurity threats, which nowadays affect both software and hardware in modern SoC applications, especially in critical areas such as medicine, communication or smart grids. Secondly, the usual justification for using ASEP designs is their excellent performance and power efficiency characteristics which are comparable or can exceed those of dedicated hardware. Results presented in this paper show that automated ASEP designs can be more than an order of magnitude smaller, and therefore more power efficient, than equivalent general-purpose application-specific embedded processors. The small size results from the fact that both the architecture and instruction set of an ASEP are tailored to the unique needs of a particular application within the embedded system. The automation approach presented in this paper helps to reduce the high design costs and necessity to use highly skilled design engineers.

Keywords—application-specific embedded processors, picoRISC, RISC-V, NIOS, ultra-low power, embedded systems, automated digital synthesis

I. INTRODUCTION

The growing importance of complex System-on-Chip (SoC) applications is accompanied by rapid advances in the embedded processor design techniques. In a typical SoC there is at least one processor, frequently with multiple cores, to perform control operations, digital signal processing, network operations or advanced peripheral data transfers. Generally, SoC designers face the difficult challenge of having to minimize energy consumption as many of these powerful, single-IC systems are used in battery powered or energy-harvester powered devices. For this reason, many of the embedded processors in SoC designs are application-specific embedded processors (ASEPs), with customised instruction sets and customised hardware. ASEPs can be customised to obtain extremely small silicon areas or minimal use of FPGA resources and therefore ultra-low power consumption. The results pre-

sented in Section IV show that. There are many embedded applications of standard general-purpose processors such as MIPS [1][2], ARM [3][4], SPARC [2][5] or, in the case of FPGA designs, soft-cores such as NIOS [7][8]. In recent years there has been an exponential growth of RISC-V embedded applications [6][9]. The popularity of RISC-V is driven by its very flexible ISA, which allows for a high degree of customisation, hardware flexibility and power consumption efficiency [11][12]. RISC-V is the most recent of the major RISC processors and has been specifically designed to support both high-performance implementations as well as low-complexity embedded designs [10][13]. The obvious advantage of using general-purpose processors is a standardized ISA and software support in the form of compilers, cross-compilers, extensive debugging, software applications and libraries. Designers of ASEPs do not have the comfort of standardized software support because the instruction sets are customised hence ASEP designs are frequently performed manually. Nevertheless, in recent years there has been a growing interest in ASEPs which can be explained by the following two factors. Firstly, in contrast to general-purpose processors, ASEPs are particularly resilient to cybersecurity threats, which nowadays affect both software and hardware in modern SoC applications, especially in critical areas such as medicine, communication or smart grids. Secondly, the usual justification for using ASEP designs is their excellent performance and power efficiency characteristics which are comparable or can exceed that of dedicated hardware. Because the architecture and instruction set of an ASEP is tailored to the unique needs of a particular application within the embedded system, the standard software support has to be sacrificed. The manual design approach works to an extent because ASEPs are small and simple but still leads to high design costs and necessity to use highly skilled design engineers. Due to their simplicity and freedom to make a variety of hardware choices, ASEPs are akin to both dedicated hardware designs generated by high-level synthesis (HLS) tools [14], as well as No Instruction Set Computers (NISCs) which have been around for some years and nowadays enjoy a renewed interest [15]. This paper proposes a semi-automated approach to ASEP design and verification for the architecture and instruction set of an ASEP by applying a combination of architectural libraries prepared in advance with population-based multi-objective optimisation methods that can be implemented on parallel computing platforms. The overall objective is to replace the current time-consuming design flow by a methodology that can automate much of the generation and verification of both hardware and software of an ASEP. This approach has the potential to make use of artificial intelligence (AI) and machine learning (ML) to use the available expert knowledge and support the automation. The approach presented here is demonstrated by a case study where an 8-bit variable-architecture ASEP is synthesised with a limited amount of manual intervention. The application used in this case study performs affine transformation of an image represented as a stream of 8-bit pixel coordinates. The semi-automated design flow of the software tool named SYNAPSE, which supports the automation, is outlined in Section II, details of the application used to test SYNAPSE are in Section III. Section IV presents sample designs optimised by SYNAPSE for the smallest size and Section V outlines the conclusions and further work.

II. PROPOSED SYNTHESIS APROACH - SYNAPSE

SYNAPSE generates multiple designs and optimizes them for power consumption or performance as specified by the configuration data. The SYNAPSE flow is shown in Figure 1.

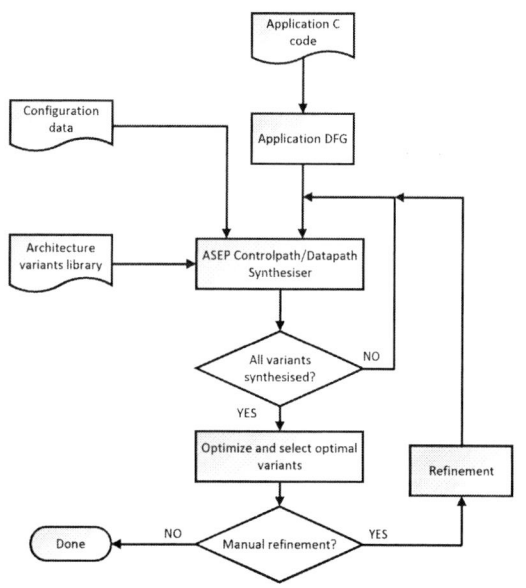

Fig. 1. SYNAPSE ASEP generation flow

The process is similar to that in High-Level Synthesis (HLS) [14] but SYNAPSE does not follow the classical HLS flow that generates dedicated hardware in the form of a controller and data path. Instead it generates the hardware and software for application specific processors. The current, experimental version, composes multiple designs as No Instruction Set

Computers [15] or ASEPs based on the picoRISC variable architecture concept [16]. There is a close relationship between HLS designs, NISC and ASEP. All three approaches rely on a control path which in the case of HLS is a state machine, in a NISC it is a program counter and a control word memory and in an ASEP - a program counter, instruction memory and an instruction decoder. These three control paths are presented in Figure 2.

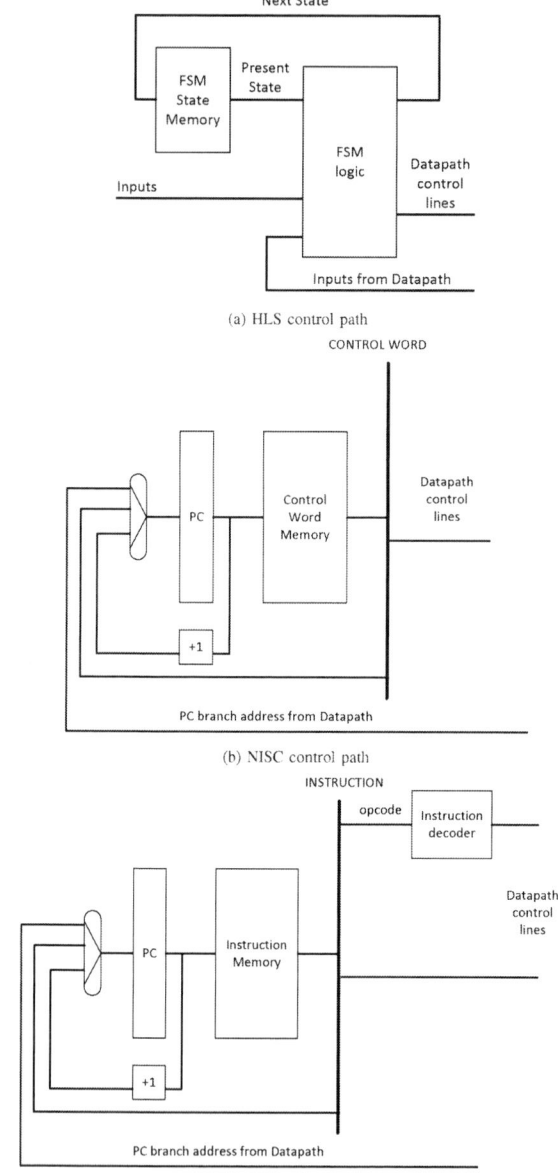

(a) HLS control path

(b) NISC control path

(c) picoRISC ASEP control path

Fig. 2. HLS, NISC and ASEP control paths

Any form of control path can be used to drive a variety of different data paths. SYNAPSE builds data paths using all combinations of architectural variations present in its architecture library. As an example, Figure 3 shows a sample picoRISC [16] datapath variant which features an 8-bit data bus, a RISC-V style ALU and branching logic, with support for a JALR instruction (Jump and Link with Register), a two bus architecture with a three-register General Purpose Register file (GPR) with two source registers Rs1, Rs2 and a destination register Rd.

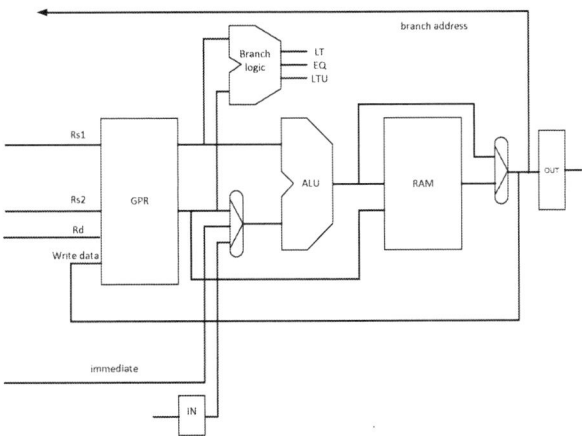

Fig. 3. A dual-bus picoRISC datapath variant

Figure 4 shows an alternative variant of picoRISC datapath with a single-bus architecture, a two-register GPR, a MIPS style ALU which generates four condition codes: Carry (C), Zero (Z), Negative (N) and 2's complement overflow (V) for conditional branches and a SPARC-style RAM with a load/store unit. Many other variants are possible.

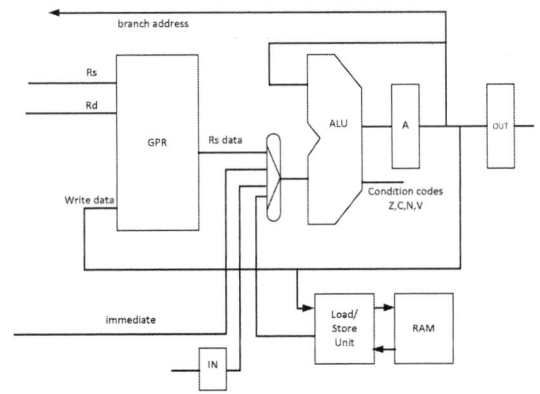

Fig. 4. A single-bus picoRISC datapath variant

SYNAPSE will also generate equivalent NISC designs and compare them for size and performance. The simplest NISC variant can be generated from a picoRISC design where the Instruction Memory (IM) and Instruction Decoder are replaced with a Control Word Memory where the instruction code is expanded to include the decoded signals that control the data path of the processors. Many other configurations of the NISC data path are also possible [15].

Both NISC and ASEP can have customized datapaths to suit the application. In ASEP custom instructions must be generated to reflect the data processing hardware capabilities and limitations. Unlike in a NISC, an ASEP design must also implement an instruction decoder which is usually built of combinational logic. This may reduce the program memory size compared with an equivalent NISC, as NISC has to store all the control bits necessary to drive the datapath. The difference in memory usage is expected to be in favor of ASEP, especially if the program is large. In our case study SYNAPSE generated similar memory sizes, slightly in favor of the ASEP designs as the programs in the case study were not large as outlined in Section IV.

III. CASE STUDY - AFFINE TRANSFORMATION ALGORITHM

An affine transformation [18] is a geometrical transformation that preserves co-linearity, i.e. all points lying on a line will also lie on a line after the transformation and distance ratios are preserved. For two-dimensional images, the general affine transformation can be expressed as:

$$\begin{bmatrix} y_1 \\ y_2 \end{bmatrix} = \begin{bmatrix} a_{11} & a_{12} \\ a_{21} & a_{22} \end{bmatrix} \begin{bmatrix} x_1 \\ x_2 \end{bmatrix} + \begin{bmatrix} b_1 \\ b_2 \end{bmatrix} \quad (1)$$

where $[x_1, x_2]$ are the original pixel coordinates, $[y_1, y_2]$ are the resulting coordinates after the affine transformation and matrix A and vector B represent the transformation coefficients. Figure 6 shows the Data Flow Graph (DFG) of the affine transformation operations from which SYNAPSE can generate ASEP program instructions.

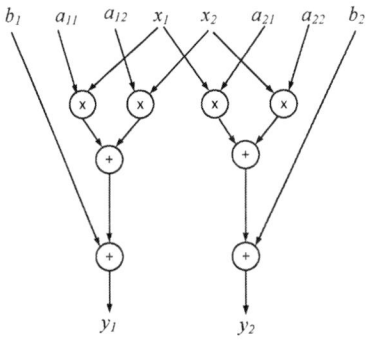

Fig. 6. Affine transformation DFG

A. Application Specific Arithmetic

Pixel coordinates are 2's complement 8-bit integer numbers in the range $-128.. + 127$ but coefficients of matrix A are 2's complement signed fixed-point fractions in the range $-1...+1-2^{-8}$, i.e. they are 2's complement fractional numbers with the radix point positioned after the most significant bit. Therefore the weights of the individual bits are as shown in Table I. Table I also shows examples of how the matrix coefficients and pixel coordinates are represented in this arithmetic.

137

TABLE I.
APPLICATION SPECIFIC ARITHMETIC: (A) WEIGHTS FOR MATRIX A COEFFICIENTS; (B) REPRESENTATION OF -0.25 AS A MATRIX A COEFFICIENT; (C) REPRESENTATION OF 20 AS A PIXEL COORDINATE.

(A)

Bit position	Weight
7	-2^0
6	2^{-1}
5	2^{-2}
4	2^{-3}
3	2^{-4}
2	2^{-5}
1	2^{-6}
0	2^{-7}

(B)

-0.25	Weight
1.	-2^0
1	2^{-1}
1	2^{-2}
0	2^{-3}
0	2^{-4}
0	2^{-5}
0	2^{-6}
0	2^{-7}

(C)

20	Weight
0	-2^7
0	2^6
0	2^5
1	2^4
0	2^3
1	2^2
0	2^1
0.	2^0

The double-length 16-bit product has the radix point positioned after the 9-th bit. The product is truncated and the result $[y_1, y_2]$ is rounded to an 8-bit 2's complement whole number as it is a pixel coordinate.

IV. RESULTS

The ASEP (picoRISC) and NISC designs generated by SYNAPSE and selected for optimal speed and size are shown in Table II. The structure of the smallest design generated by SYNAPSE, named picoRISC '1011' is shown in Figure 5.

TABLE II.
SYNAPSE GENERATED PICORISC AND NISC DESIGNS; OPTIMAL SELECTION FOR SPEED AND SIZE

Name	ALM (Cyclone V)	Fmax@85 °C (MHz)	Fmax@0 °C (MHz)
picoRISC '0000'	111	131.67	131.15
picoRISC '1011'	94	129.47	128.53
NISC '0111'	100	129.17	129.7
NISC '1011'	120	126.44	126.77

For comparison, Table III shows similar data for state-of-the-art embedded processors based on RISC-V and NIOS architectures. In a recent work [20], a number of open source processors were selected for comparative analysis: Amber, LatticeMicro32 (LM32), S1, Altor32, OpenRISC 1200 (OR 1200), LEON2 and LEON3 and three versions of NIOS II cores. They are all economic, standard and fast cores. In Table III for the purpose of our comparison we quote their results for NIOS II/e [20], the smallest of the three NIOS cores to provide a fairer comparison with the ASEPs and NISCs

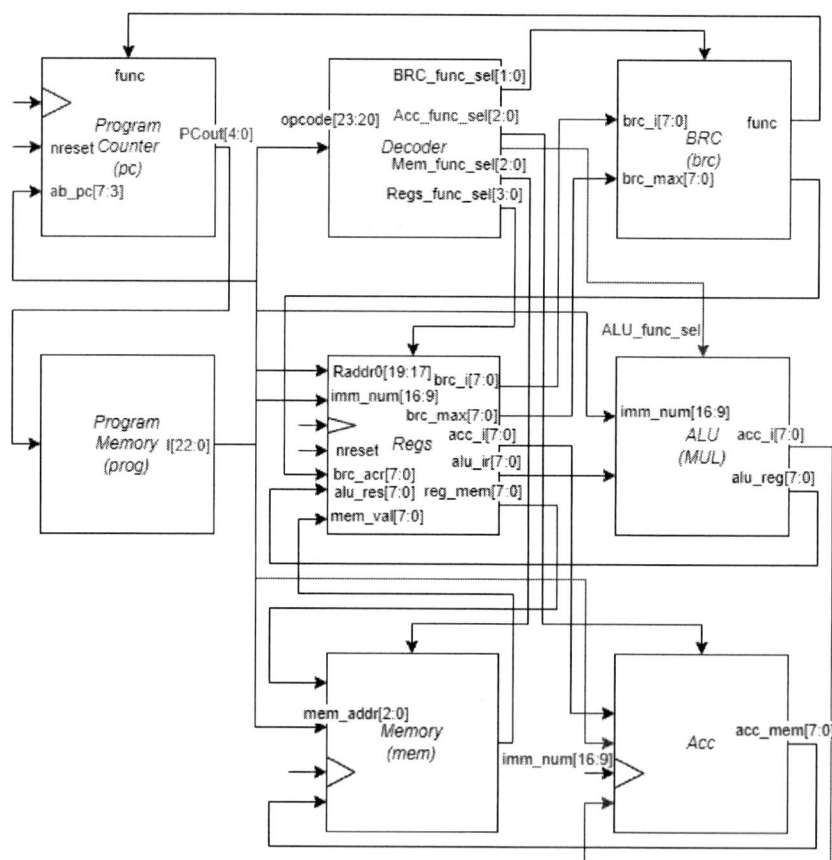

Fig. 5. Design "1011", synthesised by SYNAPSE, with the lowest count of ALMs (94) and the maximum clock rate of 129MHz at 85 °C.

generated by SYNAPSE. In this comparison, SYNAPSE generated ASEPs and NISCs win by more than one order of magnitude in terms of size. The differences in maximum clock frequencies are due to the different Intel FPGA models the designs were implemented on. SYNAPSE results were synthesized for Cyclone V and those presented in Table III relating to the survey paper [20], for Stratix V which is a much more powerful FPGA. The extremely small hardware which can be obtained by SYNAPSE makes a very good case for an automated synthesis of ASEPs and NISCs. The attraction of general-purpose reconfigurable embedded processors based on modern platforms such as RISC-V and NIOS lies in the fact that machine code can be easily generated by freely available compilers and debugging tools. No such tools are generally available for ASEPs and NIOS as the hardware and software is custom generated. Generally, automated generation of embedded processors is still largely limited to general-purpose reconfigurable architectures [17] and fully automated tools for ASEP designs are lagging.

TABLE III.

STATE-OF-ART RISC V AND NIOS DESIGNS SHOW A MUCH GREATER USE OF FPGA RESOURCES (ALMS) COMPARED WITH SYNAPSE GENERATED ASEPS AND NISCS.

Name	ALM (Cyclone V)	$Fmax@85°C$ (MHz)	$Fmax@0°C$ (MHz)
RISC-V Full OCD[19]	4615	76	-
RISC-V Small UART1[19]	1691	89	-
Name	LUT (Stratix V)	$Fmax@100°C$ (MHz)	$Fmax@-40°C$ (MHz)
Amber23[20]	3073	84.7	83.1
LM32[20]	2141	179.9	171.1
S1[20]	39519	52.3	56.2
LEON3[20]	1239	212.3	210.0
NIOS II/e[20]	409	367.5	337.0

V. CONCLUSION

A novel approach to semi-automated design and optimisation of Application-Specific Embedded Processors has been proposed. The synthesis of ASEP architectural and software variants is based on a combination of architectural libraries prepared in advance with optimisation methods that can be implemented on parallel computing platforms. The overall objective is to replace the current time-consuming design flow by a methodology that can automate much of the generation and verification of both hardware and software of an ASEP. An experimental tool named SYNAPSE has been developed and experiments presented in this paper indicate that extreme savings in terms of hardware resources can be achieved, yielding results better by more than one order of magnitude in terms of size, and therefore also power consumption, compared

with state-of-the-art synthesis of reconfigurable general-purpose embedded architectures such as those based on RISC-V and NIOS. The proposed approach has the potential to make use of artificial intelligence (AI) and machine learning (ML) to use the available expert knowledge and support the automation.

REFERENCES

[1] J. Hennessy, N. Jouppi, S. Przybylski et al., "MIPS: A microprocessor architecture," *ACM SIGMICRO Newsl.*, vol. 13, pp. 17–22, Dec. 1982.

[2] D. A. Patterson, "Computer Organization and Design", Elsevier, 2018, ISBN 978-0-12-407726-3.

[3] S. Bibi, M. Asghar, M. U. Chaudhry, N. Hussan, and M. Yasir, "A review of ARM processor architecture history, progress and applications," *IEEE Access*, vol. 10, pp. 171–179, Aug. 2021.

[4] T. Martin, "Designer's Guide to the Cortex-M Processor Family", Newnes, 3rd Ed; 2022; ISBN 978-0323854948.

[5] A. Agrawal and R. B. Garner, "SPARC: A scalable processor architecture," *Future Gener. Comput. Syst.*, vol. 7, no. 2, pp. 303–309, 1992.

[6] A. Sanchez-Flores, L. Alvarez, and B. Alorda, "A review of CNN accelerators for embedded systems based on RISC-V," in *Proc. IEEE Int. Conf. Omni-layer Intell. Syst. (COINS)*, Barcelona, Spain, 2022, pp. 1–6.

[7] J. Ball, "The Nios II family of configurable soft-core processors," in *Proc. IEEE Hot Chips XVII Symp. (HCS)*, Stanford, CA, USA, 2005, pp. 1–40, doi:10.1109/HOTCHIPS.2005.7476599.

[8] "Intel NIOS II Processor Reference Guide", 2019, [Online] Available: https://cdrdv2-public.intel.com/666887/n2cpu-nii5v1gen2-683836-666887.pdf [Accessed: Mar. 15, 2025].

[9] "Outlook 2025: The Role of RISC-V in Shaping the Future", March 6, 2025, [Online] Available: https://riscv.org/ecosystem-news/2025/03/outlook-2025-the-role-of-risc-v-in-shaping-the-future/ [Accessed: Mar. 15, 2025].

[10] D. Emil, M. Hamdy, and G. Nagib, "Development of an efficient AXI-interconnect unit between customized peripheral devices and a dual-core RISC-V processor," *J. Supercomput.*, vol. 79, no. 15, pp. 17000–17019, Oct. 2023.

[11] E. Choi, J. Park, K. Lee, J.-J. Lee, K. Han, and W. Lee, "Day-night architecture: Development of an ultra-low power RISC-V processor for wearable anomaly detection," *J. Syst. Archit.*, vol. 152, p. 103161, 2024.

[12] B. Satyajit, R. Paily, "A high-performance core micro-architecture based on RISC-V ISA for low power applications", *IEEE Transactions on Circuits and Systems II: Express Briefs*, v.68, 6, 2020, pp.2132–2136.

[13] K. Asanović, D. A. Patterson, "Instruction Sets Should be Free: The Case for RISC-V", [Online] Available: https://www2.eecs.berkeley.edu/Pubs/TechRpts/2014/EECS-2014-146.pdf [Accessed: Mar. 15, 2025].

[14] G. de Micheli, " Synthesis and optimization of digital circuits", McGraw-Hill Higher Education, 1994,

[15] B. Gorjiara, M. Reshadi, D. Gajski, "Low-power design with NISC technology", Chapter 2. in *Designing Embedded Processors: A Low Power Perspective*, Eds. J. Henkel, S Parameswaran, 2007, Springer, ISBN 978-1402058684.

[16] R. Zhu, "FPGA Image Processing with 8-bit picoRISC Core", MSc Dissertation, University of Southampton, 2022.

[17] S. Wang, C. Xiao, "Reinforcement Learning for Selecting Custom Instructions Under Area Constraint," in *IEEE Transactions on Artificial Intelligence*, vol. 5, no. 4, pp. 1882-1894, April 2024.

[18] G. E. Martin, "Affine Transformations", in *Transformation Geometry: An Introduction to Symmetry*, Springer, New York, 1982, pp. 167–181, ISBN 978-1-4612-5680-9.

[19] J. Op den Brouw, "The THUAS RISC-V 32-bit processor in VHDL," GitHub repository, v1.1.1.0 (Commit: 880cdf2), Jan. 2025. [Online]. Available: https://github.com/jesseopdenbrouw/thuas-riscv [Accessed: Mar. 12, 2025].

[20] R. Jia, C. Y. Lin, Z. Guo, R. Chen, F. Wang, T. Gao, and H. Yang, "A survey of open source processors for FPGAs," in *Proc. 2014 24th Int. Conf. on Field Programmable Logic and Applications (FPL)*, pp. 1-6, 2014.

Analysis and Modelling of ICs and Microsystems

Fractional Spurious Tones Analysis
of the Space-Time Averaging PLL

Radosław Wiliński, Paweł Gryboś

AGH Univeristy of Krakow

Krakow, Poland

Emails: wradoslaw@student.agh.edu.pl, pawel.grybos@agh.edu.pl

Abstract—This article presents a step-by-step process for modeling and analyzing the Space-Time Averaging Phase-Locked Loop architecture, which enables fractional frequency synthesis while significantly reducing the quantization error caused by fractional division. This reduction is achieved through the use of spatial averaging implemented as array of dividers, phase-frequency detectors, and charge pumps. A tree-structured, switching-block-based digital encoder is employed to generate the control signals for the dividers. The critical part of the design from the loop dynamics point of view is implemented at the transistor level using TSMC 40nm CMOS technology, while the reminder is modeled in the Verilog-A Hardware Description Language. The analysis focuses primarily on the fractional spurious tones originating from the quantization error. The Discrete Fourier Transform is used to obtain the output frequency spectra of the space-time and time averaging PLL to evaluate the effectiveness of spatial averaging in reducing quantization noise.

Keywords—fractional spurious tones, fractional phase-locked loop, phase-locked loop, space-time averaging, time-averaging, quantization error reduction.

I. INTRODUCTION

Phase-locked loops (PLLs) play a critical role in modern systems, spanning telecommunications, frequency synthesis, high-speed data transfer, and clocking applications. Depending on the use case, the requirements for frequency synthesis vary and may include high-frequency resolution, a wide frequency range, high speed, low cost, low power consumption, and more. The Time-Averaging Phase-Locked Loop (TA PLL) is a PLL architecture that significantly increases frequency resolution. However, this improvement comes at the cost of introducing a quantization error which induces fractional spurious tones (frac spurs) in the output spectrum.

One way of reducing the frac spurs is implementation of the Space-Time averaging architecture described for the first time in [1]. This paper presents the process of modeling and designing a Space-Time Averaging PLL (STA PLL) from the ground up to a complete system. Section II outlines the design strategy and the modeling procedure of individual building blocks. Section III presents the simulation methodology and results, while Section IV concludes the work.

A. Fundamentals of Fractional Spurious Tone Formation

TA PLL, which block diagram is shown in Fig. 1 (PFD – phase-frequency detector, CP – charge pump, LPF – low-pass filter, VCO – voltage controlled oscillator, N-DIV – feedback

divider, $\Delta\Sigma$ – delta-sigma modulator), enables fractional frequency synthesis by using a Delta-Sigma Modulator (DSM) located in the feedback path. DSM operates by varying the divider value between two or more integers based on the DSM order, such that the average of these values corresponds to the desired fractional division ratio. Consider a reference frequency of 40 MHz and a desired output frequency of 130 MHz. A traditional Integer-N PLL is limited to generating output frequencies that are integer multiples of the reference frequency, such as 40 MHz, 80 MHz, 120 MHz and 160 MHz. In contrast, the TA PLL can generate a non-integer output frequency, such as 130 MHz, by alternating between two frequency values, 120 MHz and 160 MHz, through dynamic adjustment of the divider between 3 and 4, as illustrated in Fig. 2. Divider switching results in the Voltage Controlled Oscillator (VCO) output containing a time-averaged frequency that incorporates the desired fractional portion of the reference frequency, hence the name Time-Averaging. On the downside, fractional operation inherently introduces quantization error. At any given moment, the instantaneous output frequency never equals exactly 130 MHz. This results in a persistent quantization error between the feedback and reference clocks. This error introduces additional frequency components into the VCO control voltage (VCTRL), which directly translate into fractional spurs in the output spectrum. These spurious tones occur at frequency offsets that are integer multiples of the reference frequency, divided by the denominator of the irreducible divider fraction, and contribute to the overall output phase noise.

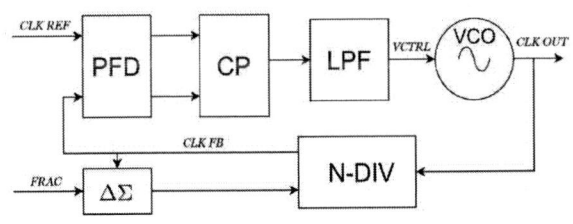

Fig. 1. TA PLL block diagram.

B. STA PLL architecture

The Space-Time Averaging PLL [1] utilizes spatial averaging to reduce quantization noise, offering a significant improvement over the TA PLLs. Unlike TA PLL that rely only

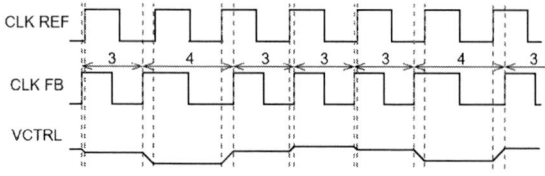

Fig. 2. TA PLL selected waveforms.

on time averaging, the STA PLL employs spatial averaging in addition to time averaging in the form of an array of dividers, phase frequency detectors (PFDs), and charge pumps (CPs), which work in parallel as shown in Fig. 3 (DVC – divider vector controller) to achieve instantaneous fractional division ratios. For example, if 130 MHz is the required output frequency with 40 MHz reference, then the division ratio is set to 3.25. In such a case, to achieve this fraction at any given time, three dividers are set to divide by 3 and one by 4 as shown in Fig. 4. As a result, this causes the instant output frequency of an STA PLL to correspond to the desired frequency. Furthermore, the STA PLL architecture can handle any arbitrary fractional division ratio by combining both spatial and time averaging. This results in more precise fractional frequency synthesis with significantly reduced quantization noise.

Central to the spatial averaging process is the Divider Vector Controller (DVC), which translates the fractional division ratio and the Delta-Sigma Modulator (DSM) output signal into a set of control signals. These control signals modify dividers value between N and N+1 while ensuring that feedback clocks do not drift excessively apart in phase.

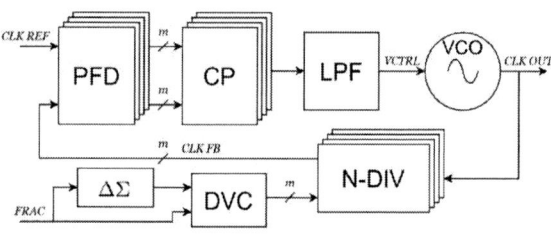

Fig. 3. STA PLL block diagram.

II. MODELING STA PLL

Before beginning the design process, careful consideration must be given to the simulation strategy. Simulating a PLL entirely at the circuit level is an extremely time-consuming process. Therefore, to accelerate simulation, the phase-frequency detector, charge pump, and low-pass filter are modeled at the circuit level using TSMC 40nm CMOS technology to preserve the accuracy of the loop dynamics and characteristics, while the remaining building blocks, that is delta-sigma modulator, divider vector controller, feedback divider, and voltage-controlled oscillator are implemented using the Verilog-A Hardware Description Language. Most of the

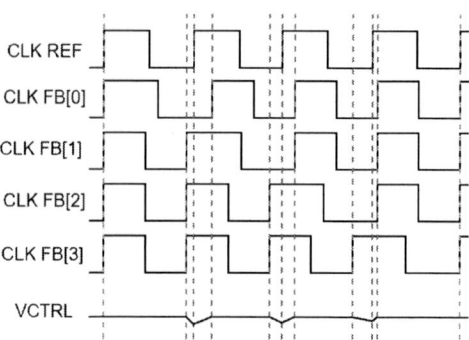

Fig. 4. STA PLL selected waveforms.

focus will be placed on designing transistor level models and the divider vector controller algorithm.

Secondly, the design requirements and core parameters must be established before proceeding with the design. Selected main system parameters are listed below in the Tab. I.

TABLE I.
MAIN SYSTEM PARAMETERS.

Reference frequency (ω_{ref})	$100MHz$
Loop bandwidth	$\frac{\omega_{ref}}{10} = 10MHz$
VCO nominal frequency (ω_{nom})	$300MHz$
VCO gain	$2.83GHz/V$
Spatial averaging array size	4
Total / individual CP current	$150\mu A$ / $37.5\mu A$
Supply voltage	$1.1V$

A. Phase-frequency detector

In this work, phase and frequency detection is realized by the conventional linear PFD presented in Fig. 5, implemented using a design centered around two NOR-based latches.

The UP control signal leaving the PFD must be inverted because it drives the PMOS devices inside the CP. Adding an inverter to the *up* path introduces a mismatch in propagation delays between the UP and DN signals. As a result, the UP signal is consistently delayed relative to the DN signal, leading to dominance of the sink correction in CP. Consequently, this issue introduces additional spurs into the output frequency spectrum at the reference frequency as the control voltage undergoes periodic fluctuations with each PFD cycle. To minimize delay, a buffer is inserted in the *down* path. Then using the logic gate delay matching technique described in [2], the sizes of the *down* path inverters are selected to achieve the smallest delay. The output gate of the NOR-based latch is increased in size to provide better drive capability. Additionally, during calculations, branch effort was not considered because the NAND gate is of minimal size and is therefore expected to have a negligible impact on the final results. The PFD load capacitance was derived from the parasitic capacitance of the CP input transistors. During the following calculations, all values are normalized to the parameters of a unit inverter ($L_{N,P} = 120nm$, $W_P = 240nm$, $W_N = 120nm$). First, to

144

achieve the smallest delay of the *down* path each stage has to bear the same effort $f = g \cdot h$, where g is a ratio of the input capacitance (C_{in}) of a gate to the input capacitance of an unit inverter that delivers the same output current, and h is a ratio of the gate load capacitance (C_{Load}) to its C_{in}. Therefore, by calculating the gate effort f and applying the following Eq. 1, individual gates C_{in_n} are acquired and optimal gate sizes are derived.

$$C_{in_n} = \frac{g_n C_{Load_n}}{f} \qquad (1)$$

Next normalized *down* path delay $D = p + f$, where p is a parasitic delay, is calculated and is equal to 9.86. Finally, using the *down* path delay, the size of the INV_UP inverter is determined to match the delays. Table II shows calculated gate sizes (N and P are the size of the NMOS and PMOS transistor, respectively), and Tab. III the PFD output C_{Load} used for the calculations, normalized to a unit inverter.

TABLE II.
NORMALIZED PFD OUTPUT PATHS GATE SIZES.

Instance	P size	N size
Unit inverter	2	1
NOR_LATCH_UP/DN	8	2
INV_DN1	5.16	2.58
INV_DN2	6.56	3.28
INV_UP	19.72	9.86

TABLE III.
NORMALIZED PFD OUTPUT CAPACITANCE.

Down path output C_{Load}	$25C$
Up path output C_{Load}	$12.5C$

To verify the accuracy of the gate size calculations, simulation results are provided for the scenario where the rising edges of both the **CLK REF** and **CLK FB** signals occur simultaneously. This allows for an assessment of how closely the UP and DN signals are matched. Fig. 6a presents the simulation results of an unmatched PFD delay, where a clear mismatch is evident. In contrast, Fig. 6b shows the results after delay matching, with the signals significantly better aligned.

B. Charge pump

Depending on the application, a CP may need to meet various requirements, including low power consumption, minimal current mismatch, compact area, high switching speed, reduced clock feedthrough, among others. However, in this work, the main focus is on the three parameters:

- High output resistance - max 10% current mismatch due to Channel Length Modulation (CLM) effect,
- Switching speed,
- Output voltage range.

The CP architecture is implemented using self-cascode transistors (CP SC) [3], as shown in Fig. 7. The reference current source is used as a means of biasing the CP. Transistor sizes are detailed in the Tab. IV.

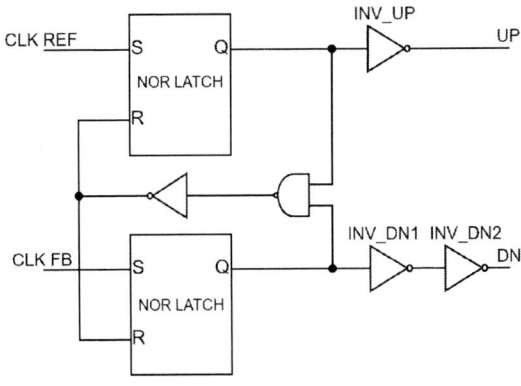

Fig. 5. PFD schematic.

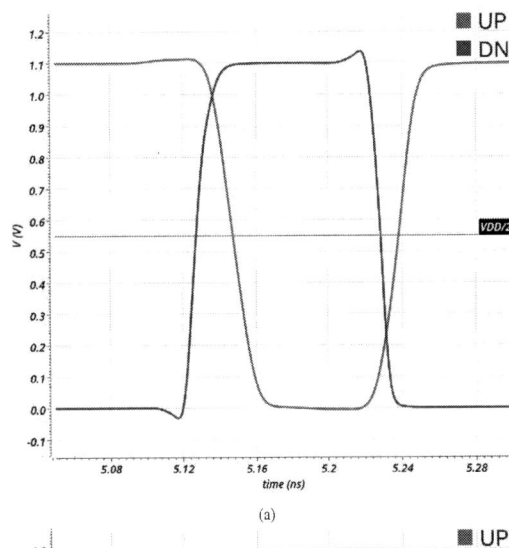

(a)

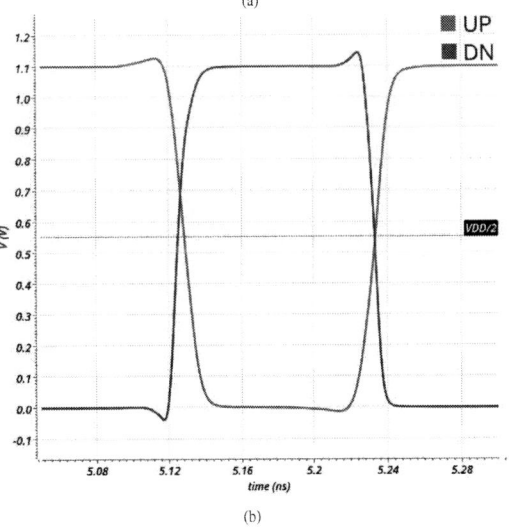

(b)

Fig. 6. PFD a) unmatched and b) matched propagation delay of UP and DN signals.

145

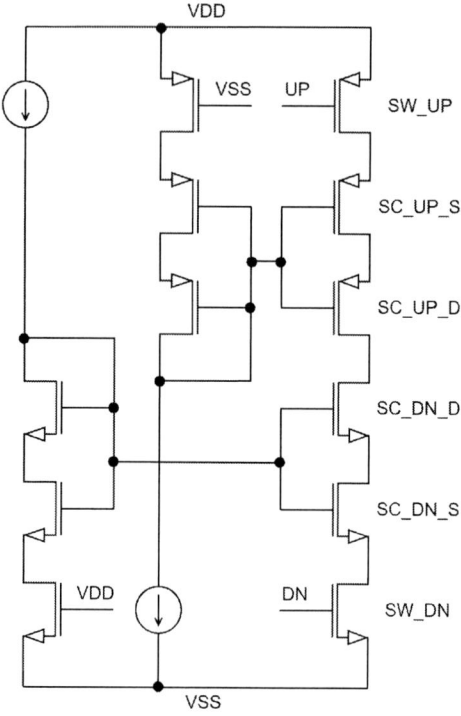

Fig. 7. CP SC schematic.

TABLE IV.
CP SC TRANSISTOR SIZING.

Instance	Width	Length
SW_UP	$3\mu m$	40nm
SC_UP_S	$6.66\mu m$	60nm
SC_UP_D	$20\mu m$	60nm
SC_DN_D	$10\mu m$	60nm
SC_DN_S	$3.33\mu m$	60nm
SW_DN	$1.5\mu m$	40nm

As the voltage on the loop filter varies with the output frequency of the PLL, the operating point of the CP inevitably changes. Variations in CP output voltage result in discrepancies between the *up* and *down* currents due to the CLM effect. These mismatches can introduce ripples in the control voltage and degrade the overall performance of the PLL. To mitigate this issue, the CP requires a high output resistance. To achieve this, self-cascode transistors are employed in the design based on [3]. Although increasing the output resistance minimizes the issue of CLM it reduces the voltage headroom available for the transistors functioning as current sources, which, in turn, decreases the CP output voltage range and directly influences PLL output frequency range. Therefore, when designing a CP, it is crucial to account for this relationship and make sure that the design requirements are met.

The CP output voltage sweep simulation presented in Fig. 8 shows how the output current of a CP utilizing self-cascode

transistors and a conventional CP changes with the output voltage. Conventional CP transistors are sized such that it achieves significant output resistance while ensuring that the area occupied by them is the same as those in the CP SC architecture to provide a fair comparison. It can be observed that the CP SC architecture exhibits less severe current modulation compared to the conventional CP. Additionally, its output voltage range (regarding +/- 10% diversion from the ideal current requirement) is wider, as anticipated, due to the smaller influence of CLM and optimized transistor sizing.

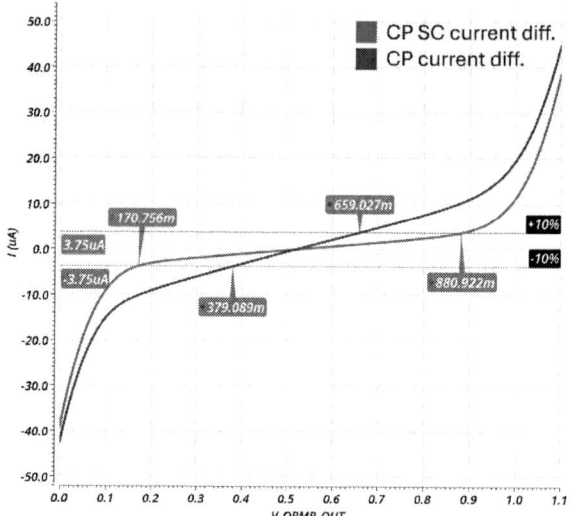

Fig. 8. PFD unmatched propagation delay of UP and DN signals.

Another critical factor to consider when designing a charge pump is its switching speed. Due to the relatively large size of the CP transistors, they exhibit significant parasitic capacitance, which can slow the switching. To address this issue and improve switching performance, the sizes of the transistors that act as switches (SW_DN for the *down* branch and SW_UP for the *up* branch) are minimized. This minimization is done carefully to maintain the required output voltage swing, resistance, and that they remain in saturation. The following simulation illustrates the response of the charge CP in the scenario where the PLL is locked and both the UP and DN signals are asserted simultaneously. This represents the most challenging operating condition for the charge pump, as the pulse width is minimal and is equal to the PFD reset time. This simulation is provided to investigate the mismatch between the source and sink switching speed. Fig. 9 presents CP sink (I_DN) and source (I_UP) current. It is observed that the I_DN current dominates (switches faster). Also, the I_UP current does not reach the expected $37.5\mu A$ and peaks around $31u\mu A$ right before the reset. This behavior is primarily attributed to the differences in capacitances between the *up* and *down* paths within the CP, where the PMOS transistors are twice as wide as the NMOS, therefore experiencing more parasitic capacitance. In an ideal scenario, when both the UP and

DN signals are asserted simultaneously, there should be no change in the control voltage. However, due to the mismatch in switching speed, a non-zero net charge is generated, leading to ripples. These periodic ripples introduce spurs into the PLL output spectrum, effectively increasing the deterministic phase noise of the output signal. Although reducing the transistor's size could minimize this issue, it would come at the cost of disturbing other performance parameters of the CP.

and using predetermined parameters from Tab. I, individual components values are calculated: $C_1 = 25.6pF$, $R1 = 7.75k\Omega$, and $C_2 = 0.2C_1 = 5.12pF$.

$$H(s) = \frac{2\zeta\omega_n s + \omega_n^2}{s^2 + 2\zeta\omega_n s + \omega_n^2} \quad (4)$$

$$\zeta = \frac{R_1}{2}\sqrt{\frac{I_{CP}K_{VCO}C_1}{2\pi N}} \quad (5)$$

$$\omega_n = \sqrt{\frac{I_{CP}K_{VCO}}{2\pi N C_1}} \quad (6)$$

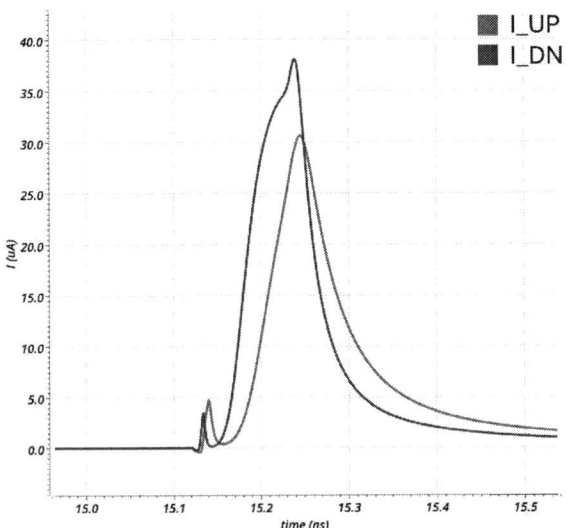

Fig. 9. CP SC *up* and *down* paths current response.

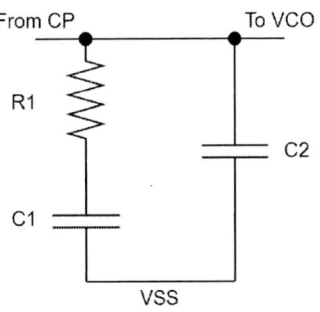

Fig. 10. LPF schematic.

C. Loop filter

Selected loop filter topology is a second-order low-pass filter consisting of two capacitors, C_1 and C_2 and one resistor R_1 as presented in Fig. 10. Although the designed system is of the third order as shown in Eq. 2, where I_{CP} is the CP current, K_{VCO} is the gain of VCO and N is the divider value, the influence of the third pole on loop stability becomes negligible when C_2 is relatively small, that is, $C_1 \leq 0.2C_2$ [4]. In this case, the third pole is located beyond the gain crossover frequency, which means that its impact on the loop dynamics is minimal. Therefore, individual values of the LPF components are derived based on the second-order PLL transfer function depicted in Eq. 3.

$$H(s) = \frac{I_{CP}}{2\pi}\frac{R_1 C_1 s + 1}{R_1\frac{C_1 C_2}{C_1 + C_2}s + 1}\frac{1}{(C_1 + C_2)s}\frac{K_{VCO}}{Ns} \quad (2)$$

$$H(s) = \frac{\frac{I_{CP}K_{VCO}}{2\pi C_1}(R_1 C_1 s + 1)}{s^2 + \frac{I_{CP}K_{VCO}R_1 s}{2\pi N} + \frac{I_{CP}K_{VCO}}{2\pi C_1 N}} \quad (3)$$

Then, the control theory equation for the second-order system, presented in Eq. 4, where ζ is a damping factor and ω_n is a natural frequency of the system, is used to resolve the equation. Finally, selecting $\zeta \approx 1$ to achieve the critically damped system, the maximum N-DIV value based on the CP output range $N = \lfloor(\omega_{nom} + VCTRL_{max} \cdot K_{VCO})/\omega_{ref}\rfloor = 26$,

D. Verilog-A models

The Verilog-A models for the VCO, DSM and feedback divider are straightforward and therefore not discussed in detail. The primary focus is on the divider vector controller, as its algorithm is crucial to preventing feedback clocks from drifting apart in phase during PLL operation. DVC is implemented as a tree-structured digital encoder which is often used in mismatch shaping algorithms as described in [5]. DVC architecture is presented in Fig. 11 where the input value is the requested fractional divider value, DIV_N[3:0] depicts control signals for the dividers, and SW_L*_S* represents switching block instances and its position in the tree structure. Switching block model is presented in Lst. 1.

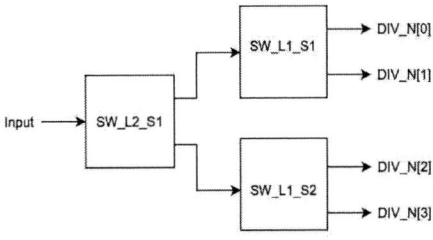

Fig. 11. DVC architecture.

```
1   prbs = symbol_type > 0 ?
2       abs($random(prbs_seed) % 2) : prbs;
3   if(data_in % 2) begin
4       symbol_head = (prbs ? 1 : -1);
5       sw_sequence = symbol_type > 0 ?
6           symbol_head : symbol_head * -1;
7       symbol_type = symbol_type * -1;
8   end else begin
9       symbol_head = symbol_head;
10      sw_sequence = 0;
11      symbol_type = symbol_type;
12  end
13  data_out1 = (data_in + sw_sequence)/2;
14  data_out2 = (data_in - sw_sequence)/2;
```

Listing 1. Switching block Verilog-A algorithm.

Crucial part of the DVC operation is generation of the switching sequence (sw_sequence), which is responsible for high-pass shaping of the output response to prevent aforementioned excessive phase drift in the feedback clocks. This is achieved by composing the switching sequence based on two symbol types: (1, -1) and (-1, 1), which are selected according to the input value and a Pseudo-Random Binary Sequence (PRBS).

Fig. 12 shows DVC output signals for the 0.75 fractional ratio. It can be observed that at any given time three of the output signals are high, which directly correlates with the selected fractional ratio.

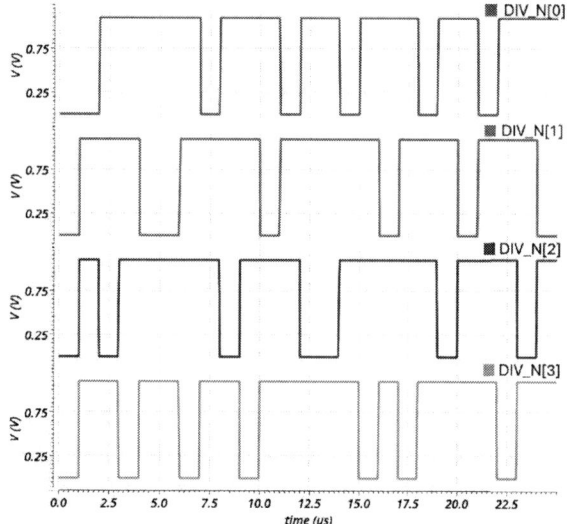

Fig. 12. DVC output signals for 0.75 fractional ratio.

III. SIMULATION RESULTS

A fast and commonly used way to simulate oscillatory systems is periodic steady state (PSS) analysis. While PSS simulation is highly effective for analyzing circuits such as high-Q filters and oscillators, it fails to achieve convergence in certain PLL architectures which exhibit varying loop dynamics during the operation. Fractional PLLs are such systems due to the fractional division process and, as a result, lack a periodic solution, making PSS analysis unsuitable [6]. Due

to the reasons outlined above, PSS analysis cannot be employed in the design described in this paper. Consequently, the behavior of the design is analyzed through a transient simulation. Discrete Fourier Transform (DFT) is employed to analyze the output frequency spectrum containing frac spurs. However, DFT requires careful setup as described in [7]. Even after ensuring that the DFT is configured correctly, additional challenges persist. For instance, achieving high precision in the DFT, such as 10 Hz, leads to unacceptably long simulation times. Consequently, the DFT frequency step must be set to a higher value. Nevertheless, reduced precision introduces further issues: spectral leakage and a lack of side-lobe symmetry. Therefore, in the following simulations, a simple fraction is chosen, as its frequency spectrum remains relatively clean even with a larger frequency step value, allowing for an acceptable simulation time. Simulation results are provided for both the TA and STA architectures for a divider value of 20.2 and reference frequency of 100 MHz which makes output frequency equal to 2.02 GHz.

Fig. 13 presents DFT output spectra of the TA and STA PLL with a frequency resolution of 1 MHz. The simulation results are normalized to the magnitude of the PLL output frequency bin (carrier frequency) according to the equation: $20 \log_{10} \frac{M_n}{M_c}$, where M_c is the magnitude of the carrier frequency (2.02 GHz), and M_n is the magnitude of the nth frequency bin. Results below $-70dB$ are not shown for readability. Effectively only frac spurs and carrier frequency are presented. Magnitude of the frac spurs are compared in Tab. V.

In addition, time-domain waveforms of the output frequency of the TA and STA PLL are extracted and presented in Fig. 14. It can be clearly observed that the STA PLL output signal (FOUT STA) exhibits significantly lower quantization error which manifests itself in a smaller frequency fluctuation compared to that of the TA PLL (FOUT TA).

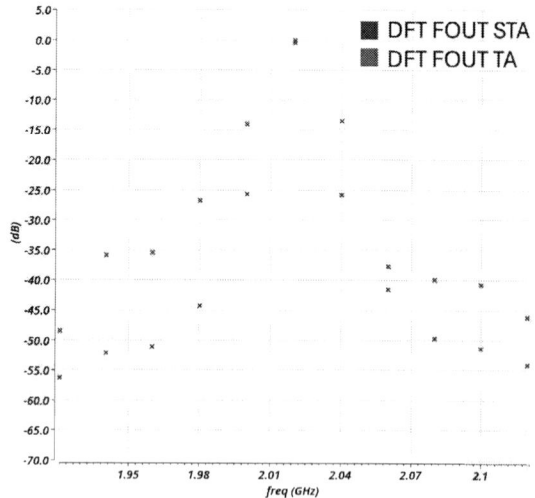

Fig. 13. DFT of STA and TA PLL output signal for 20.2 division ratio.

TABLE V.
MAGNITUDE OF FRACTIONAL SPURIOUS TONES.

Frequency offset from carrier	STA	TA
-100 MHz	$-56.5\ dB$	$-48.4\ dB$
-80 MHz	$-52.2\ dB$	$-35.9\ dB$
-60 MHz	$-51.1\ dB$	$-35.5\ dB$
-40 MHz	$-44.23\ dB$	$-26.7\ dB$
-20 MHz	$-25.6\ dB$	$-14\ dB$
+20 MHz	$-25.7\ dB$	$-13.5\ dB$
+40 MHz	$-41.5\ dB$	$-37.7\ dB$
+60 MHz	$-49.7\ dB$	$-40\ dB$
+80 MHz	$-51.3\ dB$	$-40.8\ dB$
+100 MHz	$-54.1\ dB$	$-46.1\ dB$

IV. CONCLUSION

This article presented a detailed analysis of the space-time averaging phase-locked loop, showcasing its ability to suppress fractional spurious tones effectively. Using spatial averaging with an array of dividers, phase-frequency detectors, and charge pumps, the space-time averaging architecture achieves precise fractional frequency division with reduced quantization noise. The self-cascode transistor-based charge pump design further improves performance by ensuring high output resistance and minimizing current mismatch due to the channel length modulation effect. A divider vector controller enhances this technique by maintaining phase coherence across feedback paths. Simulation results for a 20.2 division ratio with a 100 MHz reference frequency demonstrate a significant reduction in fractional spurs compared to time-averaging PLLs.

ACKNOWLEDGMENT

This research is supported by the National Science Center, Poland, project no. 2023/51/B/ST7/01782

REFERENCES

[1] Y. Zhang, A. Sanyal, K. Wen, X. Tang, X. Yu, X. Quan, G. Jin, L.Geng and N. Sun, "A Fractional-N PLL With Space–Time Averaging for Quantization Noise Reduction", *IEEE Journal of Solid-state Circuits*, 100:633–638, 2020

[2] Neil H. E. Weste and David Money Harris, "CMOS VLSI Design a Circuits and Systems Perspective Fourth Edition", *Pearson*, 2015

[3] Mohsen Karbalaei Mohammad Ali and Omid Hashemipour, "A simple and high performance charge pump based on the self cascode transistor", *Analog Integrated Circuits and Signal Processing*, 31:602–614, 2019

[4] Behzad Razavi, "Design of CMOS Phase-Locked Loops From Circuit Level to Architecture Level", *Cambridge University Press*, 2020

[5] Jared Welz, Ian Galton and Eric Fogleman, "Simplified Logic for First-Order and Second-Order Mismatch-Shaping Digital-to-Analog Converters", *IEEE Transactions on Circuits and Systems II: Express Briefs*, 48:1014—1027, 2001

[6] Ken Kundert, "Predicting the Phase Noise and Jitter of PLL-Based Frequency Synthesizers", https://designers-guide.org/, 2015

[7] RFInsights, "Transient to Spectrum", https://www.rfinsights.com/insights/simulations/cadence/fft-in-cadence/

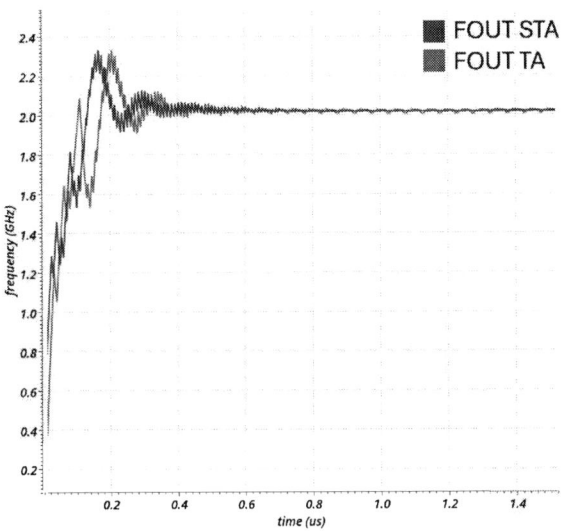

(a)

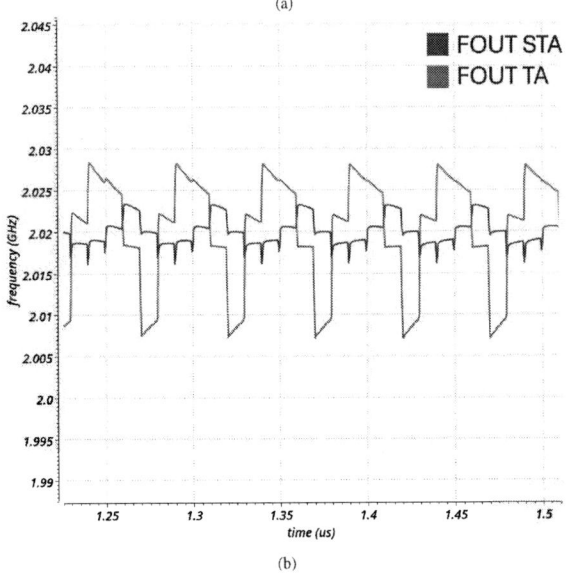

(b)

Fig. 14. TA and STA PLL a) full and b) zoomed in output signal frequency waveform for 20.2 division ratio.

High-Level Modeling of RF Power Amplifiers and Antenna Arrays for Efficient Over-the-Air Power Combination in RF Transceivers

Marius Diacu[1], João Guerreiro[1,2], João P. Oliveira[1,3], Paulo Montezuma[1,2,4], Pedro Viegas[4]

[1] DEEC, NOVA School of Science and Technology, NOVA University of Lisbon, Portugal
[2] IT, Instituto de Telecomunicações, Av. Rovisco Pais, Lisboa, Portugal
[3] UNINOVA-CTS and LASI, Caparica, Portugal
[4] Koala Tech, Portugal

Abstract—**Efficient RF amplification and power combination are key challenges in 5G/6G transceivers. In particular, Power Amplifier (PA) designs have to balance linearity and efficiency requirements to accommodate high Peak-to-Average Power Ratio (PAPR) waveforms, and power combining networks such as Wilkinson combiners introduce insertion losses, limiting the achievable energy efficiency of the RF Front-End (RFFE). This work explores Power Combination Over-the-Air (PCOA), where four sinusoidal components are individually amplified by an RF PA operating at near saturation, driving a 4×4 patch antenna array. Though beamforming, the components are spatially power combined, in the far-field. PCOA allows more efficient PAs to be used, reduces interference to other systems, and avoids circuit-based power combiner disadvantages. A high-level system model is developed to analyze the influence of per-component power control, PA sizing, and antenna array configuration on PCOA efficiency and directivity. Nonidealities such as noise, PA distortion, impedance mismatches, path loss, and mutual coupling are considered, providing insights into optimal PA output power levels, beamforming strategies, and array design trade-offs. Simulation results show that PCOA can decrease RFFE losses, making it a promising technique for next-generation transceivers.**

Keywords—**Antenna arrays, beamforming, mutual coupling, Power Amplifiers, Power Combination Over-the-Air, RF, transceiver.**

I. Introduction

Energy consumption and efficiency in telecommunication systems have become critical concerns, particularly in 4G, 5G [1], [2] and upcoming 6G mobile networks. The demand for higher data rates and improved connectivity has led to challenges in optimizing energy consumption while maintaining high-performance communication. This stems from the trade-off between spectral and energy efficiency [3], which is exacerbated by the signal processing techniques employed.

In 4G and 5G networks, Orthogonal Frequency Division Multiplexing (OFDM) is the predominant signal modulation scheme, which offers high spectral efficiency and resilience against multipath fading, however, the resulting waveforms have an inherently high Peak-to-Average Power Ratio (PAPR) [4]. This results in substantial signal envelope variations, leading to inefficient power amplification, attending to the

strict linearity requirements imposed for low bit error rates (BER).

This work introduces Power Combination Over-the-Air (PCOA), implementing spatial combination of signal components, under any input amplitude distribution. This allows highly non-linear, switched-mode Power Amplifiers (PAs) to be used, which are much more efficient [5], offering a promising solution to the PAPR-related trade-offs by eliminating PA linearity requirements. A four-component, constant-envelope scenario was considered, each amplified by a switched-mode PA, operating near saturation, transmitted by a 4×4 patch microstrip antenna array. By allocating a per-component subarray, the amplified components can be combined at the channel level [6], [7] through beamforming. In this scenario, the overall efficiency is limited by the switched PAs, which have Power Added Efficiencies (PAEs) $\geq 65\%$ [8], since power combiner insertion losses [9], [10] are avoided with spatial combination.

Section II shows the relevant theoretical background for the input signals II-A, PAs and filters II-B, antenna arrays II-C and phase-shift beamforming II-D. Section III describes the model and its configuration, Section IV and Section V present the simulation results and conclusions respectively.

II. Theoretical Analysis

Fig. 1 shows a simplified diagram of the transmitter architecture, which assumes a decomposition of the analog input signal into the four sinusoidal components presented.

A. Input Signals

As can be noted from Fig. 1 an $L = 4$ architecture was considered, where L is the number of branches. The sinusoidal input signal for each lth branch is then given by

$$s_l(t) = A_l \sin\left(2\pi f_c t + \phi_l\right), \quad l = \{1, 2, 3, 4\}, \quad (1)$$

where s_l, A_l and ϕ_l are the per-branch input signal, amplitudes and initial phases respectively. A_l allows the definition of different per-branch amplitudes, and ϕ_l is used to model phase mismatch between branches. Equal amplitudes for all branches and binary scaling with respect to power are considered.

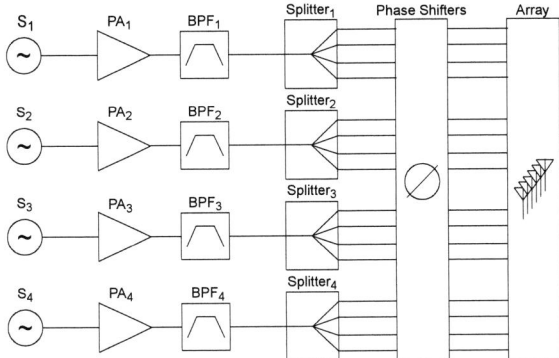

Fig. 1. Simplified diagram of the proposed transmitter architecture.

B. Power Amplifiers and Filters

For the PAs, a third-order, memoryless polynomial model was considered, with a voltage transfer function given by

$$v_{out}(t) = \alpha_1 v_{in}(t) + \alpha_2 v_{in}^2(t) + \alpha_3 v_{in}^3(t), \quad (2)$$

according to [5], where v_{out} is the output voltage, v_{in} is the input voltage, a_1 is the small-signal voltage gain, and a_2, a_3 model second and third-order distortion respectively. This model allows the PAs to be sized at the system level by defining the relationship between a_1 and a_3, with a_2 having a lower impact. Since each branch has a single frequency input, there is no intermodulation distortion, regardless of the generated second and third order harmonics. Moreover, the third-order distortion results in gain compression. Evaluating (2) with a sinusoidal input signal leads to the fundamental component voltage gain G_{v,f_c} being expressed as

$$G_{v,f_c} = \alpha_1 + \frac{3\alpha_3 A_{v_{in}}^2}{4}, \quad \alpha_3 \leq 0, \quad (3)$$

according to [5], where $A_{v_{in}}$ is the input signal amplitude. Gain compression is commonly deemed as having a significant impact on linearity at the 1-dB compression point for power, P_{1dB}, so signals which require high linearity need their whole dynamic range below this point. As shown in [5], this results in low PA efficiency, since it scales with output power. In PCOA, the components in each branch have a constant amplitude, allowing highly non-linear, very efficient switched PAs to be used. They are thus sized to operate at close to their saturation output power P_{sat} (beyond P_{1dB}), for maximum efficiency. This can be expressed as

$$P_{sat} - 1 = 20 \log_{10} |G_{v,f_c}| + 10 \log_{10}\left(\frac{Z_{in}}{Z_{out}}\right) + P_{in}, \quad (4)$$

where P_{in} is the input power and Z_{in}, Z_{out} are the PA input and output impedances, respectively. P_{sat} can be approximated with respect to an n-dB compression point, allowing for α_3 to be determined for a given α_1. With this sizing, the output signal includes harmonics, which are filtered by the output matching network acting as a high-Q Band-Pass Filter (BPF).

C. Antenna Array Factors

The $M \times N$ Uniform Rectangular Array (URA) with $M = N = 4$, equal horizontal (δ_x) and vertical (δ_y) element spacing, along with the considered coordinate system are shown in Fig. 2. From the complete array, each component has a 2x2 URA subarray allocated, represented by the index in Roman numerals. An URA was preferred to an Uniform Linear Array (ULA) for compactness and three-dimensional beamforming capabilities. Far-field beam patterns can be determined through pattern multiplication theory [11], which states that

$$B_{PP,array}(\theta, \varphi) = |AF(\theta, \varphi)|^2 B_{PP,element}(\theta, \varphi), \quad (5)$$

where $B_{PP,array}$ and $B_{PP,element}$ are the array and element power beam patterns respectively, and AF is the array factor, expressed in amplitude. Thus, the determination of the array factor allows the array beam pattern to be found for any type of antenna element. This theory disregards mutual coupling between elements, and serves as an ideal basis of comparison for the simulated results. Determination of the URA $|AF|$ can be done through ULA analysis when the excitations $I_{m,n}$ are separable, namely when

$$I_{m,n} = I_m I_n = A_m A_n e^{j((m-1)\phi_x + (n-1)\phi_y)}, \quad (6)$$

meaning that the excitation for each element m, n (corresponding to x, y indices in Fig. 2) is the product of two functions I_m, I_n which only depend on m and n respectively. This is trivially fulfilled for equal amplitude excitations across all elements, however it is not for binary power scaling between subarrays, so the radiation patterns obtained will be different.

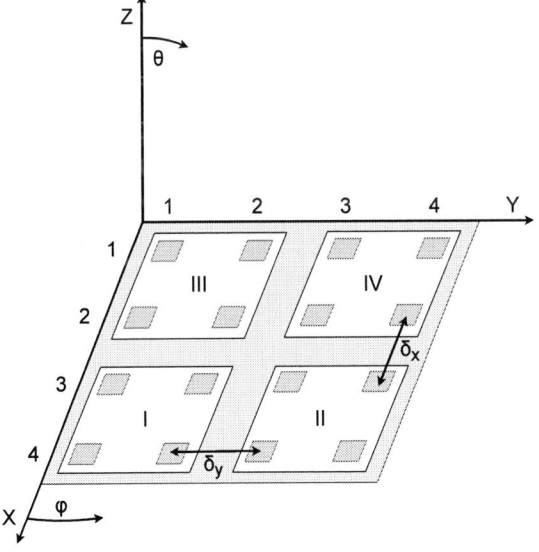

Fig. 2. Transmitter array diagram with the considered coordinate system.

For an N-element ULA with spacing δ, and considering excitations with a common amplitude and a progressive phase shift ϕ from the reference antenna, $|AF_{\text{ULA}}|$ is given by

$$|AF_{\text{ULA}}(\psi)| = \left| \frac{\sin\left(\frac{N\psi}{2}\right)}{\sin\left(\frac{\psi}{2}\right)} \right|, \quad \psi = k\delta\cos(\gamma) + \phi, \quad (7)$$

according to [11], where k is the free-space wavenumber, γ is the angle to the ULA plane, and ψ is a variable that allows for a more compact expression. $|AF_{\text{ULA}}|$ has its maximum value when $\psi = 0$ and it follows that $|AF_{\text{ULA}}(0)| = N$.

Considering an $N \times N$ URA with $\delta_x = \delta_y = \delta$, in the xy plane, with equal amplitudes for all elements, $|AF_{\text{URA}}|$ can be expressed as

$$|AF_{\text{URA}}(\psi_x, \psi_y)| = \left| \frac{\sin\left(\frac{N\psi_x}{2}\right)\sin\left(\frac{N\psi_y}{2}\right)}{\sin\left(\frac{\psi_x}{2}\right)\sin\left(\frac{\psi_y}{2}\right)} \right|, \quad (8)$$

according to [11], where

$$\psi_x = k\delta\left(\sin(\theta)\cos(\varphi)\right) + \phi_x, \quad (9)$$

$$\psi_y = k\delta\left(\sin(\theta)\sin(\varphi)\right) + \phi_y, \quad (10)$$

where φ, θ are the spherical coordinate azimuth and elevation angles (shown in Fig. 2). Similarly to before, $|AF_{\text{URA}}|$ has its maximum value when $(\psi_x, \psi_y) = (0,0)$ and it follows that $|AF_{\text{URA}}(0,0)| = N^2$.

It is important to note that the previous results do not provide a full theoretical description of the array in Fig. 2 for two reasons, namely:

- The URA results assume equal amplitudes and separability, which the binary power scaling considered in this work doesn't fulfill;
- Branches can have zero amplitude, effectively removing the respective subarrays from the geometry.

However, it will be briefly shown that these properties do not pose significant difficulties in implementation.

D. Phase Shift Beamforming

Only beamforming at the phase shift level will be considered, and amplitude alteration for array factor optimization will not be explored. For the ULA, since ψ depends on ϕ, the direction which results in the maximum, γ_{max}, can be controlled by setting ϕ to

$$0 = k\delta\cos(\gamma_{max}) + \phi \Leftrightarrow \phi = -k\delta\cos(\gamma_{max}), \quad (11)$$

however, this only allows horizontal or vertical adjustment for the maximum's position, depending on ULA orientation. For the URA, simultaneous horizontal and vertical adjustment is possible, by setting the progressive phase shifts to

$$\phi_x = -k\delta\left(\sin(\theta_{max})\cos(\varphi_{max})\right), \quad (12)$$

$$\phi_y = -k\delta\left(\sin(\theta_{max})\sin(\varphi_{max})\right). \quad (13)$$

An important property can be derived when observing the generic array factors for both cases. For the ULA, it can be expressed as

$$AF_{\text{ULA}}(\psi) = \sum_{n=1}^{N} A_n e^{j(n-1)\psi}, \quad (14)$$

according to [12]. Since this function does not alter the excitation amplitudes A_n, the maximum is obtained when all of the components sum in-phase, which is only possible when $\psi = 0$ since n changes for each term. This result is independent of A_n, thus, the main beam direction will not change with amplitude alterations. For the URA, the generic array factor is given in [12], after algebraic manipulation,

$$AF_{\text{URA}}(\psi_x, \psi_y) = \sum_{n=1}^{N}\sum_{m=1}^{M} A_{m,n} e^{j((m-1)\psi_x + (n-1)\psi_y)}, \quad (15)$$

which also does not modify $A_{m,n}$, leading to the maximum condition being given by $(\psi_x, \psi_y) = (0,0)$ for analogous reasons, which is also independent of $A_{m,n}$. This means that the main beam direction will remain unaltered, whether equal amplitude components or binary power scaling is considered.

The conclusion is then that the beamforming progressive phase shifts in (12) and (13) will orient the main beam to the target direction for both amplitude cases considered, and also when there exist branches with zero amplitude. However, the array factors do change, and only the direction of the main beam remains constant.

The required progressive phase shifts are applied to the array input signals with ideal phase shifters. The splitters are not a necessary part of the architecture and a completely splitterless analog implementation is possible, it serves as a model shorthand to avoid N^2 input signal, filter and PA blocks.

III. HIGH-LEVEL TRANSCEIVER MODEL

Having summarized the relevant theoretical background, a schematic of the model utilized, implemented in MATLAB [13] Simulink, can be found in Fig. 3. Table I summarizes the main design parameters considered for the model. The operating frequency f_c was chosen to be compliant with the latest 5G Radio Access Network (RAN) requirements [14].

TABLE I
MAIN MODEL PARAMETERS FOR TRANSCEIVER SIMULATION.

Parameter	Value
Operating frequency, f_c	5.3 GHz, "n46" band [14]
Characteristic impedance, Z_0	50 Ω
Temperature	300 K
Sector central angle	60° (±30°)
Transmission impairments	Noise (channel and antennas), path loss
Total power at PA outputs	250 mW / 23.98 dBm
Types of input	Equal amplitudes, binary power scaling
Channel noise PSD	10^{-17} W/Hz / -140 dBm/Hz

This model expands Fig. 1 by adding combined splitter and filter insertion loss modeling, channel modeling with parameters and Power Spectral Density (PSD) described in

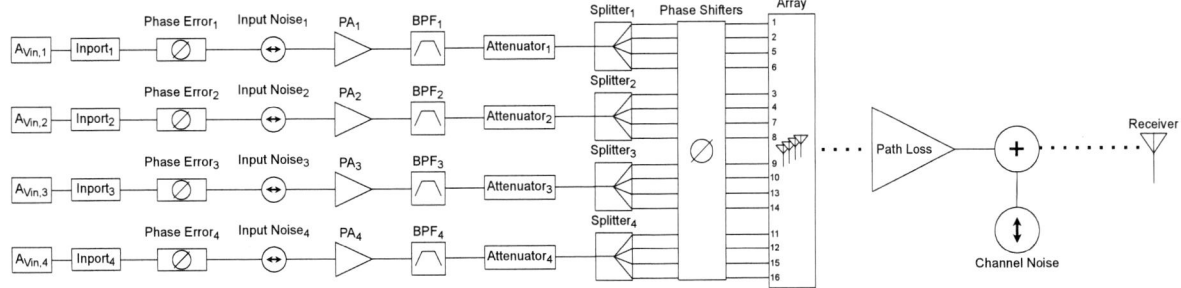

Fig. 3. Schematic of transceiver model implemented in MATLAB [13] Simulink.

Table I and a single-antenna receiver, simulating a Multiple-Input Single-Output (MISO) system.

Specific block configuration parameters are presented in Table II, excluding the array and antennas. A singular value means it is common to all branches, whenever applicable. All impedances were set equal to Z_0, resulting in perfect matching, except for the array and receive antenna.

TABLE II
PARAMETERS FOR THE RF BLOCKS IN FIG. 3.

Parameter	Value
PA available power gain	16 dB
PA IIP$_2$	50 dBm
PA IIP$_3$ (equal amplitudes)	10.25 dBm
PA IIP$_3$ (binary power scaling)	{13.84, 10.52, 7.08, 3.86} dBm
PA NF	3 dB
BPF topology and order	T network, 2
BPF transfer function	Butterworth, $\varepsilon = 1$
BPF centre frequency, bandwidth	5.3 GHz, 100 MHz
Insertion loss	1 dB
Input P_i (equal amplitudes)	3.19 dBm
Input P_i (binary power scaling)	{6.54, 3.54, 0.54, -2.46} dBm
Splitter type	Ideal Wilkinson power divider

Externally to the Simulink project, the antenna array was sized using the Antenna Array Designer and the receiving antenna using the Antenna Designer [13]. The array elements and the receive antenna are identical. Table III summarizes the antenna and array parameters. The Fraunhofer distance d_{FF} is given by

$$d_{FF} = \frac{2D_{TX}^2}{\lambda}, \tag{16}$$

according to [12], where D_{TX} is the largest array dimension and λ is the wavelength.

IV. SIMULATION RESULTS

The three-dimensional directivity for an isolated element is shown in Fig. 4, along with the considered orientation, coordinate systems, and maximum value. It should be noted that in MATLAB [13] simulation, θ starts in the x, y plane, (i.e. the direction $\theta = 90°$ lies in the z axis). In all the following results, only the elevation plane will be considered.

Knowing the isolated element's power beam pattern, it is possible to predict the array power beam pattern through (5),

TABLE III
PARAMETERS FOR THE ARRAY AND ANTENNAS IN FIG. 3.

Parameter	Value
Type of element / receive antenna	Patch microstrip
Element length, width, height	2.72 cm, 3.53 cm, 566 μm
Element ground plane length, width	5.66 cm, 5.66 cm
Element dielectric	Air, $\varepsilon_r = 1$
Element conductor	Ideal conductor, $\sigma = \infty$
Array δ_x, δ_y spacing	3.76 cm, 3.76 cm ($\lambda/2$)
Fraunhofer distance, d_{FF}	1.47 m
Resonant frequency	5.3 GHz (f_c)

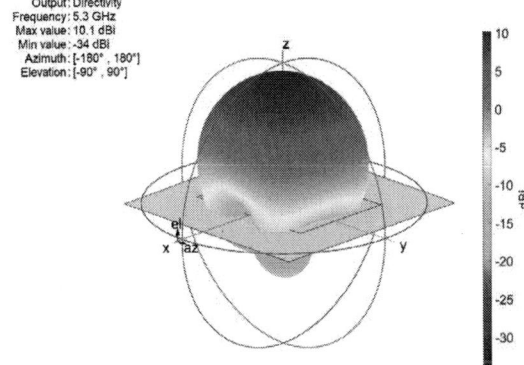

Fig. 4. Isolated element three-dimensional directivity with axes, element orientation and maximum value.

with $|AF_{\text{URA}}|$ given in (8). This, in turn allows the array directivity to be predicted, and compared against the simulated array directivity, which includes mutual coupling effects. This is shown in Fig. 5, where significant differences can be observed in the simulated scenario, namely a higher maximum value, along with differing positions and number of sidelobes. However, the direction where the maximum is found remains unchanged, meaning that the previously described progressive phase shifts can still be used.

Besides the directivity, it is important to represent the impedance behavior of the array under mutual coupling, which dictates how it interacts with the RF chain. This was done in Fig. 6 where all S-parameters are shown. Despite the influence of mutual coupling, the $S_{1,n}$ values are similar, which allows for good matching across the whole array. The array has a

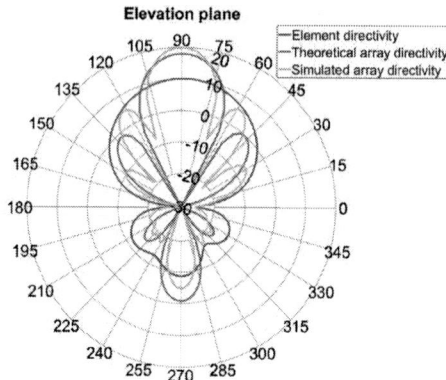

Fig. 5. Element directivity, theoretical prediction for array directivity and simulated array directivity, in dBi.

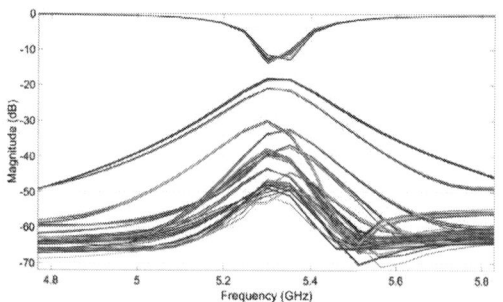

Fig. 6. Complete representation of the array S-parameters, showing the influence of mutual coupling.

narrowband characteristic, suppressing harmonics and out-of band spurs. Lastly, the process of phase-shifting the incoming signals to perform beamforming does influence the active impedances [12], which effectively leads to a varying degree of mismatch depending on the target direction, adding slight losses by reflection.

Having explored the array and element properties through simulation, the transmitter as a whole can be evaluated. The default scenario considered assumed perfect phase alignment between the input signal branches, all components enabled, and no beamforming phase shifts applied. Examples of the transient input (transmitter) and output (at receiver) signals as well as the signal power spectrum for various signals in the first transmitter branch are shown in Figs. 7 and 8 respectively, for binary power scaling. Perfect receiver alignment is always assumed.

Table IV summarizes the power levels at the PAs and the transmitting array's Effective Isotropic Radiated Power (EIRP) with and without beamforming. Considering the input powers from Table II clear gain compression can be observed, and most notably the EIRP differs between setups, even with the same total PA output power, indicating different array behavior. The power levels also change significantly between both beamforming scenarios which shows the impact of the array's elements active impedance variation.

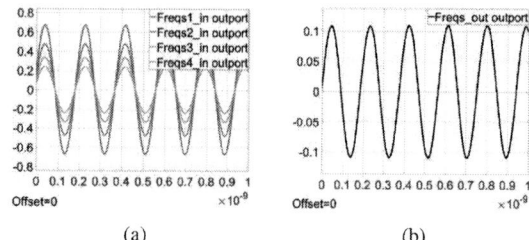

(a)   (b)

Fig. 7. Transient input (a) and output (b) signals for binary power scaling, default conditions, for a distance of 5 meters.

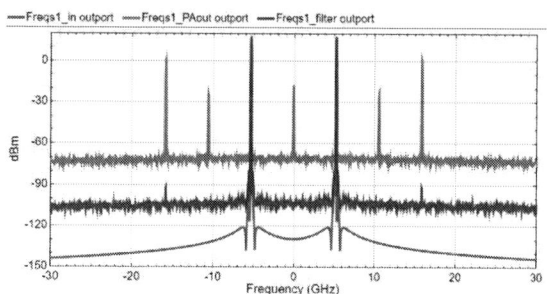

Fig. 8. Signal power spectrum for the first branch, considering binary power scaling and default conditions.

TABLE IV
PA OUTPUT POWERS AND ARRAY EIRP FOR BOTH INPUT SCALING MODES AND BEAM DIRECTIONS.

Input	Main beam $\theta = 90°$		Main beam $\theta = 120°$	
	Output power		Output power	
Equal Amplitudes	PA$_1$	17.97 dBm	PA$_1$	16.66 dBm
	PA$_2$	17.96 dBm	PA$_2$	16.63 dBm
	PA$_3$	17.96 dBm	PA$_3$	16.42 dBm
	PA$_4$	17.96 dBm	PA$_4$	16.46 dBm
	EIRP	42.10 dBm	EIRP	41.00 dBm
	Output power		Output power	
Binary scaling	PA$_1$	21.24 dBm	PA$_1$	20.04 dBm
	PA$_2$	18.24 dBm	PA$_2$	17.08 dBm
	PA$_3$	15.53 dBm	PA$_3$	13.20 dBm
	PA$_4$	12.24 dBm	PA$_4$	10.29 dBm
	EIRP	41.52 dBm	EIRP	40.50 dBm

To further show how the RF chain and array interaction affects the transmission, directivities were obtained through transmitter simulation, and were compared to the simulated array directivity presented in Fig. 5. These results are shown in Fig. 9 for default conditions and target main beams at $\theta = 90°$ (a) and $\theta = 120°$ (b) respectively. An example where only two branches (I and IV) are enabled is shown, to validate the theoretical prediction. It is possible to observe that binary power scaling alters the directivity, however all cases have their maximum for the same value of θ, as predicted. Furthermore, it is possible to verify once again that the process of phase shifting the signals alters the active impedances of the elements, which results in mismatch with the RF chain. Slight error is noticed in Fig. 9 (b) in the beam direction, since the phase shifting strategy does not account for mutual coupling.

154

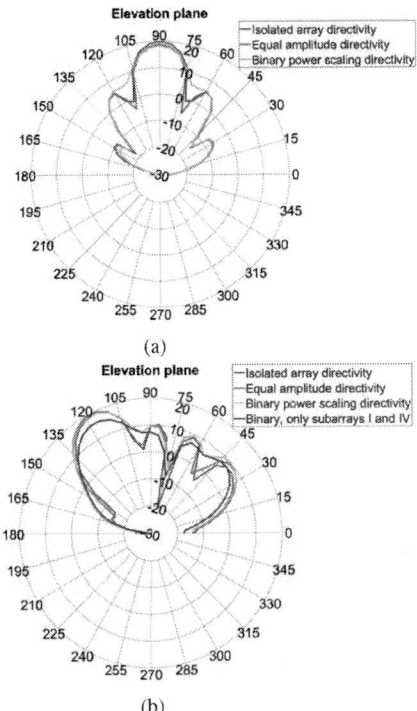

(a)

(b)

Fig. 9. Directivities obtained for $\theta = 90°$ (a) and $\theta = 120°$ (b) in dBi. Results are only valid for $\theta \in [0°, 180°]$.

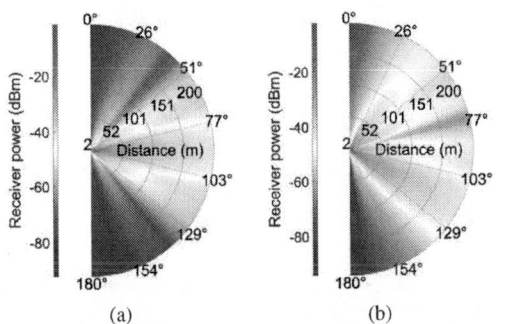

(a)   (b)

Fig. 10. Binary power scaling heatmaps, $\theta = 90°$ (a) and $\theta = 120°$ (b).

It is possible to show how the results in Fig. 9 affect the received power, through a spatial heatmap. Considering the free-space path loss model [12], Fig. 10 shows the receiver power heatmaps, considering propagation distances ranging from 2 to 200 meters and binary power scaling. As intended, the vast majority of the transmitted power is pointed towards the receiver, offering little interference to neighboring systems.

V. CONCLUSION

The proposed PCOA architecture increases transmission efficiency and offers low interference to neighboring systems. This is achieved by minimization of insertion losses and addition of array gain, along with a directive transmission respectively. It is shown to be resilient to a wide variety of nonidealities and configurations, all while utilizing maximally

efficient blocks and without sacrificing signal quality. Furthermore, it relies on simple-to-implement logic and analog circuitry.

The high-level architecture and results presented serve as a strong starting point for future optimization and synthesis works, to bridge the gap between the proposed solution and full next-generation mobile network compliance.

ACKNOWLEDGMENT

This work is funded by the Portuguese Foundation for Science and Technology (FCT) and Ministry of Education (MECI) through national funds and, when applicable, joint EU funds under UID/50008: Instituto de Telecomunicações. The authors would also like to thank FCT within the scope of UNINOVA-CTS Multiannual Strategic Funding under Grant CTS/00066, project EU-FCT LISBOA2030-FEDER-00816400 and project EU-FCT LISBOA2030-FEDER-00917900.

REFERENCES

[1] E. Kolta. (2024) Going green: measuring the energy efficiency of mobile networks. [Online]. Available: https://data.gsmaintelligence.com/api-web/v2/research-file-download?id=79791160&file=270224-Measuring-energy-efficiency-of-mobile-networks.pdf

[2] G. Y. Li, Z. Xu, C. Xiong, C. Yang, S. Zhang, Y. Chen, and S. Xu, "Energy-efficient wireless communications: Tutorial, survey and open issues," *IEEE Wireless Communications*, vol. 18, no. 6, pp. 28–35, 2011.

[3] Y. Chen, S. Zhang, S. Xu, and G. Y. Li. "Fundamental trade-offs on green wireless networks," *IEEE Communications Magazine*, vol. 49, no. 6, pp. 30–37, 2011.

[4] A. Behravan and T. Eriksson, "Some statistical properties of multicarrier signal and related measures," in *2006 IEEE 63rd Vehicular Technology Conference*. IEEE, 2006, pp. 1854–1858.

[5] M. K. Kazimierczuk, *RF Power Amplifiers*, second edition ed. John Wiley & Sons, 2015.

[6] V. Astucia, P. Montezuma, R. Dinis, and M. Beko, "On the use of multiple grossly nonlinear amplifiers for highly efficient linear amplification of multilevel constellations," in *2013 IEEE 78th Vehicular Technology Conference (VTC Fall)*. IEEE, 2013, pp. 1–5.

[7] P. Montezuma, D. Marques, V. Astucia, R. Dinis, and M. Beko, "Robust frequency-domain receivers for a transmission technique with directivity at the constellation level," in *2014 IEEE 80th Vehicular Technology Conference (VTC2014-Fall)*. IEEE, 2014, pp. 1–7.

[8] P. Montezuma, R. Madeira, H. Serra, P. Viegas, R. Dinis, J. Oliveira, and J. Guerreiro, "Quantized digital amplification with combination over the air - achieving maximum efficiency on communication links between long range uavs and satellites," in *MILCOM 2021 - 2021 IEEE Military Communications Conference (MILCOM)*. IEEE, 2021, pp. 802–807.

[9] K. H. An, O. Lee, H. Kim, D. H. Lee, J. Han, K. S. Yang, Y. Kim, J. J. Chang, W. Woo, C.-H. Lee, H. Kim, and J. Laskar, "Power-combining transformer techniques for fully-integrated cmos power amplifiers," *IEEE Journal of Solid-State Circuits*, vol. 43, no. 5, pp. 1064–1075, 2008.

[10] A. Wentzel, V. Subramanian, A. Sayed, and G. Boeck, "Novel broadband wilkinson power combiner," in *2006 European Microwave Conference*. IEEE, 2006, pp. 212–215.

[11] H. L. V. Trees, *Optimum Array Processing: Part IV of Detection, Estimation, and Modulation Theory*, first edition ed. John Wiley & Sons, 2002.

[12] W. L. Stutzman and G. A. Thiele, *Antenna Theory and Design*, third edition ed. John Wiley & Sons, 2013.

[13] T. M. Inc., "Matlab version: 24.1.0.2537033 (r2024a)," Natick, Massachusetts, United States, 2024. [Online]. Available: https://www.mathworks.com

[14] 3GPP, "3rd generation partnership project; technical specification group radio access network; nr; base station (bs) radio transmission and reception (release 18)," 3GPP, Tech. Rep. TS 38.104 V18.8.0 (2024-12), 2024.

Reliability Analyses of Ultra-Low Voltage Analog Spiking Neurons

Grégoire Brandsteert, Léopold Van Brandt, Denis Flandre

ICTEAM Institute, UCLouvain

Louvain-la-Neuve, Belgium

Email: {gregoire.brandsteert, leopold.vanbrandt, denis.flandre}@uclouvain.be

Abstract—We investigate the behaviour of a bio-inspired artificial neuron, implementing a simplified Morris-Lecar model under downscaled supply voltages in the subthreshold regime towards ultra-low-power performance. The neuron spike characteristics, including amplitude, period, and typical shape, are analysed as the supply voltage is reduced. We observe a dramatic loss of the typical linear-exponential behaviour in the rising edge of the spike related to a drastic reduction in spike amplitude and an increase in spike period. These are used to define figures-of-merit of spiking operation for ad hoc neuromorphic computations. Subsequently, through Monte Carlo variability analyses, we highlight that ultra-low-voltage neurons are highly sensitive to process variations and prone to statistically likely failures. Our study evidences the challenges of maintaining stability and performance in low-power neuromorphic circuits and the importance of reliability assessments.

Keywords—neuromorphic; spiking neural network (SNN); analog neuron; subthreshold , reliability; VLSI; ultra-low voltage

I. INTRODUCTION

Computing technologies based on traditional CMOS technology and Von Neumann architecture are currently experiencing significant bottlenecks and approaching fundamental physical limits [1]. Limitations, related to processing speed, power consumption, and scalability, are prompting researchers to explore new design paradigms to overcome these challenges for information processing systems. One promising approach that receives considerable attention lies in spiking neural networks (SNNs) that are inspired by the operation of biological brains [2]. Specifically, SNNs use time-based signals, known as spikes, to process and transmit information, mimicking the temporal nature of neural communication in biological organisms.

Spiking neurons are key building blocks of SNNs that, similarly to their biological counterparts, integrate incoming spikes from other neurons over time. When the neuron internal potential reaches a critical threshold, it emits a spike, thereby transmitting information to other connected neurons through synapses [3]. This process, also known as spike-based communication, is substantially different from the continuous activation schemes used in conventional artificial neural networks (ANNs). The design and operation of these spiking neurons can be based on a variety of mathematical models, each with their own strengths and trade-offs [4].

In the context of hardware implementation, one of the main challenges is to ensure the scalability of the neurons in SNNs. As the number of neurons increases, it becomes essential to maintain a balance between computational capacity and power consumption. This is critical when neurons are to be integrated at a very large scale (VLSI) [5], similarly to the scale of biological neural networks, or for applications with strict power requirements. As a result, considerable attention has been paid to optimizing the energy efficiency of the hardware that supports these neural networks. Purely analog designs seem capable of much lower power operation compared to existing digital [6] or mixed-signal [7] circuits investigated for small scale bio-inspired SNNs.

A particularly noteworthy analog neuron architecture, presented in Fig. 1, approximating the Morris-Lecar (ML) neuronal model has been implemented in 65 nm CMOS technology in [8]. This design combines biophysical plausibility with high energy efficiency, operating in the order of a few femtojoules (fJ) per spike reported from measurements on a 65 nm CMOS prototype [8]. The architecture is highly compact, using only two capacitors in the femtofarad range and six MOS transistors, making it a promising candidate for efficient hardware implementations.

While other neuron designs have achieved even lower power consumption, down to tens of attojoules (aJ) per spike [9], these designs are often based on simpler mathematical equations such as the Leaky Integrate-and-Fire (LIF) model. Such artificial LIF neurons are unable to emulate the wide variety of dynamic behavioral characteristics of more complex biological neurons [10]. This trade-off between energy efficiency and behavioral complexity is a major challenge regarding the hardware implementation of bio-inspired neural models.

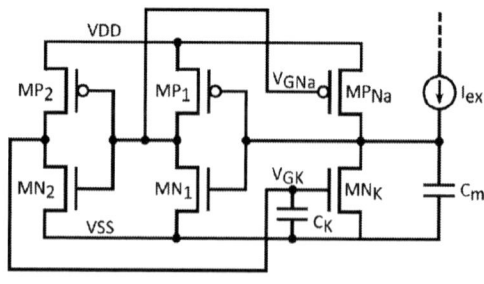

Fig. 1. Compact analog neuron design based on the Morris-Lecar model, from [8].

The massive deployment of IoT sensor is an important motivation regarding energy efficiency. While reducing the energy consumption per spike is an important goal [11], it only represents a fraction of the total energy budget, and the quiescent energy consumption during periods of inactivity has to be accounted for. To address this issue, the present work proposes an extensive analysis of the effects of supply voltage reduction on neuron behaviour, defined by several figures of merit that we present in Fig. 2: amplitude, period and shape. In addition, the study includes process variability analysis using the Monte Carlo simulation technique to assess the circuit robustness under ultra-low supply voltage, as it operates deeper in subthreshold regime. This approach aims to explore whether further power reductions can be achieved without threatening the reliable functionality of the neuron.

II. THE SIMPLIFIED MORRIS-LECAR ARTIFICIAL NEURON

The artificial analog neuron design under study is based on an approximation and simplification of the ML mathematical model appropriate in describing the dynamics of biological neurons at a relatively low computational cost [12]. In this approach, the typical response of the neuron is emulated by incorporating the exponential behaviour usually observed in biological neurons, while optimizing the design for energy efficiency. Both these goals are achieved through the subthreshold operation of MOS transistors, with gate voltages way below the transistor threshold voltages. This regime yields an exponential relationship between gate voltage and drain current, as well as low energy dissipation due to the low current and voltage values.

A brief overview of the neuron's circuit behaviour is as follows: the membrane voltage, denoted V_m, is initially set at a resting potential, typically pulled down close to the negative supply V_{SS} (set at 0 V). When a postsynaptic excitation current pulse (I_{ex}) is applied, the membrane capacitance begins to charge, causing the membrane voltage to rise linearly. When the first inverter switching voltage ($V_{switch,1}$) has been reached, the MP_{Na} transistor is activated, initiating a positive feedback loop that further increases the membrane voltage thanks to the exponential control of the drain current by the gate. This is shortly followed by negative feedback via the MN_K transistor when the C_K capacitance is fully charged. This alternance of positive and negative feedbacks creates a characteristic

linear/exponential voltage increase followed by a rapid exponential discharge (Fig. 2) that mimics the action potential (or "spike") in biological neurons.

This process of voltage rise and fall is repeated continuously as long as a sufficient excitation current (superior to 20 pA) remains present. In this architecture, the total energy consumption of the neuron is primarily determined by the supply voltage (V_{DD}) and the membrane capacitance (C_m). Hence, parameters with very low values are chosen (V_{DD} = 200 mV ; C_m = 4 fF) in order to minimize the energy requirement [8]. While both V_{DD} and C_m affect the active power consumption, the supply voltage is the major controllable parameter contributing to the static power consumption. Since the energy efficiency of a SNN is estimated to be dependent on a high spike sparsity [13], one can expect that its artificial neurons will operate with low duty cycles. The reduction of V_{DD} is therefore essential in improving the general energy efficiency of analog neurons.

III. STABILITY UNDER ULTRA-LOW SUPPLY VOLTAGE

The main focus of this work is to further reduce the supply voltage below the value of 200 mV used in [8] and observe the behaviour of the circuit operating in the deep subthreshold regime. The effects of ultra-low supply voltages on the membrane voltage (V_m) are presented in Fig. 3. Three main phenomena are observable: a reduction in peak-to-peak amplitude, an increase in spike period, and a loss of definition of the typical ML spike shape, particularly the linear-exponential rise of V_m. While these observations are somewhat expected, a question arises regarding the definition and quantification of a spike.

In the literature, an *action potential* or *spike* is described qualitatively as "an abrupt and transient change in membrane voltage that propagates to other neurons..." [14]. However, as the supply voltage decreases, the transient change becomes progressively less abrupt, as seen in Fig. 3. This is mainly due to the degradation of the inverters DC (lower gain, reduced noise margins [15]) and transient (longer delays [16]) characteristics for deep subthreshold regime operation. At very low supply voltages, the membrane voltage transition increasingly resembles a sinusoidal-like curve. These effects

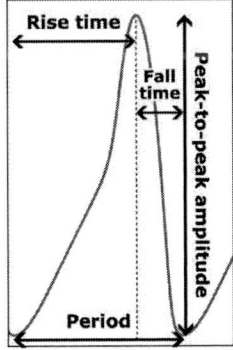

Fig. 2. Illustration of a voltage spike and definition of various the figures of merit used to characterise it.

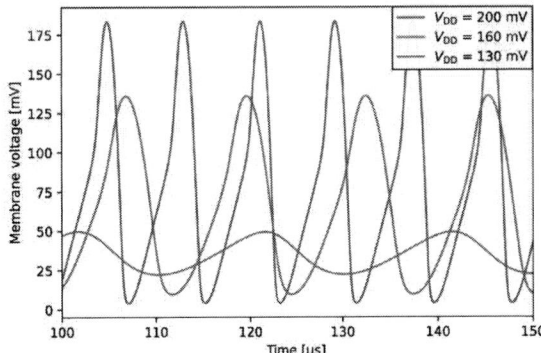

Fig. 3. Membrane voltage Vm under various supply voltages (I_{ex} = 150 pA). Illustrated case : SPICE simulation of Fig. 1, implemented in 65 nm CMOS technology with design parameters from [8].

157

also increase the spike duration, diminishing the energetic advantages related to the potential low duty-cycle operation. As a result, it is crucial in this work to establish quantitative metrics to define the shape of the generated spike.

The first metric measures the peak-to-peak voltage of the spikes relatively to V_{DD}. This metric is particularly relevant for highly interconnected neurons subject to significant fan-out issues [17]. The second metric is the spike period, defined as the time between two local minima, which quantifies the relative deviation of the spike shape from an ideal time-localised pulse. The third metric focuses on the rising and falling times of the spike, measured as the time difference between the highest and lowest voltage points, and vice versa. The results are presented in Fig. 4-6 for three different excitation currents in the range used in [8].

Regarding the membrane peak-to-peak voltage (Fig. 4), we observe a rapid degradation in the generated spike amplitude for ultra-low supply voltages, which becomes critical if the neuron is highly interconnected, making it unable to drive the next components of the SNN. The DC switching voltage (defined as the voltage such that the inverter DC transfer curve satisfies $V_{in} = V_{out}$) of the first and second inverters ($V_{switch1,2}$) are presented by the dashed and dotted lines, respectively. It is expected that the inverters become unable to operate correctly under these limits. Therefore, minimum values of supply

voltage ($V_{DD,min}$) are defined by the more restrictive dashed line for the different excitation currents, and the parts of the curves located in the greyed-out area are marked as unexploitable. The deviation from the theoretical value of $V_{DD}/2$ of the simulated switching voltages can be explained by the design parameters of the first inverter (i.e. the relative MN_1/MP_1 transistor strength) [15].

The increasing propagation delays of the inverters can be observed in Fig. 5. The spike period increases for lower supply voltages at each observed I_{ex}. The main cause seems to be the increased propagation delays of the subthreshold CMOS inverters. Note that for very low supply voltages (even below $V_{DD,min}$), the absence of spikes prevents the extraction of the period.

The rise and fall times of the membrane voltage are presented in Fig. 6. To qualitatively reproduce the biological dynamics of ion transport through the membrane, the rise time is expected to remain more important than the fall time due to the combination of linear and exponential behaviours. At voltages lower than $V_{DD,min}$, it can be observed that both times begin to merge, indicating a gradual degradation of the spike asymmetry and a loss of the biophysical plausibility of the model.

These different metrics are valuable in quantitatively comparing the generated spikes in ultra-low voltage conditions. As the spike peak-to-peak amplitude appears at this stage as the more stringent metric and has been shown to relate to the timing metrics, it will be retained in the next

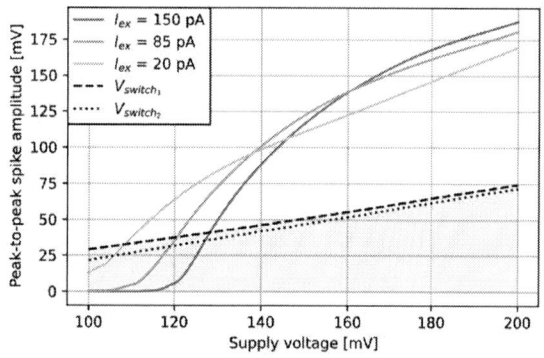

Fig. 4. Evolution of peak-to-peak spike amplitude with supply voltage for different excitation currents. The switching voltage of the two inverters are respectively shown by the dashed and dotted lines.

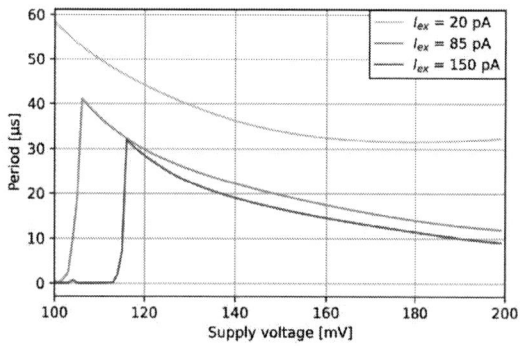

Fig. 5. Evolution of spike period with supply voltage for different excitation currents.

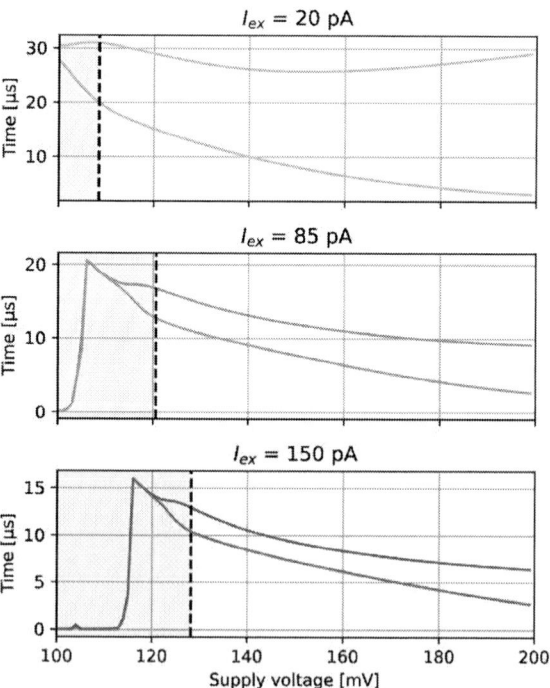

Fig. 6. Evolution of spike rise (blue lines) and fall (orange lines) times with supply voltage for different excitation currents. Respective limit supply voltages ($V_{DD,min}$) from Fig. 4 are shown by the black dashed lines.

section where we will observe that the aggressive supply voltage reduction leads to several critical variability issues for practical neuromorphic circuit applications.

IV. VARIABILITY ANALYSIS

The functionality of a prospective ultra-low voltage SNN is endangered by process variations. This motivates reliability analyses at the analog neuron level. We conducted Monte Carlo simulations on the same design to examine the effects of process variability in ML-based neurons operating in the deep subthreshold regime. In Fig. 7, simulations were performed using three distinctive supply voltages (130, 160 and 200 mV) and an intermediate excitation current (I_{ex} = 85 pA). The peak-to-peak voltage amplitude, introduced in Section III, was normalized by the corresponding supply voltage to ease comparisons. This normalized metric was analysed over 10000 trials for each case, which constitutes a statistically sufficient data set moreover consistent with the number of neurons in the brains of some small insects [18] and SNNs [19]. For each Monte-Carlo experiment, random values of transistor parameters (including threshold voltage, transistor dimensions, oxide thickness and other key factors influencing circuit performance) are drawn based on the industrial compact model to faithfully simulate the effect of process variations. The normalized spike amplitude becomes a random variable taking different realisations.

These extracted realisations, plotted in histograms (Fig. 7) highlight significant shifts from the nominal values for the three supply voltages. The general trend is that a larger fraction of spikes with lower normalized amplitudes are observed for lower supply voltages. This representation is also useful to observe the broader distribution of the spike metric, demonstrating the high variability and instability of the circuit under ultra-low voltage.

As specified in section III, we consider artificial neurons presenting peak-to-peak membrane voltages of less than $V_{DD,min}$ to be defective and establish this hard limit as criterion of functionality. Therefore, we observe in Fig. 8 the average neuron failure probability under ultra-low supply voltages for different excitation currents. The neuron variability is here

analysed over 1000 Monte Carlo trials for each supply voltage, and each tested neuron whose peak-to-peak voltage is lower than $V_{switch1}$ is classified as a failure.

We observe a rapid increase of neuron failure probability with the supply voltage reduction. Note that while neurons with low excitation current are more robust under ultra-low supply voltage, they are less reliable at higher supply voltage. These results are of high importance, since it has been shown in [20] that there is a rapid deterioration of SNNs classifying accuracy proportionally to the increasing percentage of non-functioning neurons. Moreover, the effects of intrinsic transistor noise could inflate the failure probability [21] and hence require more attention.

These findings underscore the importance of carrying out thorough stability and robustness analyses for analog neurons operating in subthreshold regimes, extending our analyses to complete process, voltage and temperature variations. It also suggests that these observations could be extended to other types of energy-efficient artificial neurons, such as capacitance-less circuits [22], where the absence of traditional capacitances may further reinforce variability and noise impact.

V. CONCLUSION AND PERSPECTIVES

This work explores the behaviour of an energy-efficient artificial analog neuron based on a simplified Morris-Lecar model, to assess its performances under ultra-low supply voltage. We observed significant degradations in spike characteristics, such as reduced amplitude, increased period, and a loss of the typical biophysically plausible spike shape, as we reduce the supply voltage. The Monte Carlo variability analysis revealed that the subthreshold regime results in a high variability and instability in the neuron operation leading to high failure probabilities at low voltages. These findings highlight the importance of considering the effects of random phenomena inherent to ultra-low-voltage devices and circuits. Reducing quiescent energy efficiency while maintaining robustness and biological plausibility of spiking neurons remains a major challenge. This work emphasises the need for further investigation of the reliability of bio-inspired neurons operating at extremely low voltages.

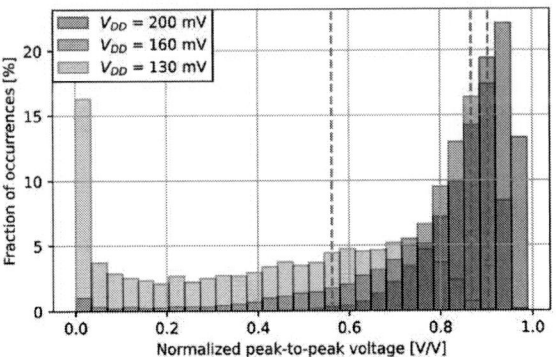

Fig. 7. Normalized spike amplitudes occurrences under various supply voltages over 10000 MC simulations. Nominal values are shown by the colored dashed lines.

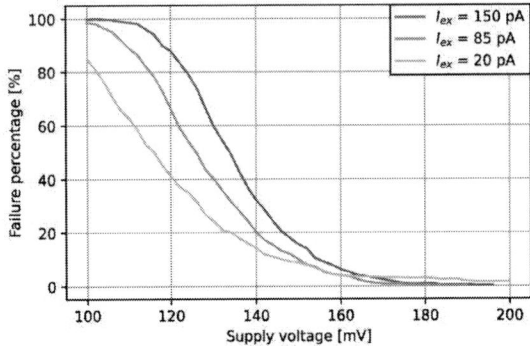

Fig. 8. Evolution of neuron failures with supply voltage for different excitation currents.

ACKNOWLEDGMENT

This work is fully supported by the European Research Council (ERC) under the European Union's Horizon AG Research and innovation program (ERC-Synergy, SWIMS, 101119062) and the FRS-FNRS 'chargé de recherche' grant supporting Dr. Van Brandt. The authors thank the Bio-Inspired Circuits and Systems (BICS) research group at the University of Groningen and Ir. Sylvain Favresse from UCLouvain for interesting discussions on this work.

REFERENCES

[1] Waldrop, M. M. (2016). The chips are down for Moore's law. *Nat. News* 530:144. doi: 10.1038/530144a

[2] N. Rathi, I. Chakraborty, A. Kosta, A. Sengupta, A. Ankit, P. Panda, and K. Roy, "Exploring Neuromorphic Computing Based on Spiking Neural Networks: Algorithms to Hardware," *ACM Comput. Surv.*, vol. 55, no. 12, Art. no. 243, Dec. 2023, pp. 1-49, doi: 10.1145/3571155

[3] Rubin, D. B. D., Chicca, E., & Indiveri, G. (2004, January). Characterizing the firing properties of an adaptive analog VLSI neuron. In *International Workshop on Biologically Inspired Approaches to Advanced Information Technology* (pp. 189-200). Berlin, Heidelberg: Springer Berlin Heidelberg.

[4] E. Izhikevich, "Which model to use for cortical spiking neurons?" *IEEE Transactions on Neural Networks*, vol. 15, no. 5, pp. 1063–1070, 2004.

[5] Vittoz, E. A. (2020). Analog VLSI implementation of neural networks. In *Handbook of neural computation* (pp. E1-3). CRC Press.

[6] Frenkel, C., Legat, J. D., & Bol, D. (2019). MorphIC: A 65-nm 738k-synapse/mm² quad-core binary-weight digital neuromorphic processor with stochastic spike-driven online learning. *IEEE transactions on biomedical circuits and systems*, 13(5), 999-1010.

[7] Moradi, S., Qiao, N., Stefanini, F., & Indiveri, G. (2017). A scalable multicore architecture with heterogeneous memory structures for dynamic neuromorphic asynchronous processors (DYNAPs). *IEEE transactions on biomedical circuits and systems*, 12(1), 106-122.

[8] Sourikopoulos I, Hedayat S, Loyez C, Danneville F, Hoel V, Mercier E and Cappy A (2017) A 4-fJ/Spike Artificial Neuron in 65 nm CMOS Technology. *Front. Neurosci.* 11:123. doi: 10.3389/fnins.2017.00123.

[9] Williams, S., Khalil, K., & Bayoumi, M. (2024, August). Penta-transistor integrate & fire (ptif) spiking neuron with an ultra-low energy consumption of 0.045 fj per spike. In *2024 IEEE 67th International Midwest Symposium on Circuits and Systems (MWSCAS)* (pp. 1060-1064) IEEE.

[10] Taherkhani, A., Belatreche, A., Li, Y., Cosma, G., Maguire, L. P., & McGinnity, T. M. (2020). A review of learning in biologically plausible spiking neural networks. *Neural Networks*, 122, 253-272.

[11] Han, J. K., Yun, S. Y., Lee, S. W., Yu, J. M., & Choi, Y. K. (2022). A review of artificial spiking neuron devices for neural processing and sensing. *Advanced Functional Materials*, 32(33), 2204102.

[12] Takaloo, H., Ahmadi, A., & Ahmadi, M. (2022). Design and analysis of the Morris–Lecar spiking neuron in efficient analog implementation. *IEEE Transactions on Circuits and Systems II: Express Briefs*, 70(1), 6-10.

[13] M. Dampfhoffer, T. Mesquida, A. Valentian and L. Anghel, "Are SNNs Really More Energy-Efficient Than ANNs? an In-Depth Hardware-Aware Study," in *IEEE Transactions on Emerging Topics in Computational Intelligence*, vol. 7, no. 3, pp. 731-741, June 2023, doi: 10.1109/TETCI.2022.3214509.

[14] Izhikevich, E. M. (2007). *Dynamical systems in neuroscience*. MIT press, pp. 2-3.

[15] Alioto, M. (2010). Understanding DC behavior of subthreshold CMOS logic through closed-form analysis. *IEEE Transactions on Circuits and Systems I: Regular Papers*, 57(7), 1597-1607.

[16] Sharroush, S. M. (2018). Analysis of the subthreshold CMOS logic inverter. *Ain Shams Engineering Journal*, 9(4), 1001-1017.

[17] Park, J., & Choi, W. Y. (2025). Ultra-Low Static Power Circuits Addressing the Fan-Out Problem of Analog Neuron Circuits in Spiking Neural Networks. *IEEE Access*.

[18] Winding, M., Pedigo, B. D., Barnes, C. L., Patsolic, H. G., Park, Y., Kazimiers, T., ... & Zlatic, M. (2023). The connectome of an insect brain. *Science*, 379(6636), eadd9330.

[19] Siddique, A., Vai, M. I., & Pun, S. H. (2023). A low cost neuromorphic learning engine based on a high performance supervised SNN learning algorithm. *Scientific Reports*, 13(1), 6280.

[20] Spyrou, T., El-Sayed, S. A., Afacan, E., Camuñas-Mesa, L. A., Linares-Barranco, B., & Stratigopoulos, H. G. (2021, February). Neuron fault tolerance in spiking neural networks. In *2021 Design, Automation & Test in Europe Conference & Exhibition (DATE)* (pp. 743-748). IEEE.

[21] Van Brandt, L., Bonnin, M., da Silva, M. B., Bolcato, P., Wirth, G. I., Flandre, D., & Delvenne, J. C. (2025). Modeling and Predicting Noise-Induced Failure Rates in Ultra-Low-Voltage SRAM Bitcells Affected by Process Variations. *IEEE Transactions on Circuits and Systems I: Regular Papers*.

[22] Kwon, D., Woo, S. Y., Bae, J. H., Lim, S., Park, B. G., & Lee, J. H. (2021). Hardware-based spiking neural networks using capacitor-less positive feedback neuron devices. *IEEE Transactions on Electron Devices*, 68(9), 4766-4772.

Power Electronics

A Thermal Behavior of Lateral (VESTIC) BJTs on SOI Substrate

Piotr Mierzwinski

Institute of Microelectronics and Optoelectronics
Warsaw University of Technology
Warsaw, Poland
piotr.mierzwinski@pw.edu.pl

Abstract—This paper analyzes the thermal behavior of lateral (Vertical Slit Transistor Integrated Circuits, VESTIC) and vertical BJTs on SOI substrates, focusing on self-heating effects, heat dissipation mechanisms, and thermal stability. The buried oxide (BOX) layer in SOI significantly impacts heat flow, leading to localized hot spots in vertical BJTs and more distributed heating in lateral BJTs. Using numerical simulations and experimental data, we evaluate thermal management strategies and their implications for complementary bipolar logic (CBip). The findings highlight the need for optimized device layouts and biasing techniques to mitigate self-heating, ensuring stable and efficient operation of SOI-based bipolar circuits.

Keywords—lateral BJT; VESTIC; SOI; self-heating; thermal management

I. INTRODUCTION

In the early years of integrated circuits, bipolar junction transistors (BJTs) dominated digital logic families like transistor-transistor logic (TTL) and emitter-coupled logic (ECL). Despite their fast switching speeds, BJTs were eventually replaced by CMOS for VLSI systems due to higher power consumption and larger device structures [2]. Additionally, bipolar processes produced excellent NPN transistors but inferior PNP devices, preventing symmetric CMOS-like logic gate designs [2].

Vertical BJTs were highly asymmetric, featuring heavily doped emitters, lightly doped collectors, and unequal junction areas, resulting in a large saturation voltage at on state [6]. This degraded logic voltage levels, making it impractical to directly replicate CMOS inverter designs using traditional vertical BJTs [6]. These power and symmetry limitations drove the industry shift from TTL/ECL to CMOS.

Nevertheless, the inherent speed advantages of bipolar devices continue to drive research into improved bipolar technologies that balance high performance with power efficiency. This paper builds on earlier works [1, 2], presenting proof-of-concept studies on using lateral bipolar transistors for digital logic gates, with a particular focus on thermal constraints.

II. LATERAL BJTs ON SOI
(VESTIC TECHNOLOGY AND OTHER)

Recent advancements in semiconductor device technology have revived using BJTs in CMOS-like digital circuits. VESTIC technology introduces the VES-BJT, a lateral BJT implemented in the thin top silicon layer of an SOI wafer [4]. Unlike traditional vertical BJTs, where current flows mainly perpendicular to the chip, the VES-BJT drives current laterally,

parallel to the wafer surface [4], [5], [Cai & Ning]. While lateral bipolar transistors were historically considered parasitic in CMOS processes, they have been implemented in BiCMOS and SOI technologies also. Various lateral transistor designs have been explored since the 1960s, including recent developments on SOI substrates. However, VES-BJT structures differ significantly from other lateral BJTs. The concept was introduced by W. Maly in a patent application and subsequent publications.

A vertical slit structure in VESTIC technology is common to several different semiconductor devices that can be manufactured in this technology. Especially after gate oxide removal from a VESFET (an insulated gate field effect transistor in VESTIC technology), the former gate regions act as the polysilicon emitter and collector, while the slit between them act as monosilicon base region. All regions are uniformly doped, and emitter and collector doping can be matched for improved symmetry [5].

In different implementation the lateral BJT is designed to be compatible with SOI CMOS process. This BJTs can be fabricated alongside CMOS transistors with minimal modifications, allowing integration of CMOS and high-speed bipolar devices on a single SOI chip [1].

Significantly, lateral BJTs enables complementary NPN and PNP bipolar devices on SOI [1]. This allowed the fabrication of complementary NPN and PNP pairs with similar characteristics, which were later optimized and integrated into experimental logic circuits, including inverters, SRAM cells, and flip-flops [1], [2].

III. THERMAL MANAGEMENT IN SOI DEVICES

One key consideration for any SOI-based device (MOSFET or BJT) is thermal management. SOI substrates include a buried oxide (BOX) layer that electrically isolates the active silicon layer from the bulk silicon wafer. While this isolation provides benefits like reduced parasitic capacitance and latch-up immunity, it severely inhibits heat flow from the device into the substrate. Silicon dioxide is a very poor thermal conductor, so any heat generated in the transistors tends to remain in the thin silicon layer, raising the device temperature [6], [8].

This self-heating effect is especially pronounced in BJTs, where power dissipation in the junctions can significantly elevate the local temperature and in turn affect transistor behavior. In fact, measurements have shown that beyond a certain bias point, BJTs on SOI wafers exhibit a sharp increase in collector and base currents compared to equivalent bulk devices, directly attributable to the inability of the BOX to conduct heat away [6]. Essentially, the transistor starts to heat itself up, which reduces the effective V_BE needed for conduction (due to the negative temperature coefficient of the base-emitter voltage), thereby driving more current – a positive feedback that can lead to thermal runaway if unchecked. Managing this self-heating requires careful attention to device geometry, biasing, and packaging (heat sinking). [8, 9]

In bulk silicon BJTs, much of the heat can spread downwards into the substrate, but in an SOI BJT the heat is largely confined to the device layer, meaning it must escape laterally or through the front-side of the chip [9]. Studies comparing different isolation schemes have found that a fully isolated SOI device relies predominantly on lateral heat flow to dissipate energy, as there is no direct path into a thermally conductive substrate [7]. For instance, 2D electrothermal simulations of SOI-based BJTs with deep trench isolation showed that significant heat spreading occurs parallel to the surface, and wider device layouts were needed to effectively distribute and remove this heat. This contrasts with a comparable bulk transistor where heat can flow vertically into the silicon substrate. Therefore, thermal design considerations (such as device spacing, the inclusion of thermal vias or heat spreaders, and limiting current densities) are crucial when implementing high-performance devices on SOI. Proper thermal management ensures that junction temperatures remain within safe limits and that device characteristics (like gain and switching speed) do not degrade due to excessive self-heating. In summary, the SOI substrate's insulating properties make thermal effects a dominant design concern: careful engineering is required to mitigate self-heating through layout and bias strategies so that the benefits of SOI (isolation and speed) can be enjoyed without compromising reliability or performance.

This makes thermal design considerations (such as device spacing, the inclusion of thermal vias or heat spreaders, and limiting current densities) crucial for implementing high-performance SOI BJTs.

IV. THERMAL BEHAVIOR: LATERAL VS. VERTICAL BJTS ON SOI

The geometry of a BJT—whether lateral or vertical—strongly influences its thermal characteristics, especially when on an SOI substrate. Lateral and vertical BJTs on an SOI substrate both experience self-heating due to the insulating buried oxide (BOX) layer. Their geometry, however, causes them to differ when it comes to heat generation and dissipation.

In contrast, a lateral BJT on SOI such as the VES-BJT distributes its active region over the plane of the silicon. Its emitter-base-collector junctions are horizontally arranged, meaning that the region of heat generation is more spread over the device's surface area, which may help reduce localized heating [9], [10]. This lateral spread could make it easier for heat to diffuse to the environment since the sources of the heat are not stacked on top of each other. In effect, the lateral BJT can have a reduced peak temperature rise for the same power, since the power is spread over a greater distance as the current flows laterally. Moreover, the base contacts for a lateral device generally lie at the base region's ends, which could be used as further sites for heat sinking (the base contact metallization serves to sink the heat, while for a vertical BJT the base contact

may be far from the region of greatest heat). It is worth noting, however, that both lateral and vertical BJTs on SOI are ultimately subject to lateral heat flow – neither enjoys the luxury of being embedded directly into the substrate. Therefore, a lateral structure does not eliminate self-heating so much as it redistributes it. Some simulation studies have indicated that, despite being fully isolated, lateral devices still enjoy being thicker (in the direction of heating spread) to reduce thermal resistance, as vertical devices on SOI enjoy being thermally shunted or containing heat-spreading dummy structures. Another difference, albeit a subtle one, lies in the manner by which the current distribution within the transistor impacts heating. In the VES-BJT, for one, the odd shape of the junctions causes most of the current to track through the center of the base (an effect of geometry current crowding [19]). This makes the center region carry the highest density of current – and hence produce most of the heat – and regions close to the base contacts carry less. In the standard planar vertical BJT, a large proportion of the current can follow closer to the region where the base contacts are implanted. The practical consequence is that for a lateral BJT the hottest location may be deeper within the device away from any contact, but for a vertical one the heat may be a little closer to the contact (but still insulated mainly by oxide beneath). These effects may impact the way each device is placed within circuits, for instance, one may place thermal trenches or straps close to a lateral BJT to draw heat away from its center. Overall, lateral BJTs on SOI have a more distributed heat profile than vertical BJTs, which have a more concentrated heat region. The lateral device can possibly sustain higher density of power before thermal runaway since the heat is distributed, but both are subject to proper design within SOI so self-heating is mitigated as far as possible. Empirically, self-heating effects for SOI BJTs occur for relatively low current levels (even for a few tens of microamps for certain cases), emphasizing that geometric details aside, designers must take thermal factors into account. Methods like the use of thicker silicon layers (partly depleted SOI), incorporation of thermal contacts, or running the devices under regimes that prevent extreme V_BE (and thus constrain power) are all techniques for controlling the thermal behavior. The lateral vs. vertical distinction provides an additional lever: one can sacrifice some transistor speed (owing to an increased lateral base transit distance) for possibly better thermal spreading using a lateral topology, while a vertical device optimizes raw speed/density but must accommodate more concentrated self-heating within a small volume. The right choice will be determined by the power and speed requirements of the application, and by how well the circuit can be engineered to dissipate heat.

In brief, lateral BJTs for SOI have a more distributed heat profile, and vertical BJTs create localized hot spots through trapping heat close to the BOX. In both scenarios, efficient thermal strategies must be employed to avoid thermal runaway and circuit degradation for SOI-based circuits [7], [9].

V. THERMALLY STABILIZED COMPLEMENTARY BIPOLAR LOGIC (CBIP)

The renewed interest in lateral BJTs on SOI is closely related to complementary bipolar logic, or what has been termed CBip. This concept, patented as early as 1976 [19], envisions logic gates built from pairs of NPN and PNP transistors analogous to

CMOS (which employs complementary N- and P-channel MOSFETs). The theoretical appeal of CBip logic is that it might offer the speed advantage of bipolar switching and the economy of power of CMOS-style operation. However, as we discussed, previous attempts at implementing bipolar logic did not work mainly due to thermal considerations and device asymmetry. Thermally stabilized CBip describes contemporary realizations of complementary bipolar gates that add design features to suppress self-heating and guarantee stable behavior. For example, a basic CBip inverter is an NPN and a PNP transistor arranged similarly to a CMOS inverter. If both transistors are conducting heavily at the same time (for instance, during a switching transition, or due to leakage currents, for example), the circuit might overheat or latch up. To avoid this, resistors for current limiting or bias networks are brought into the circuit design so the transistors never simultaneously enter a high-current regime uncontrollably. Moreover, employing SOI-based symmetric lateral BJTs means that the NPN and PNP transistors exhibit matched V_BE vs. temperature characteristics, so any thermal shifts will impact the two types of devices relatively uniformly, which will help the balance be maintained within the logic gate. A study [2] demonstrated that by biasing the complementary BJTs appropriately and adding a resistive load element, a bipolar inverter can achieve stable switching without thermal runaway, even within a DC coupled string of gates. Indeed, researchers have demonstrated that fully functional digital circuits (inverters, gates, ring oscillators) can be constructed using CBip logic on SOI if thermal stability is considered. These thermally stabilized CBip circuits could operate under a range of conditions, and significantly, they also have the property of dynamic scaling of their dynamic performance. Because the power (and thus junction temperature) within a bipolar gate can be modulated by modulating the supply voltage, the speed of CBip logic can be adjusted over orders of magnitude by modulating the supply or bias resistances. For example, increasing the supply voltage, for example, causes the BJTs to source more current and switch quicker (at the cost of more power), while reducing it makes them slow down but conserve energy. This facilitates a form of "performance-on-demand" that could be important for adaptive systems. It has been envisioned that CBip logic can thus operate over two extremes: an ultra-fast mode competitive with that of ECL for high-performance requirements, and an ultra-low-power mode competitive with that of CMOS for energy-sensitive operation, simply through modulating the bias conditions. The demonstration of successfully using thermally stabilized CBip gates is a compelling proof-of-concept that complementary bipolar technology may find niches for itself within modern circuits. From a circuit design perspective, one must simply bias each CBip gate so that it cannot self-oscillate or go into thermal runaway. This typically means adding a bias resistor or current source to the emitter or employing a partial positive feedback that squashes the VTC (voltage transfer curve) in a manner that compensates for base currents being loaded. The use of SOI benefits through supplying nearly ideal transistor isolation (so each gate's thermal and electrical behavior is relatively localized), but it also requires the detailed thermal design addressed. The presentation of ring oscillators and flip-flops constructed from CBip inverters proves that larger networks composed of such gates can function properly when each

inverter is stabilized. In the future, thermally-aware complementary bipolar logic may find use in specialized domains where CMOS is inadequate – i.e., under extreme-low-temperature conditions (where BJTs can operate reliably) or for circuits needing to span very wide supply voltage ranges. It's worth noting that CBip is not going to replace CMOS everywhere, but instead will augment it in specific areas. The quasi-CMOS behavior of CBip (so termed since the logic family behaves like CMOS but employs bipolar devices) provides another stratum for the digital designer to work with. With the twin problems of device complementarity and thermal stability solved, modern lateral BJT technologies and CBip logic prove that the eminent bipolar transistor still has a place to be found in integrated circuit development. Now that the genuinely complementary NPN/PNP pairs can be built and stabilized thermally, we are presented with the chance to investigate hybrid logic architectures that take advantage of the speed of bipolar transistors and the efficiency of design principles akin to CMOS.

This introduction has established context and the most important details of such technologies, highlighting the thermal behavior distinctions between lateral (VESTIC) and vertical BJTs on SOI. In the next sections, we will discuss the device physics and thermal modeling of lateral and vertical BJTs on SOI, and discuss design strategies (including the CBip approach) that utilize the best features of the devices for practical circuit design.

VI. THERMAL MODEL

Taking as a reference model a model presented in [5] the new coefficients for analyzing lateral bipolar transistors are needed. Considering lateral transistors two important differences in geometrical structure must be accounted. The first is the position of heat source which extends from the bottom (from BOX) to the surface of the substrate, and the second is claimed use of the pillar metal contacts which significantly change the heat distribution.

In the lateral thermal model, the heat source spans parallel to the silicon surface, resulting in a volume heat distribution (in each dimension). The lateral transistor has rather two dimensional heat distribution (a dominant heat transfer is in plane parallel to the surface). By introducing a perpendicular heat source model, we simulate scenarios in which heat generation is concentrated closer to the BOX—such as vertically extended emitter-base-collector junctions and when trench oxide structures vertically constrain the silicon island.

The role of pillar metal contacts, characteristic for VESTIC technology, becomes essential in this scenario for thermal conductivity. These vertical structures serve as alternate heat conduits, reducing vertical thermal resistance if they penetrate close to or through the BOX and serve as thermal vias. The inclusion of such features in simulations results in noticeable flattening of thermal profiles, especially under pulsed power loads.

Key innovation introduced in [5] were average heat transfer coefficients. These are empirical heat transfer coefficients h_x, h_y, h_z derived from 3D FEM simulations. These coefficients vary with trench depth, trench thickness, buried oxide thickness and silicon island footprint. General model is

$$h_{x,y,z} = \left(A_{x,y,z}e^{-a_{x,y,z}dt} + B_{x,y,z}e^{-b_{x,y,z}L} + C_{x,y,z}\right)t_{t,OX}^{-c}, \quad (1)$$

where A, B, Cs are the fitting parameters, and W is silicon island width, L is silicon island length, t_t trench thickness, d_t trench depth, t_{OX} buried oxide thickness. Unfortunately the values of coefficients were never published.

As the FEM simulation has indicated the heat from the device is blocked inside it. The trench as well as buried oxide has low thermal conductivity, as can be seen on Fig. 1 the temperature gradient in buried oxide is much less steep in comparison to the junction region. The effective thermal coupling of trench isolated SOI devices is very low, the most important issue is self-heating.

The simulation shows also that in the case of VESTIC devices the pillar contacts are crucial for effective heat dissipation, as other path for heat flux are limited by trench and buried oxide.

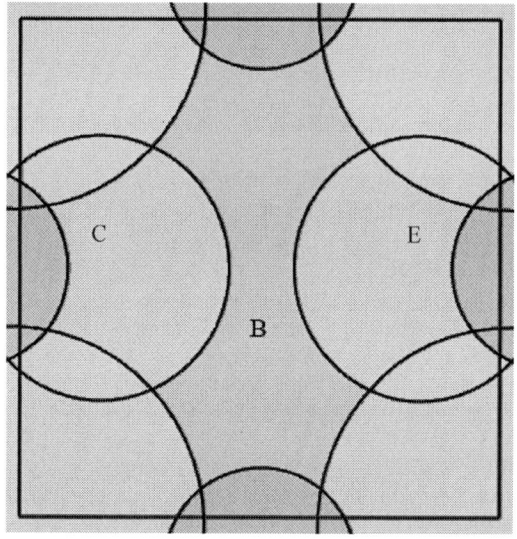

Fig. 1. The layout of VESTIC BJT in top view. Blue depicts the polysillicon region of collector (C) and emitter (E), green depicts the base (B) region made of monosilicon. Grey regions are pillar contacts (which may also work as heat sinks), yellow regions are silicon di-oxide trench isolation.

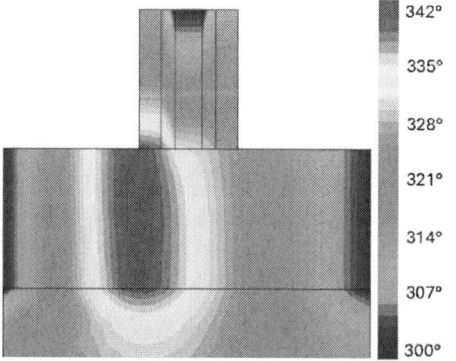

Fig. 2. Tempreature distribution at moderate current level (7 µA of collector current) in cross section of lateral transistor, the type described in [1], on the left is a emitter contact, on the left a emitter contact, and on the right is collector contact, on the bottom is the Oxide layer, W_B=50 nm. Temperature in Kelvins.

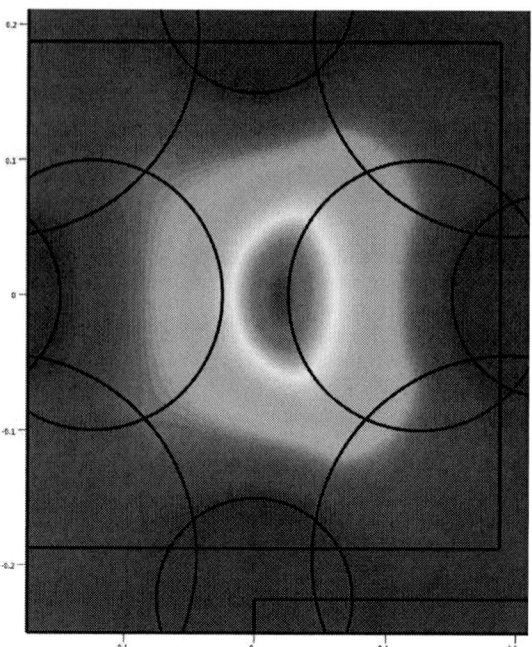

Fig. 3. Tempreature distribution at moderate current level (7 µA of collector current) in cross section of VESTIC transistor (top-view as in Fig. 2 with outlines of respective regions). As can be seen, there is significant gradient of temperature towards the contact pillars. The major heat spot remains in the middle of the structure, the cross section in perpendicular dimension from emitter to collector is similar to that from Fig. 1.

VII. CONCLUSION

The VESTIC approach to lateral BJTs on SOI offers a promising path for reintroducing high-speed bipolar transistors into logic applications. However, self-heating effects and thermal constraints remain key challenges for both lateral and vertical SOI BJTs.

This study highlights the key role of considering thermal properties in the design and operation of bipolar junction transistors (BJTs) fabricated on SOI substrates. Through the analysis of vertical and lateral BJT transistors, with particular emphasis on VESTIC-based lateral designs, it becomes clear that device geometry, heat source distribution and substrate insulation significantly affect the thermal properties of the device.

Lateral BJT transistors, especially those implemented with VESTIC technology, exhibit better heat dissipation characteristics compared to their vertical counterparts. Their horizontal current paths and distributed heat generation help mitigate localized thermal hot spots. However, lateral structures still face self-heating challenges, requiring careful consideration of layout geometry, power density and the inclusion of thermally conductive structures such as pillar contacts. Pillar contacts are a distinctive feature of transistors in VESTIC technology.

The proposed thermal stabilization circuits for complementary bipolar logic circuits (CBip) may be sensitive to the effects associated with self-heating structures. In the case of lateral transistors, self-heating is more important than mutual heating of the structures.

The key conclusion is that exploiting the advantages of SOI technology requires a holistic approach to electro-thermal design. Simulations and experimental validations confirm that lateral BJT transistors can be optimized for thermal stability, enabling new opportunities for integrating bipolar transistors into modern digital and hybrid circuits.

Future work should include improved modeling of heat sources in sideband transistors, integration of heat dissipation functions into circuit design, and further research into the temperature-dependent behavior of CBip gates.

ACKNOWLEDGMENT

P. K. M. thanks to W. Kuzmicz for the important insights for CBIC concept and theoretical background.

REFERENCES

[1] J. Cai, T. H. Ning, C. D'Emic, and K. K. Chan, "Complementary thin-base symmetric lateral bipolar transistors on SOI," 2011 International Conference on Solid-State Devices and Materials, IEEE, 2011. Available: https://ieeexplore.ieee.org/abstract/document/6131565.

[2] W. Kuzmicz, "A Thermally Stable Quasi-CMOS Bipolar Logic," Electronics, vol. 11, no. 6, pp. 1–15, 2022. Available: https://doi.org/10.3390/electronics11010006.

[3] P. Mierzwiński, "Studium wykonalności układów logicznych z wykorzystaniem stabilizowanych termicznie komplementarnych tranzystorów bipolarnych," Internal Research Report, Warsaw University of Technology, 2024.

[4] W. Maly, "VESTIC: A new approach to IC manufacturing," IEEE Transactions on Semiconductor Manufacturing, vol. 15, no. 4, pp. 359–367, Nov. 2002. DOI: 10.1109/TSM.2002.805082.

[5] I. Marano, et al., "Analytical modeling and numerical simulations of the thermal behavior of trench-isolated bipolar transistors on SOI substrates," Solid-State Electronics, vol. 52, no. 5, pp. 730-740, May 2008.

[6] P. R. Ganci, et al., "Self-heating in high-performance bipolar transistors fabricated on SOI substrates," IEEE Bipolar/BiCMOS Circuits and Technology Meeting (BCTM), 1998.

[7] J. Olsson, "Improved Thermal Performance of SOI Using a Compound Buried Layer," Microelectronic Engineering, vol. 56, pp. 339–344, 2001.

[8] A. Pacelli, et al., "Compact modeling of thermal resistance in bipolar transistors on bulk and SOI substrates," IEEE Transactions on Electron Devices, vol. 49, no. 6, pp. 1027–1035, June 2002.

[9] P. Baine, et al., "Electrothermal simulation of self-heating in SOI BJTs," IEEE Transactions on Electron Devices, vol. 61, no. 6, pp. 1999–2007, June 2014.

[10] J. S. Brodsky, et al., "Physics-based compact thermal modeling of SOI BJTs," IEEE Transactions on Electron Devices, vol. 46, no. 12, pp. 2333–2341, Dec. 1999.

[11] R. A. Wessels, et al., "Thermal behavior of lateral and vertical BJTs on SOI," IEEE Transactions on Electron Devices, vol. 45, no. 10, pp. 2218–2226, Oct. 1998.

[12] A. Gorai, et al., "Electrothermal analysis of bipolar transistors on SOI substrates," Microelectronics Journal, vol. 39, no. 4, pp. 531–539, Apr. 2008.

[13] Y. K. Li, et al., "Design considerations for thermally stable SOI bipolar circuits," IEEE Journal of Solid-State Circuits, vol. 35, no. 9, pp. 1383–1391, Sept. 2000.

[14] J. L. Prince, "Self-heating effects in SOI-based bipolar devices," IEEE Transactions on Electron Devices, vol. 48, no. 6, pp. 1157–1164, June 2001.

[15] T. H. Ning and C. D'Emic, "SOI lateral bipolar transistors for high-speed and low-power applications," IEEE Transactions on Electron Devices, vol. 45, no. 4, pp. 841–849, Apr. 1998.

[16] S. Markov, et al., "Thermal resistance scaling in lateral and vertical SOI bipolar transistors," IEEE Transactions on Electron Devices, vol. 57, no. 2, pp. 390–398, Feb. 2010.

[17] W. Kuzmicz, "Power supply circuit for digital gates, preferably composed of the complementary pairs of bipolar transistors," Polish Patent no. Pat.228465, granted by PTO of Republic of Poland, Nov. 2017.

[18] H.H. Berger, S.K. Wiedmann, "Complementary Transistor Circuit for Carrying out Boolean Functions," U.S. Patent 3,956,641, 11 May 1976.

[19] P. Mierzwiński and W. Kuźmicz, "VES-BJT: A Lateral Bipolar Transistor on SOI with Polysilicon Emitter and Collector," Electronics (Switzerland), vol. 12, Art. no. 8, 2023, doi: 10.3390/electronics12081871.

Mixed Design of Integrated Circuits and Systems – MIXDES 2025

Considerations on the Importance of Proper Modeling of Heat Transfer Coefficient Values

Marcin Janicki

Department of Microelectronics and Computer Science
Łódź University of Technology
Łódź, Poland
E-mail: marcin.janicki@p.lodz.pl

Abstract—**This paper discusses the problem of modeling the heat transfer coefficient values in electronic circuits cooled by the natural convection. Here, circuit temperature is computed for an electronic circuit containing a heat source dissipating variable amount of power and located in different locations. All thermal simulations are caried out using an analytical Green's function solver. The obtained results are compared and analyzed for cases when the value of heat transfer coefficient is constant and when it depends on the values of ambient and circuit surface temperature. The main conclusion of the analyses is that the simulated circuit temperature values are much lower if the variable value of the heat transfer coefficient is used.**

Keywords—**electronic circuit; thermal modeling; heat transfer coefficient.**

I. INTRODUCTION

Thermal analysis of electronic circuits is an essential part the design process because thermal issues are the major cause of malfunctions and failures. The temperature field is modeled by the Fourier partial differential heat equation [1]-[2]. For its solution, structure geometry, material properties and boundary or initial conditions have to be specified. Unfortunately, most engineers consider that model parameters, such as the thermal conductivity or the heat transfer coefficient, are independent from temperature. However, in reality some of these quantities strongly depend on temperature and the change of their values should be taken into account. Otherwise, obtained simulation results might bear significant errors. This problem is described in this paper based on the results of 3D thermal simulations carried out for a test structure. The following section presents the particular structure analyzed throughout this paper and its thermal model. Then, the problem of heat transfer coefficient value modeling is explained in detail. Next, thermal simulation results are provided and discussed.

II. TEST STRUCTURE AND ITS THERMAL MODEL

The test structure analyzed in thermal simulations presented in this paper resembles a real electronic, such as a power circuit manufactured in a thin substrate of high thermal conductivity. This structure has the dimensions of 10 cm x 10 cm x 1.5 mm and it contains a 1 cm x 1 cm square surface heat source, whose location is varied during thermal simulations along the circuit diagonal, as indicated in Fig. 1.

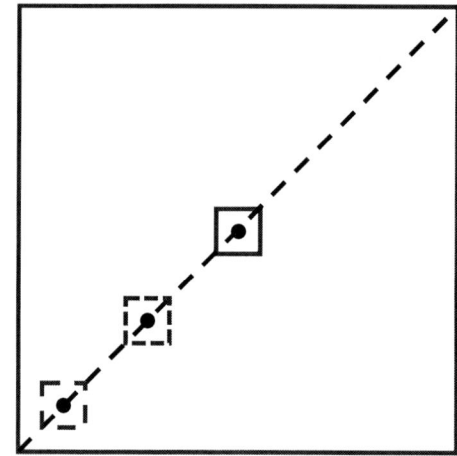

Fig. 1. Test structure layout indicating heat sources, the temperature sensing points and the temperature profile line.

The structure thermal model, presented in Fig. 2, consists of two layers. The power dissipated in the heat source is taken into account in the model, by a heat flux flowing through its top surface. Furthermore, taking into account that the structure is relatively thin in comparison to its area, it is assumed that the lateral surfaces are adiabatic and heat is exchanged with the ambient only at the two remaining surfaces, what is modelled by the heat exchange coefficient.

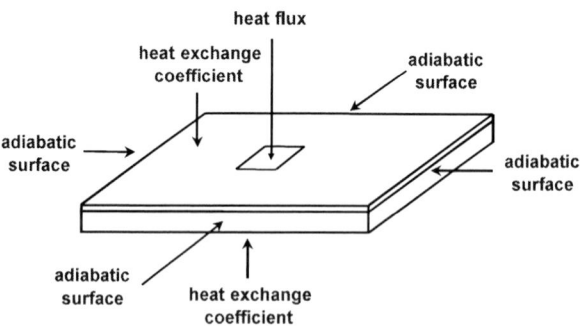

Fig. 2. Test structure thermal model.

Assuming that the thermal model should resemble a power module manufactured in the insulated metal substrate hybrid technology, as shown in Fig. 3, the base is made of an 1.5 mm thick aluminum plate covered with an 100 μm insulating resin layer. The value of the heat transfer coefficient at the top and bottom surfaces is variable and consistent with the description provided in the following section of the paper.

III. Modeling of Heat Transfer Coefficient Value

The heat transfer coefficient value represents two different physical phenomena responsible for circuit cooling: radiation and convection. The radiation occurs when surface temperature is higher than the absolute zero, whereas the convection only if there exists a temperature difference between this surface and the surrounding ambient. Since the cooling mechanisms mostly occur at the same time, they are usually represented in thermal models by a single value of the total heat transfer coefficient h_t, being the sum of the radiation heat transfer coefficient h_r and the convection heat transfer coefficient h_c.

The first component h_r, as presented in (1), depends on the circuit surface absolute temperature T and the ambient absolute temperature T_a. The proportionality coefficient σ is the Stefan-Boltzmann constant. On the other hand, the latter component h_c is proportional to the circuit surface temperature rise over the ambient temperature ΔT, as shown in (2), where the value of coefficient a is usually found empirically [3]-[4]. Here, the value of 4.06, determined in [5] for a similar kind of circuit, was used.

$$h_r = \sigma * \left(T + T_a\right) * \left(T^2 + T_a{}^2\right) \qquad (1)$$

$$h_c = a * \Delta T^{0.25} \qquad (2)$$

The values of the total heat transfer coefficient and its both components computed using the above equations for different surface temperature rise values and the ambient temperature of 20 °C are plotted in Fig. 4. From the chart, it can be seen that the value of the total heat transfer coefficient increases rapidly, mainly due to the variation of the convective component, from just 5.7 W/(m² K) at the ambient temperature to 22.2 W/(m² K) when the surface temperature rise equals 100 K. Thus, the local heat transfer coefficient values always have to be taken into account in thermal simulations so as to obtain accurate results.

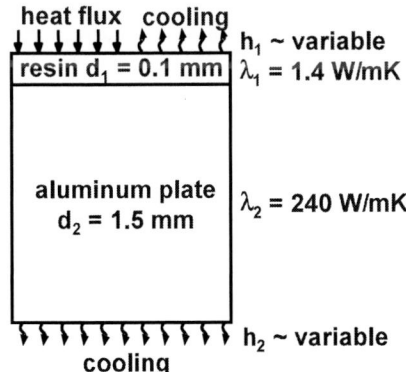

Fig. 3. Thermal model cross-section.

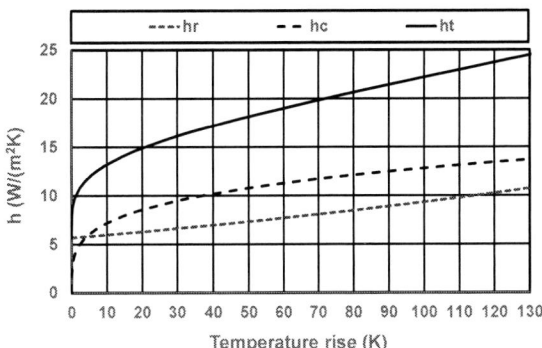

Fig. 4. Heat transfer coefficient values for ambient temperature of 20 °C.

IV. Simulation Results

The temperature values were computed using the analytical Green's function method. Green's functions are mathematical tools, which can be regarded as transient temperature responses at a location of interest caused by instantaneous heat generation at another point. Then, the temperature response in time could be computed by integrating the Green's function over the entire volume and time of heat generation. Alternatively, a Green's function is the transient temperature response at a point in time due to the initial temperature rise at another location. Then, the temperature distribution in time can be computed as an integral of a Green's function evaluated at the initial time instant over the entire domain of analysis. Here, the Green's function was determined using the Fourier separation of variables method. More information on different methods for obtaining Green's function temperature solutions can be found in [6]-[7], whereas the particular solution of the heat equation resulting from the thermal model discussed here can be found in [8].

Initially, thermal simulations were carried out varying the value of power dissipated in the heat source for three locations indicated in Fig. 1 as large dots, the structure center, the corner point distant by 1 cm from each side and an intermediate spot located right in the middle between the other ones, i.e. 5 cm from the sided. The results obtained for these locations, shown in Figs. 5-7 respectively, were obtained applying at the top and bottom surfaces either the constant heat transfer coefficient value equal to 10 W/(m² K) (lighter lines) or the variable one computed based on (1)-(2) (black lines). The solid, dashed and double lines denote in the figures respective locations where temperature values are read out.

When the heat transfer coefficient value is variable, as can be seen in the figures, the black lines from the very beginning depart from their lighter counterparts as a result of decreasing thermal resistance. Then, for 20 W of power dissipated in the heat source its temperature rise value predicted by the thermal model with the variable coefficient is almost 40% lower than in the case when the heat transfer coefficient value is constant. These differences are even more pronounced in the locations outside the source where the temperature values simulated with the variable heat transfer coefficient value are almost halved, compared to the case when it is constant. This proves that the variations of heat transfer coefficient values with temperature always should be taken into account in thermal simulations.

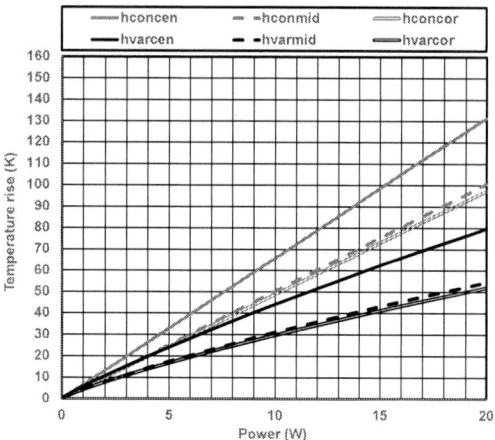

Fig. 5. Simulated temperature rise values for heat source in the center with constant and variable heat transfer coefficient values.

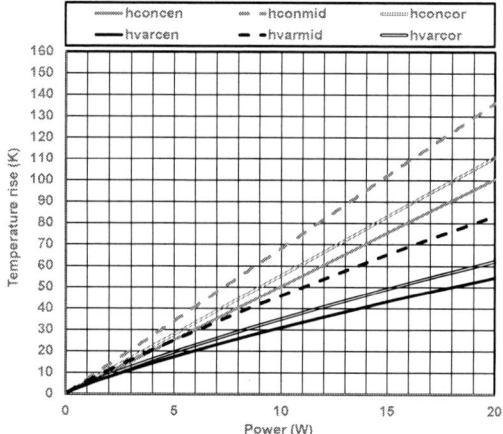

Fig. 6. Simulated temperature rise values for heat source in an intermediate location with constant and variable heat transfer coefficient values.

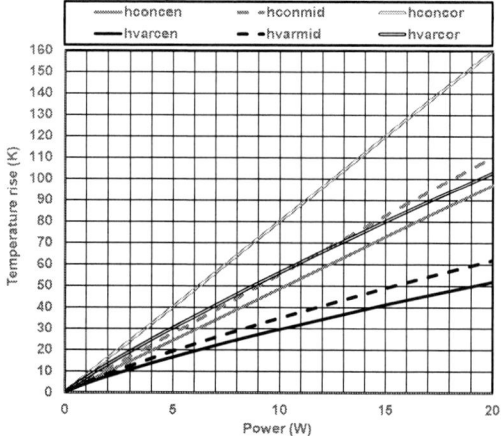

Fig. 7. Simulated temperature rise values for heat source in the corner with constant and variable heat transfer coefficient values.

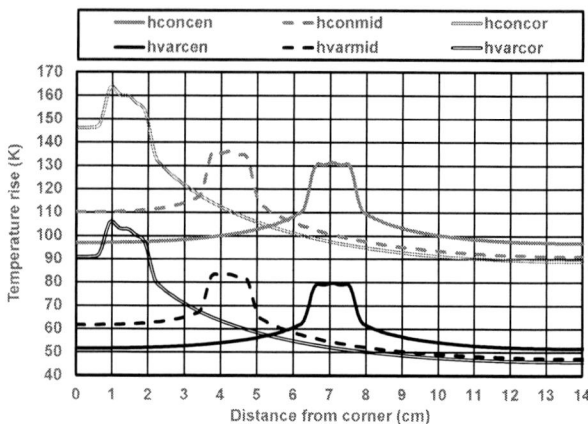

Fig. 8. Simulated temperature rise profiles with constant and variable heat transfer coefficient values for heat source in different locations.

Next, assuming the power dissipation of 20 W, temperature rise profiles were computed along the circuit diagonal for three earlier considered heat source locations, again with the constant and variable heat transfer coefficient. The simulation results are plotted in Fig. 8 using the same line marking scheme as in the previous three figures. With the constant coefficient value, the hot spot temperature varies from 131.4 K if the source is placed in the center to 163.8 K for the corner location and from 96.9 K to 88.8 K in the coolest one. With the variable coefficient value the respective temperatures values are 79.8 K, 106.1 K, 51.6 K and 45.8 K, thus they are noticeably lower, what confirms the earlier analyzed results.

Moreover, with the constant heat transfer coefficient value the temperature differences change from 34.5 K for the central heat source location to 75.0 K when the source is placed in the corner. With the variable coefficient value these values amount to 28.1 K and 60.3 K respectively. Thus, temperature profiles are much flatter when variable heat transfer coefficient values are used and then temperature differences are lower by around 20%. Regarding the circuit layout, placing the heat source close to the corner increases its temperature, even by one third when the heat transfer coefficient value is variable.

Then, dynamic simulations were performed assuming again that the power of 20 W is dissipated in the centrally placed heat source. The curves representing temperature evolution in time computed for three earlier considered locations with constant and variable heat transfer coefficient value are plotted in Fig. 9. Except for the earlier established fact that the temperatures are much lower when the coefficient values are variable, the first observation is that the thermal responses outside the source are delayed by a few seconds and that these heating curves have entirely different shape. The thermal steady states are reached after approximately half an hour. The influence of the variable heat transfer coefficient values becomes visible only after a few minutes. These results confirm that the heat transfer coefficient values have also an important influence on results of dynamic thermal simulations and their variations with structure surface temperature should be always taken into account.

170

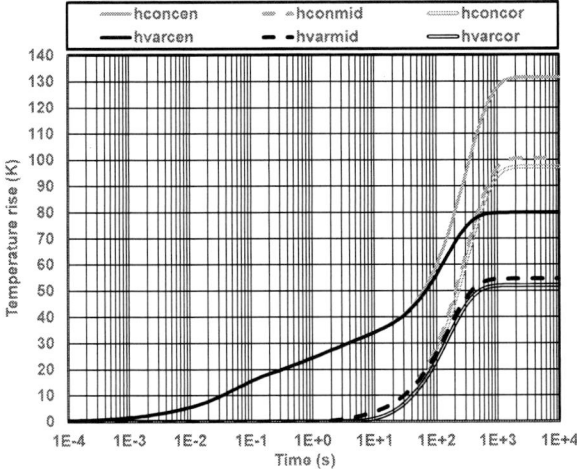

Fig. 9. Simulated temperature evolution in time with constant and variable heat transfer coefficient values for heat source located in the center.

V. CONCLUSIONS

The thermal simulations presented in this paper showed that in the temperature range close to the ambient temperature the value of heat transfer coefficient changes rapidly due to sudden appearance of air convection. Therefore, in the case of natural convection cooled circuits, the local heat transfer coefficient values always should be taken into account. In this paper, for each location the Green's function temperature solutions with such nonlinear boundary conditions were obtained iteratively, typically in 4 to 7 steps to reach the accuracy of 0.1 K.

Thermal simulations with variable heat transfer coefficient values yield much lower temperature values with visibly flatter temperature profiles than in the case of the constant coefficient value equal to 10 W/(m² K), which is often used in thermal simulations as a rule of thumb. Furthermore, the heat source temperature notably increases when it is placed closer to the structure edges.

REFERENCES

[1] M.N. Ozisik, Heat Conduction, John Wiley & Sons Inc., 1993.

[2] H.S. Carslaw and J.S. Jaeger, Conduction of Heat in Solids. Oxford: Clarendon Press, 1947.

[3] J.P. Holman, Heat Transfer. McGraw – Hill, 1985.

[4] F.P. Incropera and D.P. De Witt, Introduction to Heat Transfer, John Wiley & Sons, 1985.

[5] M. Janicki and A. Napieralski, "Analytical transient solution of heat equation with variable heat transfer coefficient," Proceedings of 8th THERMINIC, 1-4 October 2002, Madrid, Spain, pp. 235-240.

[6] J.V. Beck, K.D. Cole, A. Haji-Sheikh and B. Litkouhi, Heat Conduction Using Green's Functions. Hemisphere Publishing, 1992.

[7] M.D. Greenberg, Application of Green's Functions in Science and Engineering. Prentice-Hall, 1971.

[8] M. Janicki, G. De Mey and A. Napieralski, „Thermal analysis of layered electronic circuits with Green's functions," Microelectronics Journal, vol. 38, pp. 177-184, 2007.

Mixed Design of Integrated Circuits and Systems – MIXDES 2025

Influence of the Cooling System on Characteristics of Power LEDs in COB Packages

Krzysztof Górecki. Przemysław Ptak
Department of Power Electronics
Gdynia Maritime University
Gdynia, Poland
k.gorecki@we.umg.edu.pl, p.ptak@we.umg.edu.pl

Damian Płokarz
Faculty of Electrical Engineering
Gdynia Maritime University
Gdynia, Poland
damionas@wp.pl

Abstract—**The paper presents the results of measurements of the characteristics of selected power LEDs mounted in COB packages. The operation of this class of semiconductor devices without any additional cooling system, with a classic aluminium heat-sink and with a liquid cooling system with a heat exchanger were considered. The structures of the tested diodes and the measured current-voltage characteristics of the tested devices operating in the forward and reverse modes, the dependence of the luminous flux and thermal resistance on the forward current and the waveforms of the illuminance of the emitted light were presented. The obtained measurements results were discussed. Particularly, an influence of self-heating phenomena on the obtained characteristics was considered.**

Keywords—**power LEDs, DC and dynamic characteristics, measurements, COB, thermal phenomena, self-heating, cooling systems**

I. INTRODUCTION

Power LEDs are a basic component of modern lighting systems used both for lighting rooms and in automotive applications [1, 2]. They usually contain many diode chips mounted on a common substrate and forming LED modules emitting a desired luminous flux value [3]. In order to obtain white light, diode chips emitting blue light and covered with a yellow phosphor are typically used [2].

Classic LED modules contain from several to several dozen semiconductor chips, each of which has a separate phosphor and lens [2, 4]. A separate group of light sources are COB (Chip on Board) devices containing many diode chips placed on a common ceramic substrate, connected in series-parallel and covered with a common phosphor layer [5, 6].

The properties of power LEDs are described in the literature [1, 2, 7, 8]. In particular, many papers [2, 8, 9] are devoted to the study of the influence of thermal phenomena on the electrical and optical properties of this group of semiconductor devices and LED modules. Characteristics determined taking into account the influence of thermal phenomena (in particular self-heating) are called non-isothermal characteristics. Such characteristics are presented in the literature for various types of soli-state light sources [2, 7, 8].

Many papers [10-22] are devoted to research on the properties of LEDs in COB packages. The subject of considerations included in these papers is the analysis of thermal

properties of the luminophore used in such devices [10], design of devices with adjustable correlated colour temperature [11]. The paper [12] describes the properties of ceramics used to construct substrates for COB devices. Modeling of thermal properties of COB packages of LED modules is the subject of the paper [13]. Research on the use of graphene grease as a material improving the efficiency of natural convection of the considered packages is described in the paper [14].

In [15] a compact model of the considered devices is proposed. Studies on the distribution of light emitted by LEDs in COB packages are described in [16]. The influence of materials used for mounting semiconductor chips on the properties of the considered devices is described in [17, 18]. Microlenses for structures placed in COB packages that increase the luminous efficiency of power LEDs are described in [19].

The influence of the size of COB LEDs on their thermal properties is analyzed in [20]. The degradation processes for high-power COB LED modules were analyzed in [21]. A tunable COB LED with improved heat dissipation capability is presented in [22].

Despite the big number of papers on the study of power LEDs in COB packages, there is still a lack of information in the literature on the study of non-isothermal characteristics of these devices obtained at different cooling conditions.

The aim of this paper is to present the results of experimental studies illustrating the effect of self-heating on the electrical characteristics and optical and radiometric parameters of power LEDs in COB packages. Two components with different sizes and different permissible values of forward current were selected for the studies. Devices operating under different cooling conditions were considered.

Section 2 describes the tested devices. Section 3 describes the applied measurement set-up. Section 4 contains the obtained measurement results.

II. TESTED DEVICES

Two power LEDs in COB packages manufactured by Cree were selected for the tests: CXB2540-0000-000N0ZU4L5A (hereinafter referred to as CXB2540) and CXB1507-0000-000F0ZG2L5A (hereinafter referred to as CXB1507). The values of the most important parameters of these devices are given in Table I.

TABLE I.
VALUES OF SELECTED PARAMETERS OF THE TESTED DEVICES [23, 24].

Parameter	CXB2540	CXB1507
Maximum forward current [mA]	2100	750
Maximum reverse current [mA]	0.1	0.1
Forward voltage V_F [V] @ I_F = 1.1 A, T_j = 85 ºC	34.1	16.9
Viewing angle [º]	115	115
Typical luminous flux Φ_V [lm] @ T_j = 85 ºC	4451 @ 1.1A	818 @ 400mA
CCT [K]	4000	4000
CRI	95	95

As can be seen, both devices under consideration are characterized by the same value of correlated colour temperature CCT = 4000 K and colour rendering index CRI = 95. The CXB2540 diode allows to obtain more than 5 times higher luminous flux, 3 times higher forward current and 2 times higher forward voltage. The view of the packages of the tested devices are shown in Fig. 1 and Fig. 2.

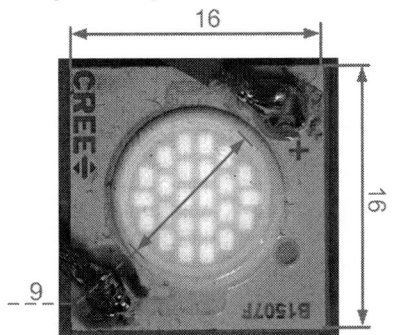

Fig. 1. View of the CXB1507F device with pointed LED chips.

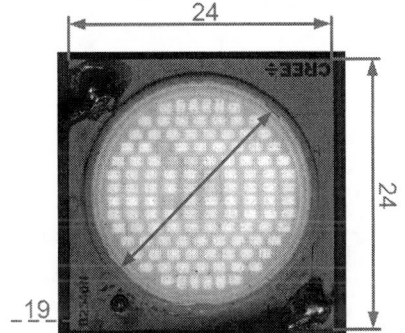

Fig. 2. View of the CXB2540 device with pointed LED chips.

The tested diodes differ in the size of the substrate. They are 16x16 mm for the CXB1507F diode and 24x24 mm for the CXB2540 diode. Diode chips are mounted in a circle with a diameter of 9 and 19 mm, respectively. The CXB1507F diode contains 24 semiconductor chips connected in 4 parallel chains of 6 chips each. In turn, the CXB2540 diode contains 120 semiconductor chips connected in 10 parallel chains of 12 chips each. The ceramic substrate of each of the tested diodes is 1 mm thick.

III. MEASUREMENT SET-UP

The measurements of the characteristics of the tested diodes were performed using the measurement set-up with the block diagram shown in Fig. 3.

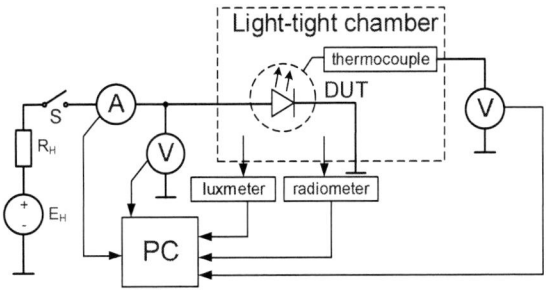

Fig. 3. Set-up to measure DC and dynamic non-isothermal characteristics of tested power LEDs.

In the considered measurement set-up, the device under test (DUT) is placed in a light-tight chamber. It is powered from the voltage source E_H via the resistor R_H. The switch S is used for dynamic measurements, while it is turned on during measurements of DC characteristics. The coordinates of the operating point of the device under test are measured using a voltmeter and ammeter. A thermocouple with a voltmeter is used to measure the temperature of the soldering point T_{SP}. The optical parameters are measured using a luxmeter, and the radiometric parameters are measured using a radiometer. The values measured using each of the above-mentioned instruments are recorded in a computer (PC).

IV. RESULTS

Using the measurement set-up described in Section 3, measurements of non-isothermal characteristics of the diodes described in Section 2 were performed. The measurements were performed for three variants of the cooling system of the tested diodes. In the first one, the tested diodes operated without any additional cooling system. In the second variant, the diodes were attached to an aluminum heat-sink with dimensions of 175 x 118 x 10 mm. In the third variant, the diodes were placed on a coldplate with dimensions of 150 x 70 x 10 mm connected to the forced cooling system described in [25].

The rest of this section presents the measured non-isothermal characteristics of the devices under consideration. All presented characteristics were measured at an ambient temperature of 20.5 ºC.

Fig. 4 shows the non-isothermal current-voltage characteristics of the CXB1507 (Fig. 4a) and CXB2540 (Fig. 4b) diodes.

Due to the different number of semiconductor chips contained in the tested devices, the range of changes in the forward voltage of both diodes is different. The deterioration of the heat dissipation efficiency when changing the cooling system causes a shift of the considered characteristics to the left and a limitation of the maximum forward current value, at which the soldering point temperature does not exceed the maximum allowable value. The most effective cooling is provided by coldplate. For a diode placed on a heat-sink, the maximum value of the measured current decreases twice, and for a diode operating without a heat-sink – even 6 times. For the CXB1507 diode, the V-I characteristics obtained for each of the considered cooling conditions are unambiguous, while for the CXB2540 diode operating without any heat-sink, a fragment of the considered characteristic with a negative slope is observed.

173

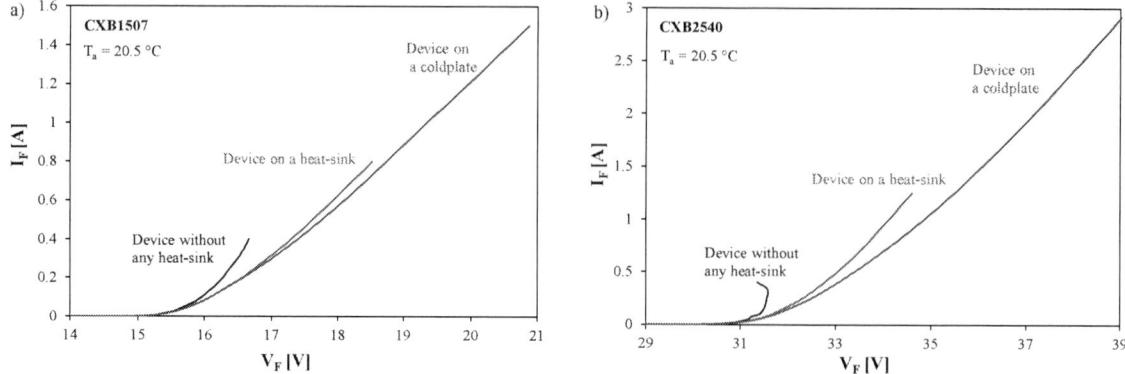

Fig. 4. Measured non-isothermal V-I characteristics of diodes CXB1507 (a) and CXB2540 (b) operating in the forward mode

Fig. 5 presents the reverse characteristics of the diodes under consideration.

Comparing the characteristics presented in Fig. 5a and Fig. 5b, it can be seen that their shapes differ significantly. This indicates the use of different system solutions for the blocks protecting the LEDs under consideration against reverse polarity of these devices. For the CXB1507 diode, the protective component is a p-n junction connected antiparallel to the terminals of the device under consideration. In turn, the characteristics measured for the CBX 2540 diode show ambiguity of the characteristic typical for devices containing a thyristor structure. It effectively blocks the flow of reverse current up to a voltage close to 9 V. After exceeding the switch-on point of the thyristor system, the reverse current of the tested device increases and the voltage decreases. If the external circuit does not limit the reverse current value below 0.75 A, the tested device may be damaged. For the characteristics under consideration, no significant effect of the cooling system on their course is observed.

Fig. 6 illustrates the dependence of the soldering point temperature T_{SP} on the forward current for both tested diodes operating in the forward mode.

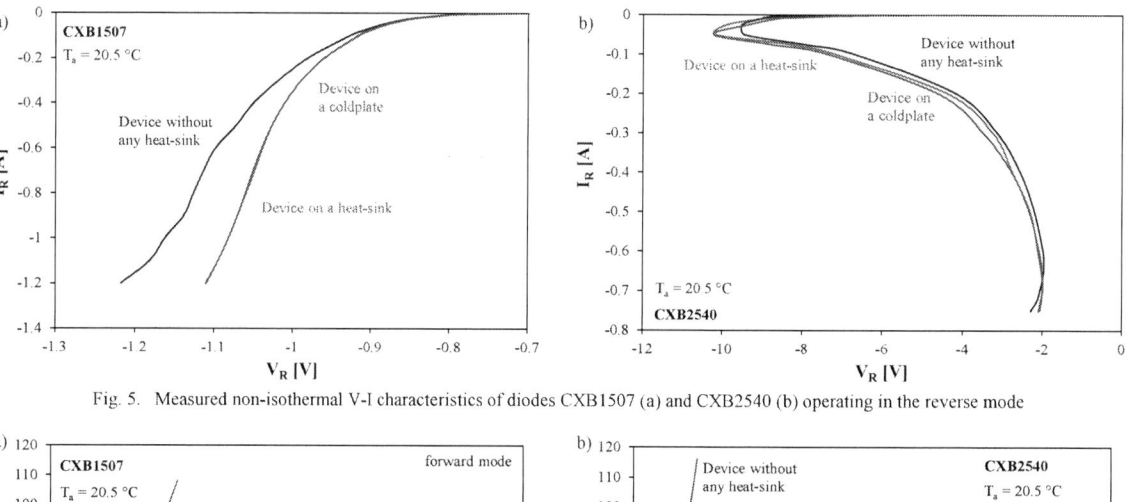

Fig. 5. Measured non-isothermal V-I characteristics of diodes CXB1507 (a) and CXB2540 (b) operating in the reverse mode

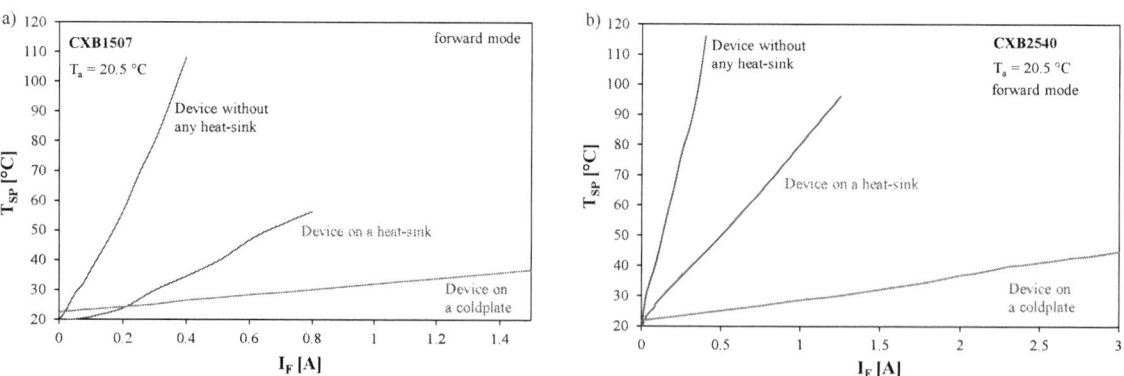

Fig. 6. Measured dependences of the temperature of the soldering point on the forward current for the diodes CXB1507 (a) and CXB2540 (b) operating in the forward mode

For both the devices under consideration, a monotonically increasing $T_{SP}(I_F)$ dependence was obtained. It is approximately linear. A strong influence of the cooling system on the obtained T_{SP} temperature values at a fixed I_F current value is visible. The T_{SP} temperature increases above the ambient temperature for a fixed I_F for a diode placed on a coldplate are even 30 times smaller than for the same diode operating without any heat-sink. Based on the presented measurement results and the forward voltage values of the tested devices, the thermal resistance values R_{thSP-a} were determined using the formula

$$R_{thSP-a} = \frac{T_{SP} - T_a}{V_F \cdot I_F} \qquad (1)$$

where T_a denotes the ambient temperature.

The R_{thSP-a} values obtained for the tested diodes are summarized in Table II.

As can be seen, for both the tested diodes, the lowest R_{thSP-a} value was obtained for the diode mounted on a coldplate, and the highest for the diode operating without any heat-sink. These values differ by more than 35 times for the CXB2540 diode and by more than 25 times for the CXB1507 diode. It is also worth noting that the R_{thSP-a} values determined for the CXB2540 diode are even twice as small as for the CXB1507 diode. This is due to, among other things, the different surface on which heat exchange between the diode and the ambient takes place.

TABLE II.
VALUES OF THERMAL RESISTANCE R_{THSP-A} MEASURED FOR THE TESTED DEVICES.

Cooling conditions	LED without any heat-sink	LED on the heat-sink	LED on the coldplate
CXB2540	7.07 K/W	1.77 K/W	0.20 K/W
CXB1507	12.2 K/W	2.48 K/W	0.48 K/W

Fig. 7 shows the dependence of the luminous flux Φ_V produced by each of the tested diodes as a function of their forward current. In this figure, the solid lines refer to the CXB1507 diode, while the dashed lines refer to the CXB2540 diode. The Φ_V measurements were performed using the method described in [26].

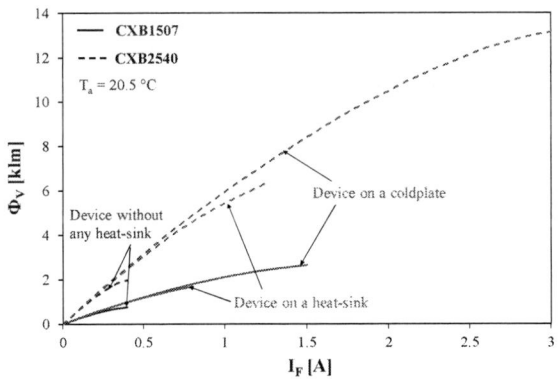

Fig. 7. Measured dependences of the produced luminous flux on the forward current for the tested diodes

It is clearly visible that the maximum value of the luminous flux obtained for the CXB2540 diode is even 6 times higher than for the CXB1507 diode. This discrepancy results, among others, from the difference in the active surface of the devices under consideration and the maximum permissible value of the

forward current. At the current value of $I_F = 1.5$ A, the Φ_V values for both diodes differ four times. The deterioration of the cooling conditions of the diodes under consideration causes a limitation of the maximum permissible value of the I_F current, and consequently – a limitation of the maximum value of Φ_V. Additionally, due to self-heating, the value of the luminous flux corresponding to the set value of the I_F current decreases. This decrease reaches as much as 20%.

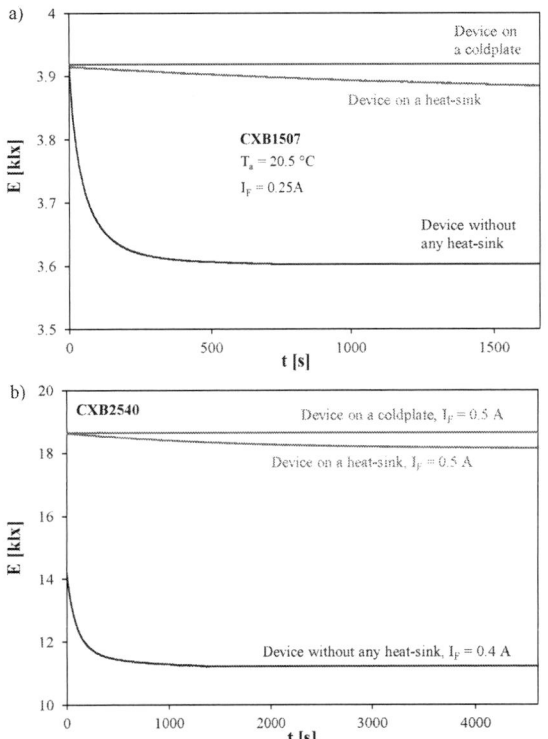

Fig. 8. Measured waveforms of illumination for the diodes CXB1507 (a) and CXB2540 (b) powered by the current step of the value I_F

To illustrate the dynamics of thermal processes and their effect on optical parameters, Fig. 8 shows the waveforms of the illumination intensity E recorded when the tested diodes were supplied with a step-shaped forward current of I_F value.

As can be seen, for each of the tested diodes operating under all considered cooling conditions, a decreasing E(t) curve was obtained. The decrease in the illuminance value is the result of the self-heating phenomenon causing an increase in the junction temperature of the tested devices. The biggest decrease in the E value is visible for diodes operating without any heat-sink. It reaches 10% for the CXB1507 diode and as much as 20% for the CXB2540 diode. In turn, for both diodes placed on a coldplate, the decrease in the E value as a function of time is practically invisible.

V. CONCLUSIONS

The paper presents the results of measurements illustrating the influence of the cooling system selection on the DC and dynamic characteristics of power LEDs in COB packages. The studies were carried out for two LEDs containing different numbers of semiconductor chips and characterized by different sizes. Three different cooling systems were considered for each of the LEDs.

As a result of the measurements, it was found that the cooling system used for the diodes under consideration determines the value of their thermal resistance. It determines the increase in the junction temperature of these diodes caused by the self-heating phenomenon. This temperature limits the upper range of permissible values of the forward current of the tested diodes. For the tested diodes operating on a coldplate and without any heat-sink, the maximum permissible values of the forward current differ by as much as six times.

It was shown that the reverse characteristics of the diodes tested differ in shape. This is due to the fact that a diode was used in one device, and a thyristor circuit in the other as protection against reverse connection of the supply voltage.

It was observed that the value of the emitted luminous flux and illuminance is proportional to the number of diode chips contained in the tested device, and not to its surface covered with phosphor. The cooling system of the tested diodes determines the maximum value of the luminous flux. For the tested diodes, this flux measured for the device situated on a coldplate is even 6 times bigger than for a diode without any heat-sink.

The illuminance changes the most due to self-heating in the absence of any heat-sink. Its value decreases by up to 20%. When using a coldplate, this change is practically invisible.

The presented measurement results indicate that the design of LEDs in COB packages and the selection of their cooling system significantly affect the electrical and optical parameters of the considered class of semiconductor devices. In further studies, the authors will attempt to prepare a mathematical description of the observed relationships.

The presented research results may be useful for designers of lighting systems. They may also be useful in teaching to illustrate to students the influence of selected factors on the parameters of power LEDs in COB packages.

References

[1] C.J.M. Lasance, A. Poppe: *Thermal Management for LED Applications*. Springer Science+Business Media, New York, 2014.

[2] E.F. Schubert, Light emitting diodes. Second edition. Cambridge University Press, New York, 2008.

[3] F. S. -S. Chien, R. A. N. Khasanah, P. -T. Lin, Y. -F. Lin and Y. -W. Suen, "Impedance Elements of Significant Junctions in InGaN Light-Emitting Diodes Studied by Electric Modulus Spectroscopy," *IEEE Transactions on Electron Devices*, vol. 66, no. 8, pp. 3393-3398, 2019, doi: 10.1109/TED.2019.2921393

[4] K. Górecki, P. Ptak, "Modelling LED lamps in SPICE with thermal phenomena taken into account", Microelectronics Reliability, Vol. 79, pp. 440-447, 2017.

[5] Z. Xia, S. Liang, B. Li, F. Wang, D. Zhang, Influence on temperature distribution of COB deep UV LED due to different packaging density and substrate type, Optik - International Journal for Light and Electron Optics, Vol. 231, 2021, 166392

[6] S.-H. Moon, Y.-W. Park, H.-M. Yang, A single unit cooling fins aluminum flat heat pipe for 100 W socket type COB LED lamp, Applied Thermal Engineering, Vol. 126, 2017, pp 1164-1169, doi: 10.1016/j.applthermaleng.2016.11.077

[7] A. Poppe, "Multi-domain compact modeling of LEDs: An overview of models and experimental data", Microelectronic Journal, Vol. 46, pp. 1138-1151, 2015.

[8] K. Górecki, P. Ptak, "New method of measurements transient thermal impedance and radial power of power LEDs", in IEEE Transactions on Instrumentation and Measurement, Vol. 69, No. 1, 2020, pp. 212-220.

[9] H.-H. Wu, K.-H. Lin, S.-T. Lin, A study on the heat dissipation of high power multi-chip COB LEDs, Microelectronics Journal, Vol. 43, No. 4, 2012, pp. 280-287, doi: 10.1016/j.mejo.2012.01.007

[10] Z. Chuluunbaatar, C. Wang, E.S. Kim, N.Y. Kim, Thermal analysis of a nano-pore silicon-based substrate using a YAG phosphor supported COB packaged LED module, International Journal of Thermal Sciences, Vol. 86, 2014, pp. 307-313, doi: 10.1016/j.ijthermalsci.2014.07.013

[11] J. Fan, J. Cao, Ch. Yu, Ch. Qian, X. Fan, G. Zhang, A design and qualification of LED flip Chip-on-board module with tunable color temperatures, Microelectronics Reliability, Vol. 84, 2018, pp. 140-148, doi: 10.1016/j.microrel.2018.03.033

[12] J.-K. Sim, K. Ashok, Y.-H. Ra, H.-Ch. Im, B.-J. Baek, Ch.-R. Lee, Characteristic enhancement of white LED lamp using low temperature co-fired ceramic-chip on board package, Current Applied Physics, Vol. 12, No. 2, 2012, pp. 494-498, doi: 10.1016/j.cap.2011.08.008

[13] M. Ha, S. Graham, Development of a thermal resistance model for chip-on-board packaging of high power LED arrays, Microelectronics Reliability, Vol. 52, No. 5, 2012, pp. 836-844, doi: 10.1016/j.microrel.2012.02.005

[14] Ch.-N. Hsu, K.-W. Lee, Ch.-Ch. Chen, Using Graphene-Based Grease as a Heat Conduction Material for Hectowatt-Level LEDs: A Natural Convection Experiment, Processes, 2021, Vol. 9, No. 5, 847, doi: 10.3390/pr9050847

[15] L. Pohl, G. Hantos, J. Hegedüs, M. Németh, Z. Kohári, A. Poppe, Mixed Detailed and Compact Multi-Domain Modeling to Describe CoB LEDs, Energies 2020, Vol. 13, No. 16, 4051, doi: 10.3390/en13164051

[16] D. Czyżewski, Research on Luminance Distributions of Chip-On-Board Light-Emitting Diodes, Crystals 2019, Vol. 9, No. 12, 645, doi: 10.3390/cryst9120645

[17] Y.-F. Kong; X. Li; Y.-H. Mei; G.-Q. Lu; Effects of Die-Attach Material and Ambient Temperature on Properties of High-Power COB Blue LED Module, IEEE Transactions on Electron Devices, Vol. 62, No 7, 2015, pp. 2251 – 2256, doi: 10.1109/TED.2015.2436820

[18] X. Yu; L. Xiang; S. Zhou; N. Pei; X. Luo; Realization of Microlens Array on Flat Encapsulant Layer for Enhancing Light Efficiency of COB-LEDs, IEEE Photonics Technology Letters, Vol. 32, No 20, 2020, pp. 1315 – 1318, doi: 10.1109/LPT.2020.3022794

[19] X. Yu; L. Xiang; N. Pei; S. Zhou; X. Luo; A Simple Method to Realize Millilens Array on Encapsulant Layer for Enhancing Light Efficiency of COB-LEDs, IEEE Transactions on Electron Devices, Vol. 67, No. 9, 2020, pp. 3655 – 3659, doi: 10.1109/TED.2020.3008371

[20] S.-P. Ying; W.-B. Shen; Thermal Analysis of High-Power Multichip COB Light-Emitting Diodes With Different Chip Sizes, IEEE Transactions on Electron Devices, Vol. 62, No 3, 2015, pp. 896 – 901, doi: 10.1109/TED.2015.2390255

[21] A. Herzog; M. Wagner; S. Benkner; B. Zandi; W. D. van Driel; T. Q. Khanh; Long-Term Temperature-Dependent Degradation of 175 W Chip-on-Board LED Modules, IEEE Transactions on Electron Devices, Vol. 69, No 12, 2022, pp. 6830 – 6836, doi: 10.1109/TED.2022.3214169

[22] J. C. Camacho-Arriaga; D. Cahue-Diaz; N. D. Herrera-Sandoval; J. A. Salazar-Torres; G. D. Conejo-Magaña; New Tunable LED With Improved Heat Dissipation on Chip-on-Board, IEEE Transactions on Components, Packaging and Manufacturing Technology, Vol. 14, No 10, 2024, pp. 1737 – 1743, doi: 10.1109/TCPMT.2024.3456768

[23] XLamp® CXB2540 LED, Product family data sheet, Cree, https://downloads.cree-led.com/files/ds/x/XLamp-CXB2540.pdf

[24] XLamp® CXB1507 LED, Product family data sheet, Cree, https://downloads.cree-led.com/files/ds/x/XLamp-CXB1507.pdf

[25] K. Górecki, P. Ptak, M. Janicki, M. Napieralska: Comparison of properties for selected experimental set-ups dedicated to measuring thermal parameters of power LEDs. Energies, Vol. 14, No. 11, 2021, 3240.

[26] K. Górecki, A. Kalinowska, P. Ptak: Comparison of selected methods of measuring the luminous flux of solid-state light sources. Opto-Electronics Review, Vol. 32, No. 1, 2024, e149234.

Signal Processing

Azure Kubernetes Service Design Principles in Machine Learning Systems

Yevhen Bershchanskyi, Halyna Klym
Department of Specialized Computer Systems
Lviv Polytechnic National University
Lviv, Ukraine
yevhen.v.bershchanskyi@lpnu.ua, halyna.i.klym@lpnu.ua

Abstract—The deployment of machine learning (ML) systems at scale necessitates a robust, flexible, and well-orchestrated infrastructure. Azure Kubernetes Service (AKS) has emerged as a key platform for managing ML workloads, offering scalability, automation, and integration with cloud-native AI services. This article explores the fundamental design principles for architecting ML systems on AKS, focusing on scalability, security, cost efficiency, and operational reliability. Key architectural considerations are analyzed, including cluster resource management, model training and deployment strategies, and observability practices. Furthermore, security and governance frameworks are examined to ensure compliance and data protection in ML workflows. Real-world case studies and best practices illustrate successful implementations of ML on AKS across various industries. Finally, emerging trends and challenges are discussed, emphasizing the continuous evolution of Kubernetes-based ML infrastructures and the need for adaptive design strategies in cloud-native AI ecosystems.

Keywords—AKS, ML, Cloud-Native AI, Kubernetes Architecture, Model Deployment, ML Design Best Practices.

I. INTRODUCTION

Machine learning has become a driving force behind technological innovation, transforming industries and redefining traditional approaches to data-driven decision-making. From healthcare and finance to manufacturing and retail, ML-powered applications are enabling advancements in predictive analytics, natural language processing, computer vision, and autonomous systems. These developments have significantly improved operational efficiency, enhanced customer experiences, and unlocked new business opportunities. However, the deployment and management of ML workloads require a highly scalable and flexible infrastructure capable of handling the computational demands, data processing requirements, and real-time responsiveness of modern AI systems.

Kubernetes has emerged as a foundational technology for orchestrating ML workloads, offering containerization, automated scaling, and resource optimization. By enabling efficient workload distribution and seamless integration with cloud services, Kubernetes addresses the complexities of deploying ML models in dynamic, production-grade environments. It provides high availability, fault tolerance, and elasticity, making it an essential tool for enterprises aiming to operationalize ML at scale.

Among managed Kubernetes services, Azure Kubernetes Service stands out as a preferred choice for ML workloads due

to its deep integration with Microsoft Azure's AI and cloud ecosystem. AKS simplifies cluster provisioning [1], scaling, and security management, reducing operational overhead and accelerating ML deployment cycles. Furthermore, its compatibility with Azure Machine Learning, and GPU-powered compute nodes allows organizations to efficiently train, serve, and monitor ML models in a cloud-native environment. The ability to implement hybrid and multi-cloud ML workflows further enhances its appeal for enterprises with complex infrastructure needs.

Despite these advantages, designing and managing ML systems on AKS presents significant challenges [2]. Efficient resource allocation is crucial, as ML workloads require careful optimization of CPUs, GPUs, and memory to balance performance and cost. Security and compliance concerns must also be addressed, as ML models often rely on sensitive datasets that require stringent access controls, encryption mechanisms, and network isolation. Moreover, observability and monitoring are essential for detecting model drift, ensuring inference accuracy, and maintaining system reliability. Cost optimization remains another key consideration, as over-provisioning resources can lead to excessive cloud expenses, while under-provisioning may impact ML performance.

This article explores the essential components of ML infrastructure on AKS, examining best practices for resource management, model deployment, observability, and governance. Real-world case studies provide insights into successful AKS implementations, highlighting innovative strategies for overcoming common challenges. Finally, emerging trends and future directions in Kubernetes-based ML architectures are discussed, emphasizing the evolving landscape of cloud-native AI systems and the need for continuous innovation in ML infrastructure design [3].

II. ARCHITECTURAL CONSIDERATIONS FOR ML ON AKS

Designing machine learning systems on AKS requires careful architectural planning to ensure scalability, security, efficiency, and seamless integration with cloud-native AI services. The complexity of ML workloads, characterized by high computational demands, fluctuating resource utilization, and stringent security requirements, necessitates a well-defined infrastructure strategy.

ML workloads vary significantly, from computationally intensive training jobs requiring GPU acceleration to lightweight inference tasks optimized for CPU execution [4]. AKS supports dynamic resource management through autoscaling mechanisms that adjust compute capacity based on real-time demand. Cluster sizing should account for workload-specific requirements, balancing node pool configurations to prevent resource contention while avoiding excessive over-provisioning. Segregating workloads using namespaces, role-based access control (RBAC), and network policies provides logical separation while enabling centralized governance [5]. Workload isolation prevents resource monopolization and minimizes the risk of cross-tenant interference, ensuring that training, testing, and production environments remain independent and secure.

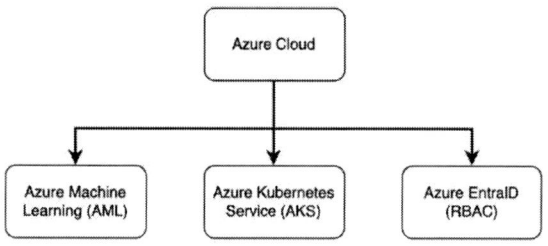

Fig. 1. Azure Cloud/ML structure

Security remains a paramount concern in ML system design, particularly when handling sensitive data and proprietary models [5]. Establishing robust security and compliance mechanisms safeguards against unauthorized access, data breaches, and regulatory violations. Secure secret management using Azure Key Vault ensures that API keys, credentials, and model artifacts are protected from unauthorized exposure (Fig.1).

Seamless integration with Azure AI and ML services enhances operational efficiency by enabling automated workflows, model versioning, and real-time inferencing capabilities. AKS serves as a scalable backend for Azure Machine Learning (AML), allowing organizations to orchestrate end-to-end ML pipelines with minimal manual intervention [6]. The adoption of Kubernetes-native tools such as Kubeflow further optimizes ML workflow orchestration, enabling distributed training, experiment tracking, and hyperparameter tuning within AKS clusters.

A well-architected ML system on AKS balances scalability, security, resource efficiency, and integration with cloud-native AI services to support dynamic and production-grade workloads. By applying these architectural principles, organizations can build resilient ML infrastructure that meets performance expectations while optimizing cost and governance in a cloud-native environment. The following sections explore best practices for model training and deployment strategies, providing deeper insights into optimizing ML pipelines on AKS.

III. MODEL TRAINING AND DEPLOYMENT STRATEGIES

Effective model training and deployment strategies on Azure Kubernetes Service are essential for optimizing ML workflows. Given the resource-intensive nature of ML training and the need for seamless model deployment, organizations must adopt scalable, automated, and efficient approaches. By leveraging distributed training techniques, automating ML pipelines, and integrating continuous integration and deployment (CI/CD) practices, AKS provides a robust environment for operationalizing ML models in cloud-native architectures.

Managing distributed training with AKS and GPU scheduling is critical for handling large-scale ML models that require significant computational power. Training deep learning models often necessitates parallelized execution across multiple nodes, efficiently distributing workloads to accelerate convergence. AKS supports GPU-accelerated node pools, enabling ML workloads for enhanced performance.

ML pipeline automation with Azure Machine Learning and Kubeflow streamlines the end-to-end lifecycle of ML models, reducing manual intervention and enhancing reproducibility. Azure Machine Learning provides a comprehensive framework for defining, orchestrating, and monitoring ML workflows within AKS. By leveraging Azure ML Pipelines, teams can automate data preprocessing, feature engineering, model training, validation, and deployment, ensuring consistent execution across multiple iterations. Kubeflow, an open-source ML toolkit optimized for Kubernetes, enhances workflow automation by enabling model tracking, hyperparameter tuning, and scalable training orchestration. Integrating Kubeflow with AKS allows organizations to manage complex ML pipelines with versioned artifacts, experiment tracking, and automated retraining cycles, improving efficiency and governance across ML operations.

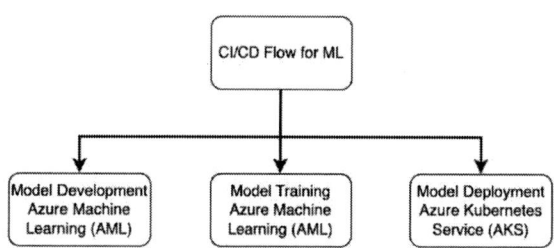

Fig. 2. Azure CI/CD for ML

CI/CD integration for ML model deployment establishes a structured approach to releasing and updating ML models in production environments. By adopting containerized model deployment practices, teams can package trained models as Docker images, ensuring consistency between development, staging, and production environments.

A well-defined training and deployment strategy ensures that ML models running on AKS remain scalable, automated, and seamlessly integrated with cloud-native development workflows (Fig.2). By combining distributed training techniques, automated ML pipelines, and robust CI/CD mechanisms, organizations can optimize resource utilization, reduce operational complexity, and enhance the agility of AI-driven applications.

180

IV. CASE STUDIES AND BEST PRACTICES

Organizations across various industries are increasingly leveraging AKS to deploy and manage ML workloads, taking advantage of its scalability, automation, and integration with Azure's AI ecosystem. This section explores real-world implementations of ML systems on AKS, key lessons learned from industry deployments, and best practices to ensure long-term maintainability.

A notable case study in the financial sector involves a global bank that implemented fraud detection models using AKS to process real-time transaction data. By deploying deep learning models for anomaly detection, the organization was able to analyze millions of transactions per second while ensuring high availability. The use of GPU-accelerated AKS node pools significantly reduced model inference latency, while Azure Monitor and Prometheus provided real-time visibility into system health. The key takeaway from this implementation was the importance of autoscaling inference workloads to balance cost and performance, ensuring that fraud detection models could handle spikes in transaction volumes without over-provisioning compute resources.

In the healthcare industry, a leading medical research institution deployed genomic data analysis pipelines on AKS to accelerate drug discovery [8]. Leveraging Kubeflow on AKS, the team automated distributed training workflows, reducing the time required for large-scale genome sequencing. The institution implemented Azure Machine Learning pipelines to orchestrate data preprocessing, feature engineering, and model training, ensuring reproducibility and compliance with regulatory requirements. One of the key lessons learned was the need for efficient resource management, as training complex models required fine-tuning GPU scheduling strategies to optimize compute usage and reduce idle resource costs.

Another case study in the retail sector highlights the use of AKS for personalized recommendation engines. A multinational e-commerce company deployed ML models on AKS to analyze customer behavior and deliver personalized product suggestions in real-time. The company adopted a GitOps-based CI/CD approach [6] to automate model versioning, testing, and deployment. Implementing canary deployments allowed the team to gradually roll out new recommendation models while monitoring user engagement and conversion rates.

From these case studies, several best practices emerge for ensuring long-term maintainability of ML workloads on AKS. Organizations should prioritize modular and scalable architectures, adopting microservices-based ML deployments to enable independent model updates and reduce operational complexity.

By learning from real-world implementations and adhering to best practices, organizations can maximize the scalability, efficiency, and reliability of their ML workloads on AKS. These insights provide a roadmap for enterprises seeking to deploy, optimize, and sustain AI-driven applications in cloud-native environments, ensuring long-term success in ML operations. The final section will summarize key takeaways and highlight future trends in ML system design on AKS.

V. RESULTS EVALUATION AND ANALYSIS

Evaluating the outcomes and performance of ML workloads on AKS requires a comprehensive look at the strategies implemented and results achieved across all sections of this article. This section brings together insights from architectural design considerations, model training and deployment strategies, observability, performance optimization, and real-world case studies to provide a holistic analysis of AKS's effectiveness for ML workloads.

In the earlier sections, key architectural considerations such as cluster sizing, resource allocation strategies, and security were discussed. For instance, GPU-accelerated AKS nodes were a consistent feature in all case studies, leading to significant improvements in training times and real-time inference latency. In the financial sector, the use of GPU nodes contributed to a 50% reduction in inference latency.

From the model training and deployment strategies, the integration of Azure Machine Learning and Kubeflow into the deployment pipeline played a vital role in streamlining operations. In the object data analysis project, the ability to automate model training and deployment via Kubeflow resulted in the processing data with faster model updates [9]. This automation, combined with CI/CD practices, helped ensure that model updates and rollouts were smooth, reducing errors and downtime.

The observability and performance optimization strategies discussed in previous sections focused on monitoring tools, which were key to tracking the performance of the ML workloads. In the financial sector, the real-time monitoring of fraud detection models led to an 18% improvement in fraud detection accuracy by enabling continuous model updates based on real-time data. These tools also allowed organizations to identify performance bottlenecks, optimize resource utilization, and fine-tune model performance, ensuring smooth operation and optimal resource allocation.

When examining the real-world case studies in more detail, the global bank that adopted AKS for fraud detection achieved impressive performance, including 50% faster fraud detection. This was due to effective GPU utilization, autoscaling policies, and continuous monitoring with Prometheus and Azure Monitor (Fig.3).

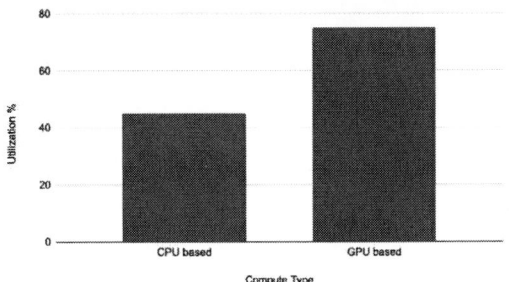

Fig. 3. Compute usage and utilization comparison for ML workflow

Similarly, the healthcare institution achieved 25% savings in compute and a dramatic reduction in model training time. These outcomes demonstrate AKS's ability to handle large, data-intensive workloads while remaining cost-effective.

The e-commerce company benefited from AKS by deploying a personalized recommendation engine that resulted in a 15% increase in customer engagement and a 10% improvement in conversion rates. Their ability to implement canary deployments and continuous integration enabled a 20% reduction in rollback incidents (Fig.4), improving overall system reliability and customer experience.

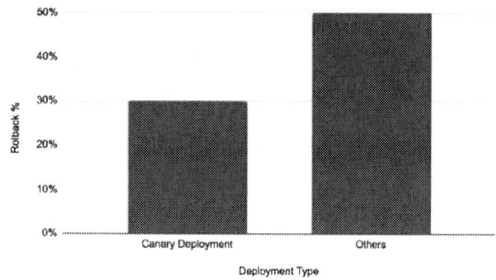

Fig. 4. CI/CD Rollback comparison for ML workflow

GPU-accelerated AKS node pools consistently delivered significant improvements in training times and inference performance. For instance, in the healthcare case study, the 80% reduction in training time was a direct result of leveraging AKS's parallel processing capabilities. Similarly, in the financial sector, the 50% improvement in latency was attributed to using GPU resources for real-time fraud detection.

In terms of cost optimization, all case studies showed positive results from autoscaling and using Azure Spot Virtual Machines for non-critical workloads. The e-commerce case study demonstrated a 30% savings in infrastructure costs, while the genomic data analysis project saw 25% savings. This highlights AKS's ability to provide cost-efficient solutions without compromising on performance.

Another key finding is the importance of monitoring and performance optimization tools. The 18% increase in fraud detection accuracy in the financial sector and improved customer engagement in the retail case study both underscore the value of continuous monitoring for real-time model optimization (Fig.5). In all cases, organizations were able to fine-tune their ML models by leveraging real-time data and performance analytics, leading to better overall performance.

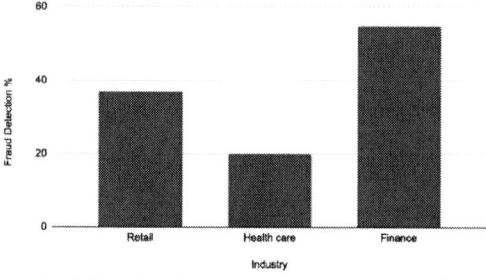

Fig. 5. Fraud detection accuracy comparison by industries

The results from architectural design, model training and deployment strategies, observability, and case studies clearly demonstrate the benefits of using AKS for deploying and managing machine learning workloads. AKS's ability to handle scalable, high-performance workloads while optimizing costs makes it a preferred choice for organizations looking to implement machine learning at scale. Key advantages include reduced training times, improved inference speeds, real-time model updates, and cost savings through efficient resource management.

Overall, AKS has proven to be a highly effective platform for ML workloads, offering robust performance, scalability, and cost optimization, while also enabling seamless integration with other Azure AI/ML services. These results reinforce the value of AKS in building and managing large-scale ML systems, paving the way for more sophisticated AI solutions in the future.

VI. CONCLUSION

In this article, a comprehensive exploration of design principles and best practices for deploying machine learning workloads on AKS was conducted, providing organizations with a clear understanding of how to effectively utilize AKS for ML applications. Recognizing the increasing importance of cloud platforms in supporting the scalability, performance, and efficiency of machine learning systems, various considerations that drive successful ML deployments on AKS were examined.

Throughout the exploration, existing strategies for managing ML workloads on AKS were analyzed, covering key topics such as cluster sizing, resource allocation, and security, which are crucial for optimizing performance and ensuring long-term maintainability. Insights from industry implementations were synthesized, demonstrating how AKS offers a flexible and cost-effective solution for containerized ML workloads, benefiting from seamless integration with Azure Machine Learning and Kubeflow to automate pipelines and streamline model deployment.

Additionally, real-world case studies across sectors like healthcare and retail were highlighted, illustrating how organizations have leveraged AKS to enhance operational efficiency, improve customer experiences, and boost data processing capabilities. These case studies underscored the scalability and performance optimization achievable through AKS, as well as the security best practices needed to safeguard sensitive ML data.

In conclusion, this research emphasizes the transformative role that Azure Kubernetes Service plays in enabling efficient, scalable, and secure ML workloads. As machine learning continues to evolve, AKS is positioned to remain a key platform in the cloud ecosystem, offering the tools and flexibility necessary to meet the growing demands of ML applications. Enterprises adopting AKS for their ML workloads will benefit from optimized performance, cost efficiency, and robust scalability, driving innovation and competitiveness in an increasingly data-driven world.

REFERENCES

[1] Ramos Apolinario, V., & Ramos Apolinario, V. (2021). Moving your containers to the Cloud with Microsoft Azure. Windows Containers for IT Pros: Transitioning Existing Applications to Containers for On-premises, Cloud, or Hybrid, 167-210.

[2] Habtemariam, Y. (2024). The Role of Cloud-Based AI and ML for Interactive Web Applications: Opportunity and challenges.

[3] Bershchanskyi, Y., Klym, H., & Shevchuk, Y. (2024). Containerized artificial intelligent system design in cloud and cyber-physical systems., Advances in Cyber-Physical Systems (ACPS) 2024; Volume 9, Number 2 pp. 151 - 157

[4] Sawyers, D. M. (2021). Automated Machine Learning with Microsoft Azure: Build highly accurate and scalable end-to-end AI solutions with Azure AutoML. Packt Publishing Ltd.

[5] Kalyva, G. (2023). Machine Learning Security with Azure: Best practices for assessing, securing, and monitoring Azure Machine Learning workloads. Packt Publishing Ltd.

[6] Milad, A., Yusoff, N. I. M., Majeed, S. A., Ibrahim, A. N. H., Hassan, M. A., & Ali, A. S. B. (2020, February). Using an azure machine learning approach for flexible pavement maintenance. In 2020 16th IEEE International Colloquium on Signal Processing & Its Applications (CSPA) (pp. 146-150). IEEE.

[7] Reddy, S., Catharine, A., & Shanthamalar, J. J. (2024, May). Efficient Application Deployment: GitOps for Faster and Secure CI/CD Cycles. In 2024 International Conference on Advances in Modern Age Technologies for Health and Engineering Science (AMATHE) (pp. 1-7). IEEE.

[8] Bershchanskyi, Y., & Klym, H. (2023, October). Information System for Administration of Medical Institution. In 2023 13th International Conference on Dependable Systems, Services and Technologies (DESSERT) (pp. 1-4). IEEE.

[9] Bershchanskyi, Y., & Klym, H. (2024, October). Development Approaches of Cloud-Based System for Object Recognition on Images. In 2024 IEEE 17th International Conference on Advanced Trends in Radioelectronics, Telecommunications and Computer Engineering (TCSET) (pp. 205-208). IEEE.

High-Accuracy ECG Signal Acquisition Using a Power-Efficient 6-bit Level-Crossing ADC

Abdollah Amini[1], Hamed Norouzi Kalehsar[2]

[1] Department of Electrical, Computer, and Biomedical Engineering, University of Pavia, Italy
[2] Microelectronics Research Laboratory, Urmia University, Urmia, Iran

Abstract—**This paper presents a novel approach using level-crossing ADCs (LC-ADCs), which provide a more power-efficient solution by sampling only when the input signal exceeds predefined amplitude thresholds. A 6-bit LC-ADC architecture that eliminates the need for an n-bit DAC and uses a single comparator, incorporating integrated sample-and-hold and logic functions, is proposed. This design drastically reduces power consumption while maintaining accuracy, making it particularly suitable for low-power biomedical applications. Simulation results in 180 nm CMOS technology with 1.8-V power supply demonstrate the system's high performance, including a Figure-of-Merit (FoM) that outperforms traditional designs. The proposed LC-ADC architecture is shown to be highly efficient, with significant reductions in power usage and FoM to 90 nW and 0.115 fJ/step, offering an ideal solution for long-term, energy-constrained biomedical signal monitoring.**

Keywords—**Level-crossing ADC, power efficient, ECG signal acquisition, high-accuracy, biomedical circuits and systems.**

I. INTRODUCTION

Power consumption remains a pivotal challenge in the design of portable and implantable biomedical systems, as well as wireless neural acquisition devices, which typically rely on batteries or wireless energy transfer. To enable extended operation without frequent recharging or battery replacement, it is critical to minimize power dissipation, especially during data transmission and processing [1–3].

Analog-to-digital converters (ADCs) are integral components in these wireless and wearable biomedical systems, yet they are a significant source of power consumption. Conventional ADCs operate on the basis of uniform sampling, adhering to the Nyquist theorem, which divides the time axis into uniform intervals and samples signal values at fixed time points. Although this approach is effective in many applications, it becomes inefficient when processing signals that are sparse in the time domain. For example, many bio-signals exhibit extended periods of low-frequency activity interspersed with brief bursts of high-frequency content. In such cases, uniform sampling leads to redundant data collection, resulting in unnecessary energy expenditure during processing, transmission, and storage [4–6].

Level-crossing ADCs (LC-ADCs) present a promising solution to mitigate these inefficiencies. Figure 1 illustrates the operational difference between Nyquist sampling (Fig. 1(a)) and level-crossing (LC) sampling (Fig. 1(b)) in the context of medical signal acquisition. In Nyquist sampling, the time axis is subdivided into fixed intervals, with signal values captured

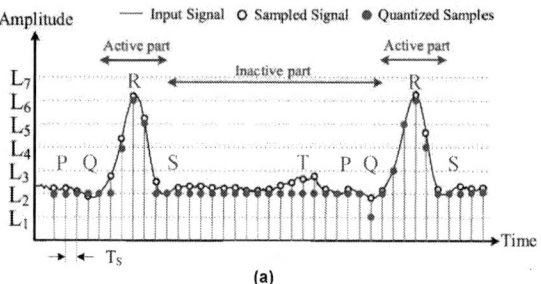

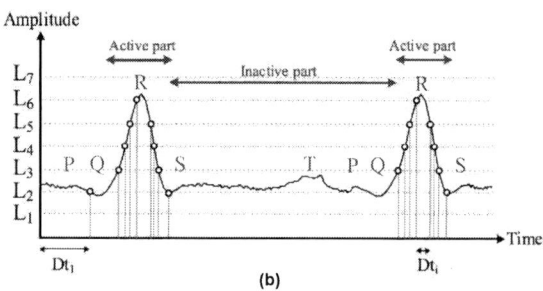

Fig. 1. A real ECG signal is sampled with both: (a) conventional synchronous sampling based on the Nyquist theorem, (b) level-crossing sampling

at each of these intervals. In contrast, LC sampling divide the input amplitude range ($2 \times AFS$) into 2^M discrete levels and sample the signal only when it crosses these predefined amplitude thresholds, which are defined by the least significant bit (LSB). The value of each threshold is determined by $2 \times AFS/2^M$.

This distinction underscores that Nyquist sampling is time-driven, with fixed intervals governing the sampling process, whereas LC sampling is event-driven, responding dynamically to variations in signal amplitude. In conventional synchronous ADCs (Fig. 1(a)), the accuracy of amplitude quantization is constrained by a limited number of discrete quantization levels, imposing a fundamental limitation on system efficiency. By contrast, LC sampling (Fig. 1(b)) selectively avoids sampling during periods of low signal variation, significantly reducing both data rate and power consumption. By focusing on significant signal transitions, LC-ADCs enhance efficiency in capturing relevant information, thereby making them

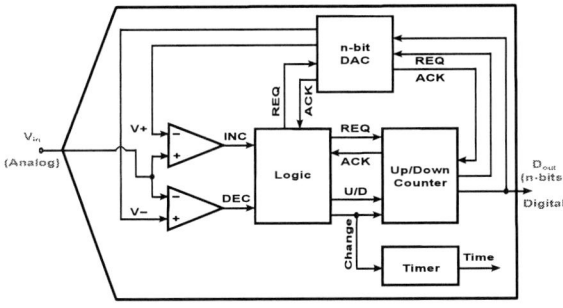

Fig. 2. Block diagram of the conventional LC-ADC using two comparators, an n-bit DAC, a logic block, a timer, and an up/down counter

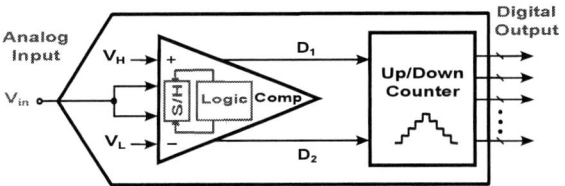

Fig. 3. Block diagram of the proposed LC-ADC, consisting of a comparator with integrated S/H and logic blocks, along with an up/down counter

ideal for various applications, particularly in medical signal acquisition, where low power consumption and reduced data bandwidth are critical [7–12].

The standard nomenclature of the PQRST components in an electrocardiogram (ECG) was introduced by Willem Einthoven, the pioneer of electrocardiography, in the early 20th century. He chose the letters P, Q, R, S, and T arbitrarily from the middle of the alphabet to label the sequential deflections of the ECG waveform, avoiding confusion with other physiological variables. This convention provides a systematic and neutral method for identifying key cardiac electrical events: the P wave represents atrial depolarization; the QRS complex reflects ventricular depolarization, with Q, R, and S denoting specific directional deflections; and the T wave corresponds to ventricular repolarization. This naming scheme has since become universally accepted in both clinical and engineering domains.

Typically, LC-ADCs consist of two comparators, an n-bit digital-to-analog converter (DAC), an up-down counter, a timer, and control logic (Fig. 2). Prior studies [13–15] show that the n-bit DAC and comparators contribute most to the ADC's power consumption. To mitigate this, some designs have eliminated the n-bit DAC or reduced the number of comparators to just one [5,6], and [16–19].

This paper presents a 6-bit LC-ADC with a modified structure that eliminates the need for an n-bit DAC and reduces the architecture to a single comparator, replacing multiple conventional blocks. Specifically, the sample-and-hold (S/H) and logic functions are integrated within the comparator, significantly lowering power consumption while preserving accuracy. This innovative design enhances energy efficiency, making it particularly well-suited for long-term biomedical signal monitoring applications.

The rest of this paper is organized as follows: The proposed LC-ADC implementation both in the block diagram and in transistor level is described in detail in Section II, while Section III reports the simulation results of the proposed LC-ADC structure and a comparison with other similar approaches found in literature. Finally, Section IV concludes the paper.

II. PROPOSED LC-ADC STRUCTURE

The block diagram of the proposed LC-ADC is shown in Fig. 3. To reduce the complexity of traditional LC-ADC architectures, the proposed system is simplified into two main blocks: a comparator and an up/down counter. The comparator itself is composed of a core current comparator, a readout circuit, a sample-and-hold (S/H) unit, and a logic block. This streamlined design not only makes the system more compact and power-efficient but also enhances the reliability of ECG signal acquisition with minimal energy expenditure.

The proposed high-performance voltage comparator employs a core current comparator with cross-coupled current mirrors, as shown in Fig. 4(a). A transconductance unit converts the differential input voltage into currents (I_1 and I_2) using two differential pairs (M_1, M_2) and (M_4, M_5) with their current sources (M_3) and (M_6), respectively. The winning current (I_1) drives the losing current (I_2) through a mirror transistor (M_7), charging the winner node (O_1) and discharging the loser node (O_2). When I_2 exceeds I_1, the roles reverse: the new winner node (O_2) is charged, turning on transistor M_8 while discharging O_1. This architecture eliminates the need for a reset switch but results in longer response times to input changes.

In high-speed applications, resetting is essential, but traditional MOS switch-based resets can introduce delays due to transistor sizing. The proposed design improves this by grounding the output nodes through NMOS devices (M_9 and M_{10}) via the reset signal (V_R), enhancing reset speed by increasing overdrive voltage and reducing output capacitance. This results in faster comparison times. The system simplifies control by requiring a single control signal (V_R) for reset, evaluation, and latching, thereby avoiding complexity and preventing clock skew.

A simplified readout circuit, including M_{11} and M_{12}, along with two inverters in a back-to-back topology (inside the blue dashed line), captures and holds the output digital bits during the clock period, extending the validity time of the outputs. The designed comparators have differential outputs, so both the inverting and non-inverting output bits are available. This design reduces power consumption by toggling output bits only when necessary, minimizing superfluous transitions. As a result, power usage is significantly reduced, especially when consecutive comparisons yield the same output.

A simple structure, including three NMOS transistors (M_S, M_{H1}, and M_{H2}), forms the S/H circuit. The fundamental operational mechanism of the S/H circuit, as illustrated in Fig. 4(b), is based on the feedback signal coming from the logic block ($V_{S/H}$). During sample period, the gate terminals of M_1 and M_5 are connected together, and the sampled V_{in} is applied to both transistors for comparison with the predefined reference voltages (higher threshold V_H and lower threshold V_L). When V_{in} crosses V_H or V_L, the output returns to the common mode value of V_{in} during the hold phase.

To provide the required control signal ($V_{S/H}$) for the comparator, a logic block is employed, as illustrated in Fig. 4(c). This block consists of an *OR* gate and an edge detector circuit, also known as a one-shot generator [20]. The output bits of the comparator serve as the inputs to the *OR* gate, and its output (D) is compared with its delayed and inverted version using an *AND* gate. Whenever a low-to-high transition occurs, the *AND* gate encounters the "11" state for a brief period, generating a pulse at the output, which acts as a control signal $V_{S/H}$.

The comparator outputs are fed into the up/down counter, determining the counting direction based on the comparison results. When the input signal crosses an upper or lower threshold, the counter increments or decreases accordingly, translating the amplitude variations of the input signal into digital output values. This simplified yet high-performance architecture reduces power consumption, improves reset speed, and efficiently captures relevant signal transitions, making it an ideal solution for energy-constrained biomedical applications.

III. SIMULATION RESULTS AND COMPARISON

This section presents the simulation results of the proposed LC-ADC, demonstrating its proper functionality under realistic conditions, including the effects of power supply noise without needing any extra circuits to adapt the resolution. The proposed LC-ADC was designed in a 180-nm standard CMOS technology with the 1.8 V power supply. To demonstrate the enhanced performance of the converter, simulations were conducted for two different inputs. Fig. 5 illustrates the simulation results for a saw-tooth input signal. For the sake of simplicity, 6 bits of the output token (Q_0 to Q_5) are depicted effectively preventing unnecessary oversampling. Similarly, Fig. 6 presents the simulation results for a semi-ECG input signal. The results show that the proposed system just taking samples of the specific part of the input semi-ECG signal (QRS complex) reduces redundant sampling, eliminating unnecessary data points and optimizing efficiency.

To evaluate the Effective Number of Bits (ENOB) of the proposed LC-ADC, a periodic sinusoidal signal is applied to the input and then the Fast Fourier Transform (FFT) is performed. The analysis reveals an ENOB of 5.7 bits.

Furthermore, the performance metrics of the LC-ADC are compared against previously reported designs, with the results summarized in TABLE I. The data highlights the significant advantages of the proposed system. The figure-of-merit (FoM) is calculated using the widely recognized formula:

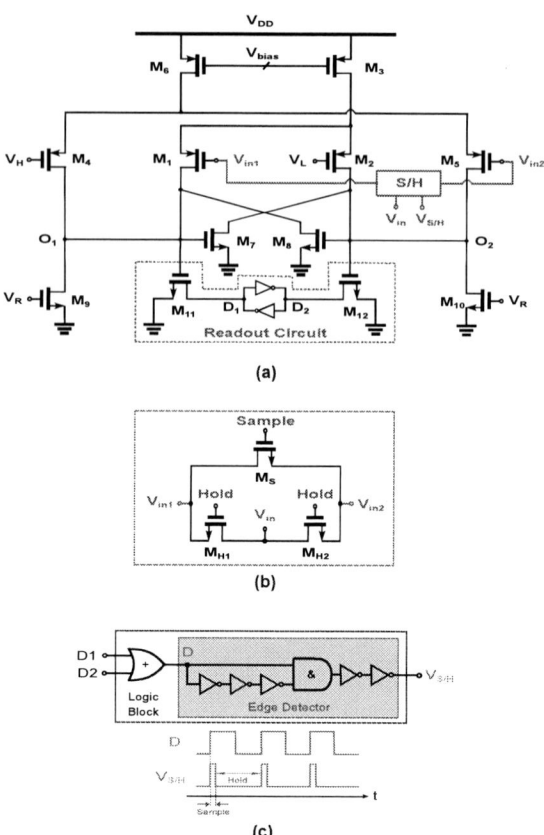

Fig. 4. Transistor level structure of the comparator block including: (a) core current comparator and readout circuit, (b) sample and hold (S/H) circuit, and (c) embedded logic block with edge detector and its input and output pulses

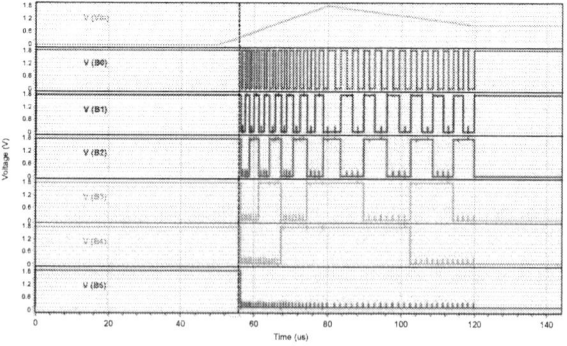

Fig. 5. Simulation results of the proposed LC-ADC using a semi-synthetic saw-tooth signal (Vin) as input and showing the extracted output bits (B0–B5).

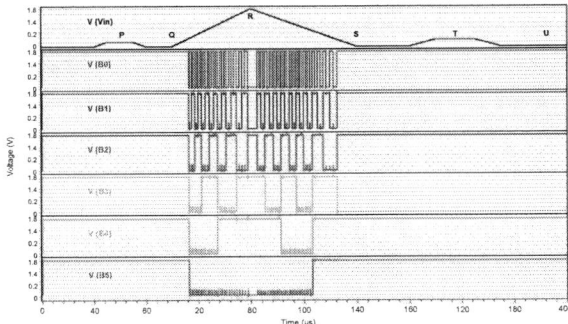

Fig. 6. Simulation results of the proposed LC-ADC using a semi-synthetic ECG signal (Vin) as input and showing the extracted output bits (B0–B5).

$$FoM = \frac{Power}{f_{\mathrm{NYQ}} \cdot 2^{\mathrm{ENOB}}} \quad (1)$$

According to this equation, a lower power consumption results in a better FoM [21]. Simulation results demonstrate that the proposed LC-ADC not only achieves low power consumption but also significantly reduces the FoM.

TABLE I
PERFORMANCE COMPARISON WITH OTHER LC-ADCs

Reference	[16]	[17]	[22]	[23]	This Work
Process (nm)	180	180	180	180	180
Supply (V)	0.55-1	0.5	1	1.8	1.8
Power (nW)	186	60-220	18	220	90
BW (kHz)	1	1	2.5	10	15
ENOB (bit)	6.2-7.9	5.6	6.2-7.7	6.8	5.7
FoM (fJ/conv.)	165	124	98	198	115
Sim./Meas.	Meas.	Meas.	Sim.	Sim.	Sim.

IV. CONCLUSIONS

A 6-bit LC-ADC design in 180 nm CMOS technology is presented to significantly reduce the power consumption in portable biomedical systems. Eliminating the need for an n-bit DAC and utilizing a simplified structure with a single comparator, integrated sample-and-hold, and logic functions, the proposed design minimizes energy expenditure without sacrificing performance or accuracy. Simulation results confirm the effectiveness of the proposed LC-ADC, showing reduced data rate and power consumption, particularly for biomedical signals that exhibit periods of low-frequency activity interspersed with high-frequency bursts. The resulting 90 nW of power consumption and 0.115 fJ/step of FoM demonstrate a significant improvement in the proposed design, making it ideal for medical signal monitoring systems.

REFERENCES

[1] M. Ghasemi et al., "An Ultra-Low Power level-crossing ADC for ECG Monitoring Application," *28th Iranian Conference on Electrical Engineering (ICEE)*, Tabriz, Iran, pp. 1-6, 2020.

[2] N. Ravanshad et al., "A fully-synchronous offset-insensitive level-crossing analog-to-digital converter," *Proc. 59th Int. Midwest Symp. Circuits and Systems (MWSCAS)*, Abu Dhabi, UAE, pp. 1–4, Oct. 16–19, 2016.

[3] N. Khiabanmanesh et al., "Diagnosis of Epilepsy Utilizing Time-Series Distribution of EEG Signals," *25th International Conference on Mixed Design of Integrated Circuits and System (MIXDES)*, Gdynia, Poland, pp. 431-435, 2018.

[4] A. Amini et al., "A Novel Online Offset-Cancellation Mechanism in a Low-Power 6-Bit 2GS/s Flash-ADC," *Analog Integr. Circuits Signal Process*, 99(2), pp. 219-229, 2019.

[5] B. Yazdani and Sh. Jafarabadi-Ashtiani, "A Low Power Fully Differential Level-Crossing ADC with Low Power Charge Redistribution Input for Biomedical Applications," *IEEE Trans. On Circuits and Syst. II*, vol. 69, no. 3, pp. 864-868, March. 2022.

[6] Y. Liet al., "A sub-microwatt asynchronous level-crossing ADC for biomedical applications," *IEEE Trans. on Biomed. Circuits Syst*, vol. 7, no. 2, pp. 149–157, April. 2013.

[7] N. Ravanshad et al., "level-crossing sampling; principles, circuits, and processing for healthcare applications," *In Compressive Sensing in Healthcare, Elsevier Academic Press Inc.*: Amsterdam, The Netherlands, pp. 223–246. ISBN 9780128212479, 2020.

[8] A. Amini et al., "Improving Accuracy of the Current Mirrors for High-Resolution Applications," *31st International Conference on Mixed Design of Integrated Circuits and System (MIXDES)*: Gdansk, Poland, pp. 114-117, 2024.

[9] A. Amini et al., "A High-Voltage TX/RX Switch with 3.3-V Supply for Ultrasound Imaging Front-End ASICs," *31st IEEE International Conference on Electronics, Circuits and Systems (ICECS)*, Nancy, France, pp. 1-4, 2024.

[10] E. Hosseini et al., "High-Speed 32*32 bit Multiplier in 0.18um CMOS Process," *25th International Conference on Mixed Design of Integrated Circuits and System (MIXDES)*, Gdynia, Poland, pp. 154-159, 2018.

[11] F. Modarresi et al., "A low-jitter, full-differential PLL in 0.18 μm CMOS technology," *International Journal of Microelectronics and Computer Science*, vol. 7, no. 4, pp. 119-122, 2016.

[12] K. Kozmin et al., "Level-Crossing ADC Performance Evaluation Toward Ultrasound Application," *IEEE Trans. Circuits Syst. I, Reg. Papers*, vol. 56, no. 8, pp. 1708-1719, Aug. 2009.

[13] B. schell et al., "A continuous-time ADC/DSP/DAC system with no clock and with activity-dependent power dissipation," *IEEE Journal of Solid-State Circuits*, vol. 43, pp. 2472–2481, Nov. 2008.

[14] M. Trakimas et al., "An adaptive resolution asynchronous ADC architecture for data compression in energy-constrained sensing applications," *IEEE Trans. Circuits Syst. I, Reg. Papers*, vol. 58, no. 5, pp. 921–934, May 2011.

[15] X. Zhang et al., "A 300-mV 220-nW event-driven ADC with realtime QRS detection for wearable ECG sensors," *IEEE Trans. Biomed. Circuits Syst.*, vol. 8, no. 6, pp. 834–843, Dec. 2014.

[16] Y. Hou, et al., "A 1-to-1kHz, 4.2-to-544-nW, multi-level comparator based level-crossing ADC for IoT applications," *IEEE Trans. On Circuits and Syst. II*, vol. 65, no. 10, pp. 1390-1394, Jul. 2018.

[17] Y. Hou, et al., "A 61-nW level-crossing ADC with adaptive sampling for biomedical applications," *IEEE Trans. on Circuits and Syst. II*, vol. 66, no. 1, pp. 56-60, Jun. 2018.

[18] T. Aspokeh, et al., "Low-Power 13-Bit DAC with a Novel Architecture in SA-ADC," *25th International Conference on Mixed Design of Integrated Circuits and Systems (MIXDES)*, Poland, 2018.

[19] W. Tang, et al., "Continuous time level crossing sampling ADC for biopotential recording systems," *IEEE Trans. on Circuits and Syst. I*, vol. 66, no. 6, pp. 1407-1418, Jun. 2013.

[20] R. Abdollahi et al, "A Simple and Reliable System to Detect and Correct Setup/Hold Time Violations in Digital Circuits," *IEEE Transactions on Circuits and Systems I*, vol. 63, no. 10, pp. 1682-1689, Oct. 2016.

[21] A. Amini et al., "On Improving Accuracy of the Resistor Strings Based on a New Design Technique," *Iran J Sci Technol Trans Electr Eng.*, (45), pp. 221–227, 2021.

[22] A. Zanjani et al., "A Power-Efficient Level-Crossing Analog-to-Digital Converter with Adaptive Resolution Based on a Signal-Dependent Sampling Mechanism," *Circuits Syst Signal Process.*, 42, 63–83, 2023.

[23] D. Makarem et al., "An Adaptive Resolution Level-Crossing Analog-to-Digital Converter for Biomedical Signal Acquisition," *30th International Conference on Mixed Design of Integrated Circuits and System (MIXDES)*, Kraków, Poland, pp. 88-91, 2023.

Mixed Design of Integrated Circuits and Systems – MIXDES 2025

Low Voltage, High Power Electronic Load Design for FPGA Current Draw Reproducing

Szymon Przybył Piotr Sarna, Zbigniew Kulesza, Mariusz Zubert

Lodz University of Technology
Department of Microelectronics and Computer Science
Łódź, Poland
szymon.przybyl@dmcs.p.lodz.pl, piotr.sarna@dmcs.p.lodz.pl, zbigniew.kulesza@p.lodz.pl, mariusz.zubert@p.lodz.pl

Abstract—**FPGA devices are complex entities which can draw extremely high currents from low voltage rails. This presents multiple problems during the design of power supplies for FPGA circuits. In this paper, an electronic load capable of replicating the current draw from FPGA circuits will be introduced, which will streamline the design of low-voltage, high-power supplies and enable robust testing of designed electrical power rails. Designing electronic loads is a complex procedure, it requires careful design of the schematic, special considerations for the layout to keep the resistance within budget, and additional thermal considerations must also be taken into account.**

Keywords—**High-power electronic, low-voltage electronic, electronic load, FPGA, power supplies**

I. INTRODUCTION

Electronic loads are one of the most important tools when it comes to testing power supplies. As the requirements for the power supplies are increasingly demanding to ensure correct operation it is necessary to test power rails. Additionally, Field-Programmable Gate Array (FPGA) devices as the technology progresses are requiring lower and lower voltages and higher currents, it creates number of challenges creates a number of challenges that come with more narrow-permitted voltage ranges.

Modern FPGA's can draw large currents, unfortunately current consumption heavily depends on what is being done on the FPGA, so each project will have different power consumption. It is possible to estimate current consumption of the FPGA using software such as AMD Power Estimator (XPE, previously Xilinx XPower) [1] or Power Play [2], but often those estimations are not accurate [3].

During DesignCon 2020 paper was presented „A Method for Dynamic Load Current Testing with a Benchtop Power Supply" [4], [5] in which using Rogowski Coil current drawn from FPGA was measured (Figure 1).

This project focused on the design of electronic load which would enable the streamline design of power supplies designated for FPGA.

To achieve that, design requirements were created:

- Voltage range of the source: 0.8V – 2.5V
- Current range: 0-100A
- Rise time of current 1µs (depending on the connection to DUT)
- Possibility of control of DC component and AC component in range 1Hz – 10kHz
- Constant Current (CC) operation
- Analog feedback loop
- Overvoltage, overcurrent, overtemperature system protections

Additionally, as the minimal voltage that device was designed is 0.8V and maximal current that this device should be able to handle is 100A, resistance of the whole device with connections should not exceed 8 mΩ:

$$R = \frac{U}{I} = \frac{0.8V}{100A} = 8m\Omega \quad (1)$$

II. DESIGN

Device first was designed as block diagram (Figure 2). The electronic load project was divided into two functional parts: firstly, the power section and secondly, the control section.

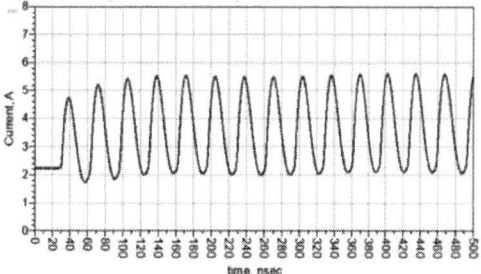

Figure 1. FPGA current [6], [7]

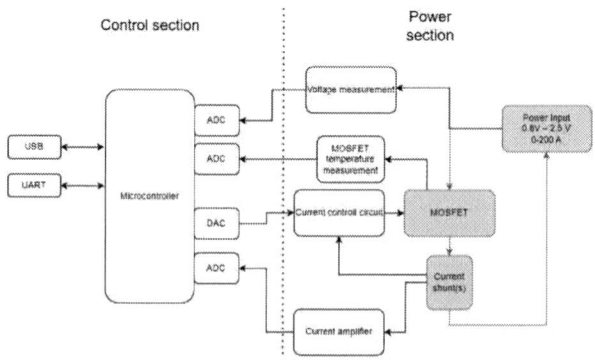

Figure 2. Block diagram of the electronic load

The power section consists of a MOSFET transistor and a control circuit. This part is responsible for generating the required current draw. The subsequent components of the power section are the voltage, current, and temperature measurement sections. These are essential for protecting the device from overpower, overvoltage, and overheating; additionally, the current measurement could be used for calibration of the device.

The control section is tasked with generating signals that will control the current draw from the device under test (DUT). Furthermore, the control section monitors all parameters and deactivates the load if overheating or overvoltage is detected. While the control section could be replaced with a generator and oscilloscope, in such cases, temperature should be monitored independently, as the overtemperature protection implemented in the microcontroller would not be operational.

A. Schematic

To create schematic Altium Designer was used. This choice was made due to rich tools included in the software, such as the Power Delivery Network (PDN) Analyser, which enabled validation of whether the PCB meets the resistance budget.

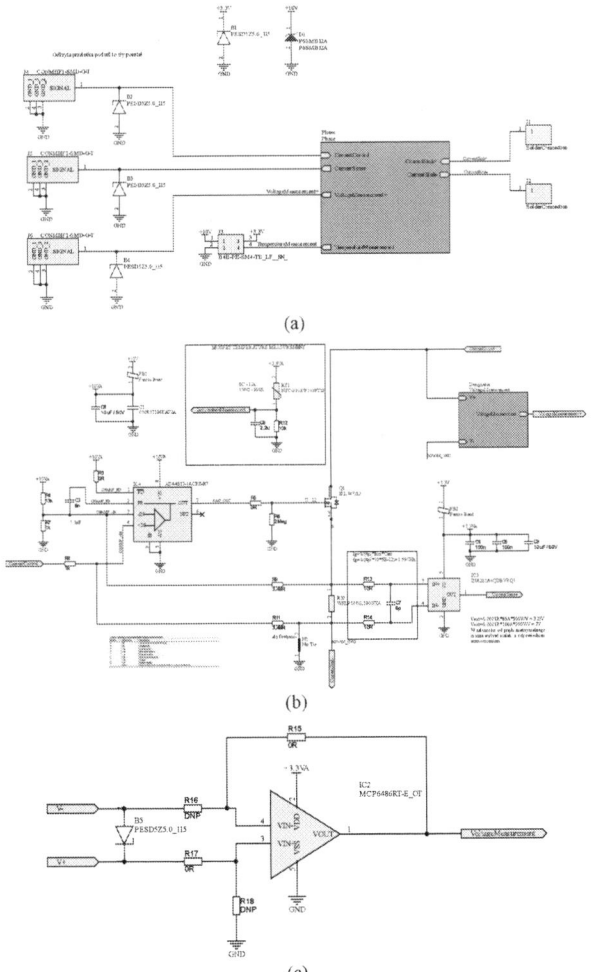

(a)

(b)

(c)

Figure 3. Schematic of the electronic load, a) connectors and ESD protections, b) main schematic, c) current voltage measurement module

Schematic was designed as hierarchical sheets the highest in hierarchy was sheet on which all connectors and ESD protection were placed (Figure 3a). On the next sheet main schematic rest of the electronic parts were placed (Figure 3b), except for the voltage measurement amplifier (Figure 3c), which was configured as voltage-follower.

B. PCB

PCB was designed using 4-layer board with 70μm thick outer layers in order to enable a 100A current flow. Owing to the low height of the transistor, most of the electronics are on the bottom side of the board; this arrangement was necessary to enable the mounting of the heatsink, which will dissipate all the power generated by the transistor. This led to only the transistor and the temperature sensor for the transistor being placed on the top layer.

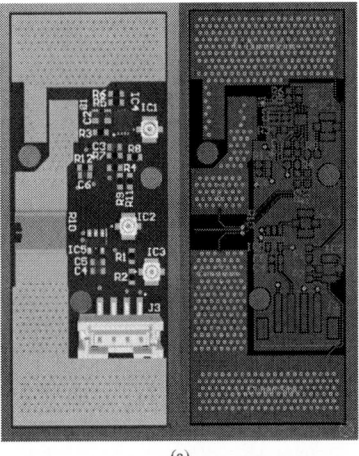

(a)

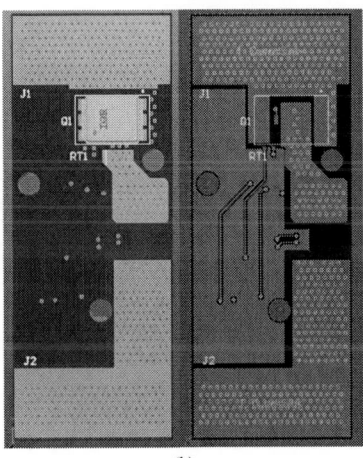

(b)

Figure 4. PCB design, a) bottom side, b) top side

C. Cooling

Device was designed to operate in range of 0.8V to 2.5V and current up to 100A. That means the maximal power that is dissipated on the electronic load is equal to:

$$P = U * I = 2.5 * 100 = 250\,W \qquad (2)$$

As the transistor on which the most power will be dissipated it was necessary to enable such high-power dissipation which led to IRL7472L1 [8] being chosen as the main transistor due to its very low junction-case resistance.

IRL7472L1 is a transistor with a metal case that enables excellent thermal specifications (see Figure 5). From these parameters, the maximal heatsink thermal resistance was calculated using the equation:

$$R_H = \frac{T_{Jmax} - T_A}{P} - R_{J-C} - R_{C-H} \quad (3)$$

As the maximal junction temperature 155°C was selected to provide safety reserve.

$$R_H = \frac{155 - 27}{250} - 0.44 - 0.0032 = 0.0688 \frac{°C}{W} \quad (4)$$

Thermal Resistance

Symbol	Parameter	Typ.	Max.	Units
$R_{\theta JA}$	Junction-to-Ambient ❶	—	40	
$R_{\theta JA}$	Junction-to-Ambient ❷	12.5	—	
$R_{\theta JA}$	Junction-to-Ambient ❸	20	—	°C/W
$R_{\theta JC}$	Junction-to-Case ❷⑤	—	0.44	
$R_{\theta JA,PCB}$	Junction-to-PCB Mounted	1.0	—	

Figure 5. IRL7472L1 thermal specification [8]

III. MEASUREMENTS

To verify that the electronic load functions properly, remains stable, and is capable of testing, a setup was created (Figure 6). Test setup comprised of laboratory power supply, signal generator Rigol DG1062Z [9], oscilloscope Rigol DS1104Z [10], evaluation board ISL68137-61P-EV1Z [11], PC power supply, Nucleo STM32L476RG [12] and electronic load tester designed by designed by the author. Additionally, heatsink RAD-A4291/80 with the dimensions of 80mm x 165 mm x 35 mm was utilised [13] to cool electronic load PCB.

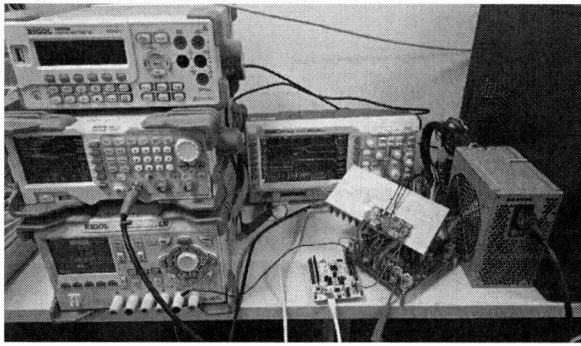

Figure 6. Test setup

A. Transfer characteristic

First, to check if current is controlled correctly, DC transfer characteristics were taken. DC transfer characteristic indicated that control of the load is mostly linear, although but on the lower currents it is entirely linear; what should be taken into account when controlling the load.

Current can be described using:

$$I_{LOAD} = 0{,}6075 * V_{CTRL}^2 + 34{,}523 * V_{CTRL} - 4{,}9001 \quad (5)$$

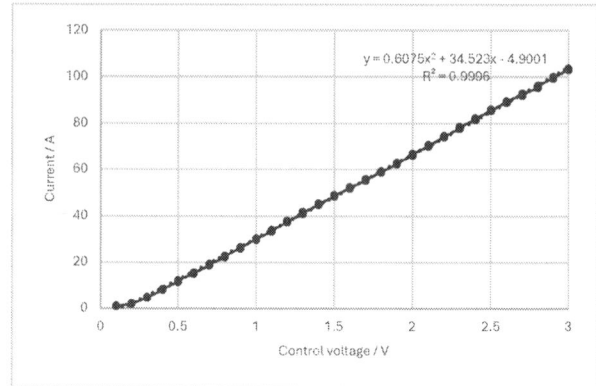

Figure 7. DC transfer characteristic

B. Rise time and signal delay

In order to ensure stable operation of the electronic load and to confirm that the electronic load can meet the required speeds, transient tests were conducted (Figure 8).

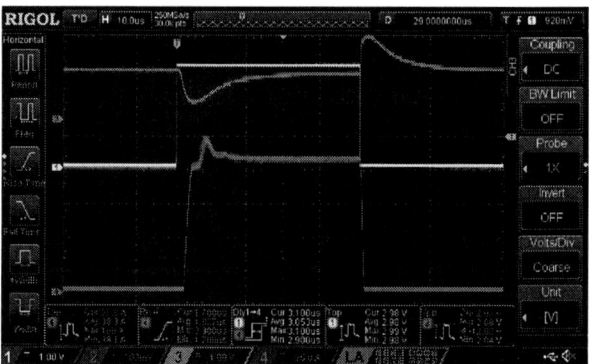

Figure 8. Transient response of the electronic load

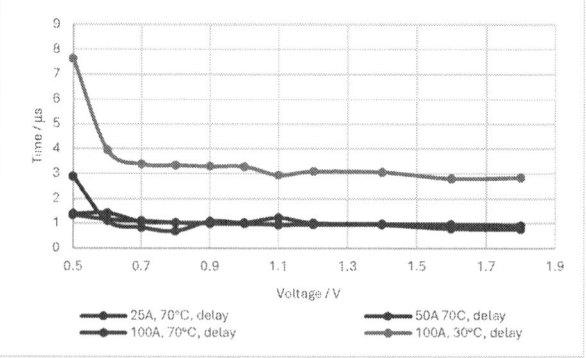

Figure 9. Signal delay characteristic

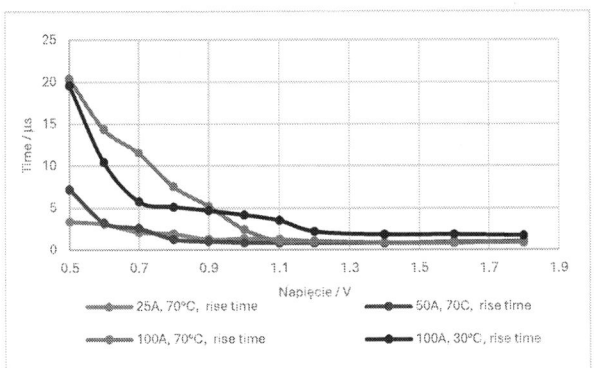

Figure 10. Rise-time characteristic

Unfortunately, ISL68137-61P-EV1Z was not able to provide high current for very low voltages, which caused the characteristics to exhibit longer times than necessary due to the module's inability to supply sufficient current, resulting in an output voltage collapse (Figure 11).

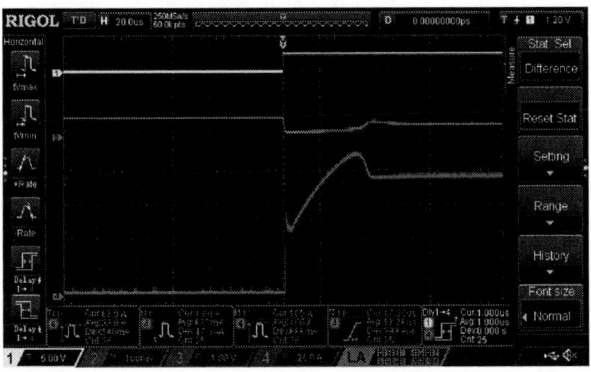

Figure 11. Operation of the electronic load with very low voltage. Signals: yellow – control voltage, magenta – input voltage, blue – current

It was discovered that the temperature of the device affected the rise-time (Figure 12) and signal delay (Figure 13). This is an effect of capacitor C3 changing its parameters due to its temperature dependency.

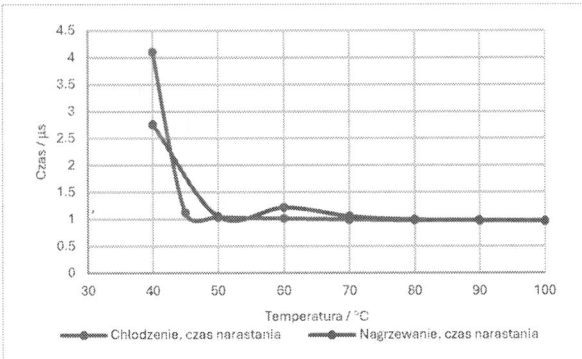

Figure 12. Rise-time characteristic vs temperature

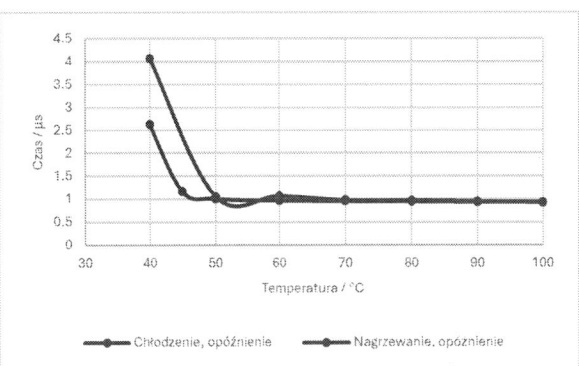

Figure 13. Signal delay vs temperature

C. FPGA current simulations

The main goal of this project was FPGA current simulation, which would enable testing of power supplies used for FPGA. In order to confirm that the designed electronic load is able to simulate FPGA, the waveform generated by FPGA (Figure 1) was reproduced using an arbitrary signal generator and the designed load (Figure 14). Additionally, the waveform was upscaled to test the full range of current load possible to generate.

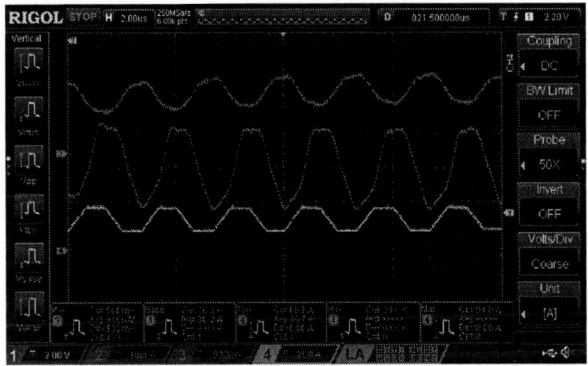

Figure 14. FPGA current simulation

IV. CONCLUSIONS

In summary, it was possible to simulate the current draw by the FPGA a using a specifically designed electronic load. However, the electronic load requires an upgrade that would enable it to operate independently of temperature. This study also demonstrates that it is feasible to generate loads analogous to the FPGA current loads using more cost-effective solutions than those commercially available [14]. For future developments, it is recommended that the control module be replaced by a combination of an oscilloscope and an arbitrary signal generator. These should be controlled via a USB or Ethernet connection, perhaps through a Python script, to streamline measurements and reduce both the cost and development time associated with the control module. In this revised configuration, the control module would only need to monitor the transistor temperature.

ACKNOWLEDGMENT

The authors would like to express their special thanks to FastLogic Sp.z o.o. and dr Kamil Grabowski for proposing this research topic, financing its implementation, and collaborating during the execution of this project

REFERENCES

[1] 'AMD Power Estimator', AMD. Accessed: Feb. 13, 2025. [Online]. Available: https://www.amd.com/en/products/adaptive-socs-and-fpgas/technologies/power-efficiency/power-estimator.html

[2] 'PowerPlay Power Analyzer Support Resources', Intel. Accessed: Feb. 13, 2025. [Online]. Available: https://www.intel.com/content/www/us/en/support/programmable/support-resources/power/sof-qts-power.html

[3] H. G. Lee, S. Nam, and N. Chang, 'Cycle-accurate energy measurement and high-level energy characterization of FPGAs', in *Fourth International Symposium on Quality Electronic Design, 2003. Proceedings.*, Mar. 2003, pp. 267–272. doi: 10.1109/ISQED.2003.1194744.

[4] 'Determining FPGA Dynamic Load Current Steve Sandler | Signal Integrity Journal'. Accessed: Feb. 02, 2025. [Online]. Available: https://www.signalintegrityjournal.com/blogs/15-extreme-measurements/post/2310-determining-fpga-dynamic-load-current-steve-sandler

[5] H. Barnes, J. Carrel, and S. M. Sandler, 'A Method for Dynamic Load Current Testing with a Benchtop Power Supply', presented at the DesignCon 2020, Jan. 29, 2020. [Online]. Available: https://www.designcon.com/en/home.html

[6] 'Determining FPGA Dynamic Load Current Steve Sandler | Signal Integrity Journal'. Accessed: Mar. 22, 2025. [Online]. Available: https://www.signalintegrityjournal.com/blogs/15-extreme-measurements/post/2310-determining-fpga-dynamic-load-current-steve-sandler

[7] Keysight Design Software, *How to Find the Elusive Dynamic Switching Current of Your FPGA Power Rail*, (Jun. 17, 2022). Accessed: Mar. 22, 2025. [Online Video]. Available: https://www.youtube.com/watch?v=UpmV5k1bVYs

[8] I. T. AG, 'IRL7472L1 - Infineon Technologies'. Accessed: Feb. 13, 2025. [Online]. Available: https://www.infineon.com/cms/en/product/power/mosfet/n-channel/irl7472l1/

[9] 'Generator Arbitralny RIGOL DG1062Z RIGOL.COM.PL'. Accessed: Feb. 02, 2025. [Online]. Available: https://rigol.com.pl/pl/p/Generator-Arbitralny-RIGOL-DG1062Z/28

[10] 'Oscyloskop Cyfrowy Rigol DS1104Z Plus', RIGOL.COM.PL. Accessed: Mar. 01, 2025. [Online]. Available: https://rigol.com.pl/pl/p/Oscyloskop-Cyfrowy-Rigol-DS1104Z-Plus/259

[11] 'ISL68137-61P-EV1Z - 6+1 Phase Digital Multiphase Controller Evaluation Board | Renesas'. Accessed: Feb. 02, 2025. [Online]. Available: https://www.renesas.com/en/products/power-management/multiphase-power/multiphase-dcdc-switching-controllers/isl68137-61p-ev1z-61-phase-digital-multiphase-controller-evaluation-board

[12] 'NUCLEO-L476RG - STM32 Nucleo-64 development board with STM32L476RG MCU, supports Arduino and ST morpho connectivity - STMicroelectronics'. Accessed: Feb. 02, 2025. [Online]. Available: https://www.st.com/en/evaluation-tools/nucleo-l476rg.html

[13] 'RAD-A4291/80 STONECOLD - Radiator: wytłaczany | żeberkowy; L: 80mm; W: 165mm; H: 35mm; surowy | TME - Części elektroniczne Polska', TME. Accessed: Feb. 12, 2025. [Online]. Available: https://www.tme.eu/pl/details/rad-a4291_80/radiatory/stonecold/

[14] 'UltraLowVoltageDCElectronicLoad 63202A-20-1000/63202A-20-2000', UltraLowVoltageDCElectronicLoad 63202A-20-1000/63202A-20-2000. Accessed: Feb. 26, 2025. [Online]. Available: https://www.chromaate.com/en/product/ultra_low_voltage_dc_electronic_load_63202a_504

Recurrent LSTM Neural Networks for Language Modelling and Speech Recognition

Piotr Kłosowski

Faculty of Automatic Control, Electronics and Computer Science
Silesian University of Technology
Akademicka 16, 44-100 Gliwice, Poland
Email: pklosowski@polsl.pl

Abstract—This paper examines interesting natural language modelling tasks, such as word-based and subword-based language modelling, where deep learning methods are making some progress. Language modelling helps to predict the sequence of recognised words or subwords and thus can be used to improve the speech recognition process. However, the field of language modelling is currently witnessing a shift from statistical methods to recurrent neural networks and deep learning techniques. This article focusses on an example of using recurrent LSTM neural networks for language modelling and speech recognition. The new research results presented in this paper, following on from previous papers, focus on how to develop word-based and subword-based LSTM language models and how to use them together. The simultaneous use of both LSTM language modelling methods allows for the development of hybrid language models that have even better properties and can further improve the speech recognition process. The results presented in this paper apply to Polish language modelling, but the results obtained and the conclusions formulated on their basis can also be applied to language modelling applications for other languages.

Keywords—artificial intelligence, neural networks, language modelling, speech recognition.

I. INTRODUCTION

Statistical language modelling enables the development of probabilistic models that can predict the next word in a sequence given the words that precede it. Language modelling is the task of assigning a probability to sentences in a language. In addition to assigning a probability to each sequence of words, language models also assign a probability for the likelihood that a given word (or a sequence of words) will follow a sequence of words [1].

A language model learns the probability of word occurrence based on examples of text. Simpler models may look at the context of a short sequence of words, whereas larger models may work at the level of sentences or paragraphs. Most commonly, language models operate at the level of words. The notion of a language model is inherently probabilistic. A language model is a function that puts a probability measure on strings drawn from some vocabulary [2]. A language model can be developed and used stand-alone, such as to generate new sequences of text that appear to have come from the corpus. Language modelling is a root problem for a wide range of natural language processing tasks. More practically, language models are used on the front- or back-end of a more sophisticated model for a task that requires language

understanding. Language modelling is a crucial component in real-world applications such as machine translation and automatic speech recognition. For these reasons, language modelling plays a central role in natural language processing, AI, machine learning and speech recognition research [1].

A good example is speech recognition, where audio data are used as input to the model and the output requires a language model that interprets the input signal and recognises each new word within the context of the words already recognised. Speech recognition is principally concerned with the problem of transcribing the speech signal as a sequence of words. From this point of view, speech is very often represented by a language model that provides estimates of $P(W)$ for all word strings W independently of the observed signal. The main goal of speech recognition is to find the most likely word sequence given the observed acoustic signal [3].

Developing better language models often results better on natural language processing and speech recognition task. It is therefore a motivation to develop ever better and more accurate language models. The main objective of the research presented in this paper is to develop language models using recurrent neural networks with a special focus on Long-Short-Term Memory (LSTM) networks.

II. NEURAL LANGUAGE MODELLING

Recently, the use of neural networks in the development of language models has become so popular that it may now be the preferred approach. The use of neural networks in language modelling is often called neural language modelling (NLM). Neural network approaches perform better than classical methods, both for autonomous language models and when the models are incorporated into larger models with demanding tasks such as speech recognition and machine translation. A major cause of performance leaps may be the ability of the method to generalise. Non-linear neural network models address some of the shortcomings of traditional linguistic models: They allow contingencies of ever-larger contexts with only a linear increase in the number of parameters, reduce the need for manual design of recall orders, and support generalisation in different contexts [1]. In particular, word embedding is adopted, which uses a real-valued vector to represent each word in the vector space of the design. This learnt representation of words based on their use allows words of similar meaning to have a similar representation. Neural language models solve the problem of the sparsity of n-gram data by parametrizing words as vectors (word embedding) and using them as input to a neural network. These parameters are set during the learning process. The words sequences obtained with NLM exhibit the property that semantically close words are similarly close in an induced vector space [4]. This generalisation is something that the representation used in classical statistical language models cannot easily achieve. True generalization is difficult to obtain in a discrete word index space since there is no obvious relation between word indices [5].

The neural network approach to language modelling can be described using the three following model properties [6]:

- Associate each word in the vocabulary with a distributed word feature vector,
- Express the joint probability function of word sequences in terms of the feature vectors of these words in the sequence,
- Learn simultaneously the word feature vector and the parameters of the probability function.

This represents a relatively simple model, where both the representation and probabilistic model are learnt together directly from raw text data. Recently, neural-based approaches have consistently started to outperform classical statistical approaches [7].

III. RECURSIVE NEURAL NETWORKS

More recently, recurrent neural networks with long-term memory like the Long Short-Term Memory Network (LSTM), allow the models to learn the relevant context over much longer input sequences than the simpler feed-forward networks [8]. Recently, researchers have been seeking the limits of these language models. In papers evaluating language models over datasets, such as a corpus of one million words, the authors find that LSTM-based neural language models outperform the classical methods Furthermore, the authors propose some heuristics to develop high-performance neural language models in general [9]:

- The best models were the largest models, specifically the number of memory units.
- Use of regularisation like dropout on input connections improves results.
- Character-level convolutional neural network (CNN) models can be used on the front-end instead of word embeddings, achieving similar and sometimes better results.
- Combining the prediction from multiple models can offer large improvements in model performance.

LSTM recursive neural networks are designed to solve sequence prediction problems and represent state-of-the-art deep learning techniques for difficult prediction problems. Recurrent neural networks are a type of neural network that adds explicit order handling to input observations. This ability suggests that the promise of recurrent neural networks is to know the temporal context of input sequences to make better predictions. This means that the set of delayed observations required to perform the forecast no longer has to be defined, as in traditional time series forecasting, or even forecasting using classical neural networks. Instead, time dependency can be learnt, and perhaps time dependency can be learned to change as well. Recently, deep neural networks are beginning to show great potential in language processing and modelling [10], [11], [12]. As a particular type of RNN, the LSTM neural network [13] has proved effective in modelling sequential data such as speech and text [14]. The fundamental problem of language modelling is the creation of increasingly efficient models and their application to better solve various NLP problems, including speech recognition.

This article describes an attempt to solve this problem by proposing more effective LSTM language models for speech recognition. Earlier publications presented the possibilities of using various models of the Polish language to improve the process of speech recognition [15], [16], [17], [18], [19]. The possibilities of using deep machine learning techniques to model the language [10], [11] were also presented. This article proposes the use of LSTM for language modelling and additionally the use of different LSTM language models simultaneously to increase the efficiency and effectiveness of the modelling process.

IV. DEVELOPED WORD-BASED LSTM LANGUAGE MODEL

In speech recognition applications, it is common practice to introduce the acoustic model $P(A|W)$ and the language model $P(W)$ when searching for the best sequence of words $\hat{W}$, where the A is the acoustic signal:

$$\hat{W} = \arg \max_W P(W|A) = \arg \max_W P(W)P(A|W) \quad (1)$$

The probability $P(W)$ of occurrence of W sequence of n words w_i, can be decomposed as [20]:

$$P(W) = \prod_{i=1}^{n} P(w_i|w_1, ..., w_{i-1}) \quad (2)$$

where $P(w_i|w_1, ..., w_{i-1})$ is the conditional probability that w_i will occur, given the previous word sequence $w_1, ..., w_{i-1}$. Unfortunately, it is impossible to compute the conditional probability of words $P(w_i|w_1, ..., w_{i-1})$ for all words and all sequence lengths in a given language. Although the sequences are limited to moderate values of i, there would not be enough data to reliably estimate all of the conditional probabilities. The conditional probability can be approximated by estimating the probability only on the preceding $N - 1$ words defined by the following formula:

$$P_N(W) = \prod_{i=1}^{n} P(w_i|w_{i-N+1}, ..., w_{i-1}) \quad (3)$$

Completely new possibilities for language modelling arise through the use of neural language modelling, in particular the use of LSTM language models. The Figure 1 shows a basic architecture ot the LSTM word-based language model.

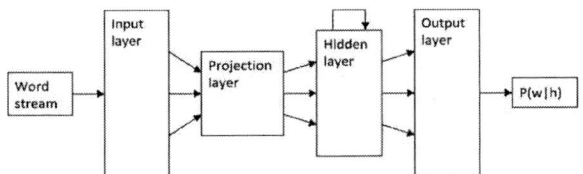

Fig. 1. Architecture ot the word-based LSTM language model

The neural network language model is composed of input, projection, hidden and output layers. Each word in the vocabulary is encoded by 1 to V coding where V is the size of the input vocabulary. In 1 to V coding, each word in the vocabulary is represented by a V dimensional sparse vector where only the index of that word is 1 and the rest of the entries are 0. The input vector is mapped into a linear projection layer. The projection layer is followed by a recurrent hidden layer and the hidden layer is connected to the output layer. Each target at the output layer corresponds to a word in the vocabulary, so that the output layer produces a probability distribution over the predicted word. Formally, given an input vector sequence $X = \{x_1, ..., x_T\}$ and an output vector sequence $Y = \{y_1, ..., y_T\}$, LSTM activations are calculated as follows:

$$h_t = tanh(W_{xh}x_t + W_{hh}h_{t-1} + b_h) \quad (4)$$

$$y_t = W_{hy}h_t + b_y \quad (5)$$

where h_t represents the hidden layer vector, W_{xh} represents the input-to-hidden-layer weight matrix, W_{hh} represents the hidden-to-hidden-layer weight matrix and W_{hy} represents the output-to-hidden-layer weight matrix. The values b_h and b_y represent the hidden and output layer biases, respectively. Note that in Equation 4, x_t represents the projection layer vector of w_t, the word at time t in the input word sequence. In LSTM language modelling, the conditional word probabilities $P(w|h)$ are calculated as follows:

$$p(w_t = i | w_{t-1}, h_{t-2}) = \frac{\exp(y_t^i)}{\sum_{j=1}^{N} \exp(y_t^j)} \quad (6)$$

where y_t^i represents the i-th element of the output vector y_t. The structure of a typical LSTM cell is presented in Figure 2. The hidden layer activations of an LSTM neural network are computed as follows:

$$i_t = \sigma(W_{xi}x_t + W_{hi}h_{t-1} + W_{ci}c_{t-1} + b_i) \quad (7)$$

$$f_t = \sigma(W_{xf}x_t + W_{hf}h_{t-1} + W_{cf}c_{t-1} + b_f) \quad (8)$$

$$c_t = f_tc_{t-1} + i_t \tanh(W_{xc}x_t + W_{hc}h_{t-1} + b_c) \quad (9)$$

$$o_t = \sigma(W_{xo}x_t + W_{ho}h_{t-1} + W_{co}c_{t-1} + b_o) \quad (10)$$

$$h_t = o_t \tanh(c_t) \quad (11)$$

The developed world-based LSTM language model has 16 inputs and 18688 outputs and uses two LSTM hidden layers with 100 memory cells each. More memory cells and a deeper network may achieve better results, but the demand for computing power will be increasing. A dense fully connected

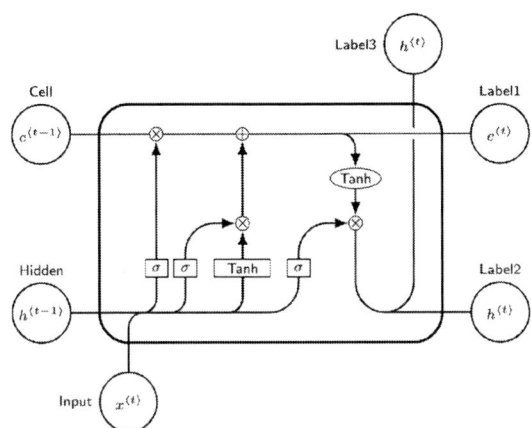

Fig. 2. Structure of a typical LSTM cell

layer with 100 neurons connects to the hidden layers to interpret the features extracted from the sequence. The output layer predicts the next word as a single vector the size of the vocabulary with a probability for each word in the vocabulary. Technical details and block diagram of the developed word-based LSTM language model are presented in Table I and in Fig. 3.

TABLE I
TECHNICAL DETAILS OF THE WORD-BASED LSTM LANGUAGE MODEL

Layer (type)	Outout shape	Number of parameters
embedding_1 (Embedding)	(None, 16, 16)	299 008
lstm (LSTM)	(None, 16, 100)	46 800
lstm_1 (LSTM)	(None, 100)	80 400
dense (Dense)	(None, 100)	10 100
dense_1 (Dense)	(None, 18688)	1 887 488
Total params:		2 323 796
Trainable params:		2 323 796
Non-trainable params:		0

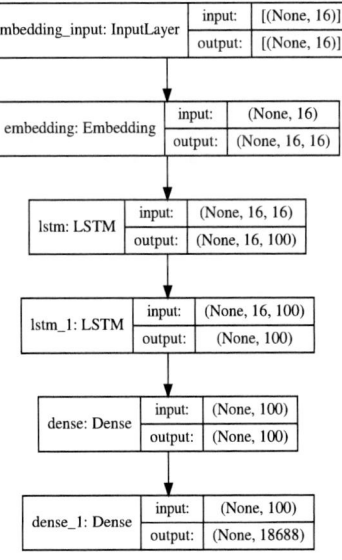

Fig. 3. Block diagram of the developed word-based LSTM language model

In the next step, the model was compiled, specifying the categorical cross-entropy loss needed to fit the model. Technically, the model is learning a multi-class classification, and this is the suitable loss function for this type of problem. Finally, the model is fit on the data for 400 training epochs with a modest batch size of 256 to speed things up. Training may take more than 3 hours on modern hardware without GPU with the computing power of approximately 26 GIPS. The model accuracy values reported during the language model training process are presented in Fig. 4. Additionally model accuracy, loss values, and perplexity as a measure of language model performance during model training are presented in Table II. The training data consist of just 68232 words (total tokens) in a clean text and a vocabulary of just 18688 words (unique tokens). This is smallish, and models fit on these data should be manageable on modest hardware. The training data allows us to build 68216 sequences consisting of 16 words (tokens).

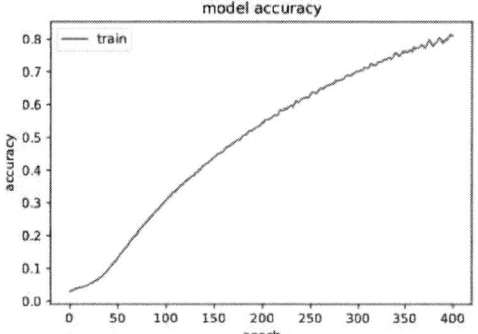

Fig. 4. Word-based LSTM language model accuracy reported during the training process

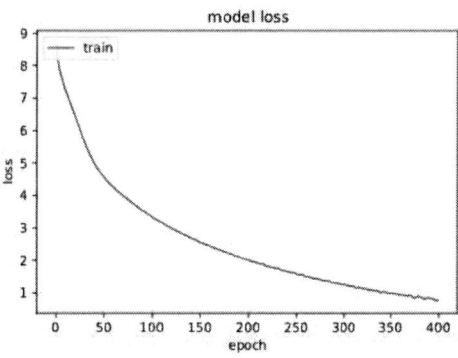

Fig. 5. Model loss reported during the language model training process

TABLE II
DETAILS OF WORD-BASED LANGUAGE MODEL TRAINING PROCESS

Training epochs	Model accuracy	Model loss	Model perplexity	Training time
50	0.129	4.604	24.323	00h25m
100	0.305	3.356	10.238	00h50m
150	0.349	2.566	5.921	01h15m
200	0.545	2.010	4.017	01h40m
250	0.629	1.598	3.027	02h05m
300	0.701	1.251	2.380	02h30m
350	0.762	0.990	1.986	02h55m
400	0.812	0.782	1.720	03h20m

V. DEVELOPED A SUBWORD-BASED LSTM LANGUAGE MODEL

The sub-word-based LSTM language model is constructed similarly. In the character-based LSTM language model, the conditional character probabilities $P(s|h)$, called Softmax, are calculated as follows:

$$p(s_t = i|s_{t-1}, h_{t-2}) = \frac{\exp(y_t^i)}{\sum_{j=1}^{N} \exp(y_t^j)} \quad (12)$$

where y_t^i represents the i-th element of the output vector y_t.

The developed character-based LSTM language model has 8 inputs, 56 outputs, and one LSTM hidden layer with 600 memory cells. More memory cells and a deeper network can achieve better performance, but the computing power requirement will increase. The parameters used in the model were determined experimentally as a compromise between efficiency and the requirements for computing power.

A dense, fully connected layer with 600 neurones interfaces with LSTM hidden layers to interpret features extracted from the sequence. The output layer predicts the next character as a single vector of dictionary-sized probability for each character in the dictionary. The technical details and block diagram of the character-based LSTM language model are shown in Table III and in Fig. 6.

TABLE III
TECHNICAL DETAILS OF THE CHARACTER-BASED LANGUAGE MODEL

Layer (type)	Output shape	Number of parameters
lstm (LSTM)	(None,600)	1 576 800
dense (Dense)	(None, 56)	33 656
Total parameters:		1 610 456
Trainable parameters:		1 610 456
Non-trainable parameters:		0

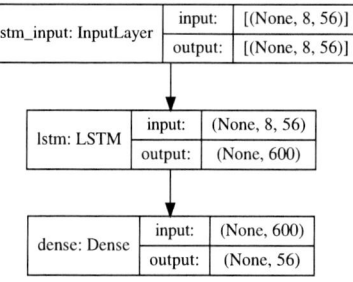

Fig. 6. Block diagram of developed character-based LSTM language model

In the next step, a subword-based language model was compiled by determining the categorical cross-entropy loss needed to fit the model. Finally, the model was fitted to the training data for 80 training epochs. Training took about 12 hours on modern non-GPU hardware with a computational power of about 26 GIPS. The model accuracy recorded during the final language model training process is shown in Fig. 7. In addition, model accuracy, loss values, perplexity as a measure of language model performance, and processing time reported during the model training process are shown in Table IV. The training data consisted of 440143 characters (sum of

196

tokens) in plain text and a dictionary of only 56 characters (unique tokens), suitable for the Polish language. The training data allow the construction of 440135 sequences consisting of 8 characters (tokens). The developed token-based language model can be used to predict token sequences based on the context of the preceding 8 tokens and can work with the developed word-based language model to predict words in speech recognition.

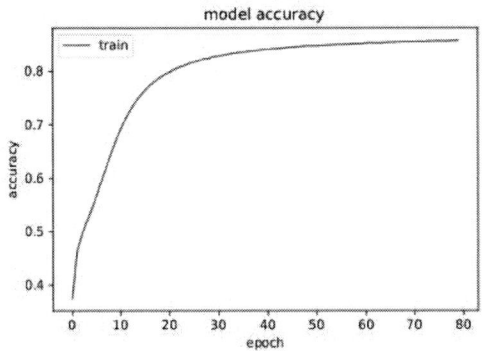

Fig. 7. The accuracy of the character-based language model reported during training process

TABLE IV
CHARACTER-BASED LANGUAGE MODEL TRAINING PROCESS DETAILS

Training epochs	Model accuracy	Model loss	Model perplexity	Training time
10	0.671	1.017	2.023	1h30m
20	0.793	0.623	1.552	3h00m
30	0.827	0.528	1.442	4h30m
40	0.840	0.478	1.393	6h00m
50	0.847	0.447	1.363	7h30m
60	0.851	0.425	1.343	9h00m
70	0.855	0.410	1.328	10h30m
80	0.857	0.395	1.315	12h00m

VI. HYBRID LSTM LANGUAGE MODEL

The simultaneous application of different language modelling methods, such as word-based and subword-based language modelling, allows the development of hybrid language models that have even better properties to improve the speech recognition process [18], [10]. The proposed idea of hybrid LSTM language models provides for the use of both developed LSTM language models: word-based and subword-based at the same time in the word prediction process in speech recognition. The hybrid LSTM language model architecture is presented in Figure 8.

An illustration of how the hybrid architecture of the LSTM language model works over time is presented in Figure 9, where the blocks labelled H represent the hidden layers of the models and the blocks labelled O represent the outputs. The input vectors of words w_1, w_2, w_3 and subwords s_1, s_{21}, s_{22} and s_{31}, s_{32}, s_{33} in the Figure 9 are fed to different recursive hidden layers. The word stream produces output at each time step, while the sub-word stream produces output only at time steps corresponding to the word boundaries. However, the sub-word hidden layer recursively gathers information

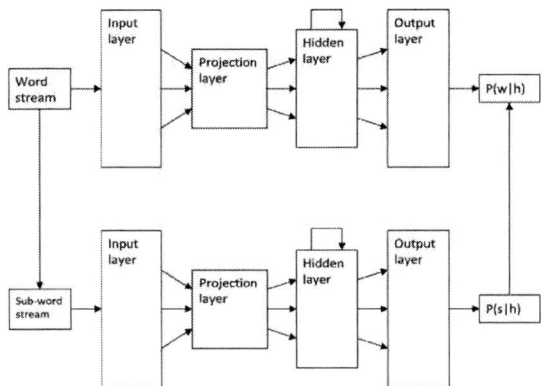

Fig. 8. Hybrid LSTM language model architecture

from previous time steps even without producing output. The proposed hybrid LSTM language model trained on a sequence of words can be interpolated with another model trained on a sequence of sub-words. In joint training, both models are trained using the same information sources together.

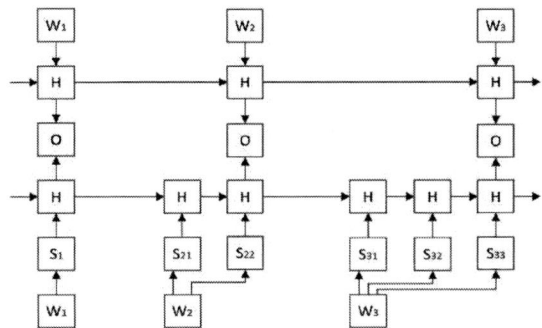

Fig. 9. Illustration of how the hybrid LSTM language model architecture works over time

The developed LSTM language models was compared and precented in Table V.

TABLE V
COMPARISON OF LSTM LANGUAGE MODELS

Training epochs	Model accuracy	Model loss	Model perplexity
word-based	0.812	0.782	1.720
subword-based	0.857	0.395	1.315
hybrid	0.880	0.202	1.113

VII. CONCLUSIONS

The results presented in this article relate to the development of an LSTM language model using word-based and subword-based methods. In addition, it was proposed to use both LSTM language modelling methods simultaneously to develop a hybrid language model, which has even better properties and can significantly improve speech recognition. A comparison of the performance of the developed LSTM language models including the hybrid model is shown in Table V. The accuracy, loss, and perplexity parameters were used as a measure of language model performance. The hybrid model performs quite well in comparison to other models developed. The performance of the hybrid model is clearly better than the word-based or subword-based language model used alone. It should be noted that the values shown in Table V are approximate and underestimated. The calculated parameters were only used to compare the developed language models. The language models were trained on a limited size test corpus to reduce the training time. Computing the real parameters of the language models would require a long learning process using corpora of very large size and sufficient quality to reflect the specifics of the language. The use of relevant language corpora as training data for developed language models provides the opportunity to simultaneously use word-based and subword-based types of models, as a hybrid language model, to predict word more effectively in the speech recognition process [18], [10], [11]. This seems to be a very promising direction for further research, and finding a suitable training corpus for the models will be crucial. Presented in this paper LSTM language models were developed with the use of Python programming language and Anaconda development environment [21].

ACKNOWLEDGEMENTS

This work was supported by Polish Ministry of Science and Higher Education funding for statutory activities.

REFERENCES

[1] Y. Goldberg, *Neural Network Methods for Natural Language Processing*, ser. Synthesis Lectures on Human Language Technologies. San Rafael, CA: Morgan & Claypool, 2017, vol. 37.

[2] C. D. Manning, P. Raghavan, and H. Schütze, *Introduction to Information Retrieval*. Cambridge, UK: Cambridge University Press, 2008. [Online]. Available: http://nlp.stanford.edu/IR-book/information-retrieval-book.html

[3] R. Mitkov, *The Oxford Handbook of Computational Linguistics (Oxford Handbooks in Linguistics S.)*. USA: Oxford University Press, Inc., 2003.

[4] Y. Kim, Y. Jernite, D. A. Sontag, and A. M. Rush, "Character-aware neural language models," *CoRR*, vol. abs/1508.06615, 2015. [Online]. Available: http://arxiv.org/abs/1508.06615

[5] H. Schwenk and J. Gauvain, "Connectionist language modeling for large vocabulary continuous speech recognition," *2002 IEEE International Conference on Acoustics, Speech, and Signal Processing*, vol. 1, pp. I-765–I-768, 2002.

[6] Y. Bengio, R. Ducharme, P. Vincent, and C. Jauvin, "A neural probabilistic language model," *JOURNAL OF MACHINE LEARNING RESEARCH*, vol. 3, pp. 1137–1155, 2003.

[7] T. Mikolov, M. Karafiat, L. Burget, J. H. Cernocky, and S. Khudanpur, "Recurrent neural network based language model," in *11th Annual Conference of the International-Speech-Communication-Association 2010 (INTERSPEECH 2010), VOLS 1-2*, 2010, pp. 1045–1048.

[8] T. Mikolov, S. Kombrink, L. Burget, J. Černocký, and S. Khudanpur, "Extensions of recurrent neural network language model," in *2011 IEEE International Conference on Acoustics, Speech and Signal Processing (ICASSP)*, 2011, pp. 5528–5531.

[9] R. Józefowicz, O. Vinyals, M. Schuster, N. Shazeer, and Y. Wu, "Exploring the limits of language modeling," *CoRR*, vol. abs/1602.02410, 2016. [Online]. Available: http://arxiv.org/abs/1602.02410

[10] P. Kłosowski, "Deep learning for natural language processing and language modelling," in *Proceedings of the 22th IEEE International Conference Signal Processing Algorithms, Architectures, Arrangements, and Applications, September 19-21, 2018, Poznan, Poland*, 2018, pp. 223–228.

[11] P. Kłosowski, "Polish language modelling based on deep learning methods and techniques," in *Proceedings of the 23th IEEE International Conference Signal Processing Algorithms, Architectures, Arrangements, and Applications, September 18-20, 2019, Poznan, Poland*, 2019, pp. 223–228.

[12] M. Auli, M. Galley, C. Quirk, and G. Zweig, "Joint language and translation modeling with recurrent neural networks," in *Microsoft Research*, Seattle, Washington, October 2013. [Online]. Available: https://www.microsoft.com/en-us/research/publication/joint-language-and-translation-modeling-with-recurrent-neural-networks/

[13] S. Hochreiter and J. Schmidhuber, "Long short-term memory," *Neural Computation*, vol. 9, no. 8, p. 1735–1780, 1997.

[14] M. Sundermeyer, H. Ney, and R. Schluter, "Fromfeedforward to recurrent lstm neural networks for language modeling," *IEEE Transactions on Audio, Speech and Language Processing*, vol. 23, no. 3, p. 517–529, 2015.

[15] P. Kłosowski, "Algorithm and implementation of automatic phonemic transcription for Polish," in *Proceedings of 20th IEEE International Conference Signal Processing Algorithms, Architectures, Arrangements, and Applications, September 21-23, 2016, Poznań, Poland*, 2016, pp. 298–303.

[16] P. Kłosowski, "Statistical analysis of Polish language corpus for speech recognition application," in *Proceedings of 20th IEEE International Conference Signal Processing Algorithms, Architectures, Arrangements, and Applications, September 21-23, 2016, Poznań, Poland*, 2016, pp. 304–309.

[17] P. Kłosowski, "Statistical analysis of orthographic and phonemic language corpus for word-based and phoneme-based Polish language modelling," *EURASIP Journal on Audio, Speech, and Music Processing*, vol. 2017, no. 1, p. 5, 2017. [Online]. Available: http://dx.doi.org/10.1186/s13636-017-0102-8

[18] P. Kłosowski, "Polish language modelling for speech recognition application," in *Proceedings of the 21th IEEE International Conference Signal Processing Algorithms, Architectures, Arrangements, and Applications, September 20-22, 2017, Poznan, Poland*, 2017, pp. 313–318.

[19] P. Kłosowski, "A rule-based grapheme-to-phoneme conversion system," *Applied Sciences*, vol. 12, no. 5, 2022. [Online]. Available: https://www.mdpi.com/2076-3417/12/5/2758

[20] F. Jelinek, *Statistical Methods for Speech Recognition*, ser. Language, Speech, & Communication: A Bradford Book. USA: MIT Press, 1997.

[21] Anaconda Inc., "What is anaconda," 2018. [Online]. Available: https://www.anaconda.com/what-is-anaconda/

Embedded Systems

A Survey and Practical Application
of Ethernet-APL, PROFINET Network and HMI

Alexandre B Lugli, Arthur S Aragão, Egídio R Neto,
Guilherme A M Vizotto, João A P Barbosa,
João Pedro M P Paiva
National Institute of Telecommunications – INATEL
Santa Rita do Sapucaí, Brazil
baratella@inatel.br

Tales C Pimenta
Institute of Engineering Systems and Information Technology
Federal University of Itajubá – UNIFEI
Itajubá, Brazil
tales@unifei.edu.br

Abstract—This paper focuses on the scenario of industrial communication protocols, with observance for the transition from fieldbus networks to protocols subsidized to the Ethernet standard. The study emphasizes the Ethernet – Advanced Physical Layer (Ethernet-APL), which enables the use of Ethernet protocols in diverse industrial environments, providing high bandwidth capacity and longer cable lengths, along with the ability to transmit both power and data through the same cabling. Thus, the work aims to develop a practical application in which there is a Process Field Network (PROFINET) network that enables the connection of a temperature sensor connected by Ethernet-APL, communicating with industrial PROFINET elements. Temporal network communication metrics will also be analyzed.

Keywords—Ethernet-APL, HMI, PROFINET.

I. INTRODUCTION

Development of industrial communication protocols has accelerated over the years, leading to increasingly diverse applications. Traditional fieldbus systems are gradually being replaced by protocols based on the Ethernet standard, due to their ability to integrate Information Technology (IT) and Automation Technology (AT) devices, which are increasingly present in automation environments [1, 2].

In this context, the Ethernet – Advanced Physical Layer (Ethernet-APL) physical medium emerged. Ethernet-APL is a physical medium that allows the use of protocols based on the Ethernet standard, such as Process Field Network (PROFINET), Ethernet/Industrial Protocol (Ethernet/IP) and Ethernet Powerlink. It allows electrical connection for severe conditions; two-wire cabling, power and data transmission through the same pair of cables and the possibility of being used in intrinsically safe hazardous areas. Its speed and bandwidth are compatible with the Ethernet standard and the cable can have up to 1,000 meters long [3, 4].

Objective of this paper is to carry out a practical study, by demonstrating an application that integrates the PROFINET network with a visualization and monitoring system using an Ethernet-APL physical layer. Utilizing PROFINET network devices, the study aims to establish the connection and monitoring of a temperature sensor. In order to achieve this, an Ethernet-APL-based medium will be used to configure a PROFINET network, program the Programmable Logic Controller (PLC) and its peripherals, and set up and program a Human-Machine Interface (HMI) to assemble the final practical application. Study will also assess the interoperability of the equipment. Additionally, it seeks to validate the communication within the PROFINET network and over the Ethernet-APL physical layer by analyzing temporal metrics and verifying the accuracy of temperature readings from the sensor via a web server embedded in the temperature transmitter.

This paper is structured into five sections. Section II provides the theoretical foundations, Section III presents the materials and methods, Section IV describes the practical applications and results, and Section V offers the conclusions.

II. THEORETICAL FOUNDATIONS

This chapter presents the definitions and theoretical foundations used in this paper.

A. Programmable Logic Controller

A PLC is defined by IEC as a digital electronic system designed to be used in an industrial environment that has a programmable memory for internal storage of user-oriented instructions. It is capable of implementing specific functions, such as logic, sequencing, timing, arithmetic and to control, through digital or analog inputs and outputs, for various types of machines or processes. The programmable controller and its associated peripherals are designed to be integrated into an industrial control system, to perform their intended functions [5].

B. PROFINET Network

PROFINET is an automation standard established by PROFIBUS International (PI) association for the implementation and integration of solutions based on industrial Ethernet. PROFINET supports the integration of field devices into real time applications, as well as the integration of automation distributed systems based on components [1].

PROFINET has three basic device types [1]:

- Controller: master controller that executes the control program (centralized system).
- Field module: remote field device, which maintains communication with a controller.
- Supervisory system: programmable graphical device that commissions and diagnostics functions in the network.

In PROFINET Input/Output (IO), communication occurs in two main forms: Non-Real-Time (NRT) and Real-Time (RT).

NRT communication is used for tasks such as configuration, diagnostics, and parameterization. It relies on standard Transmission Control Protocol/Internet Protocol (TCP/IP) and operates over the upper Open Systems Interconnection (OSI) layers (layers 4–7), making it suitable for non-time-critical data. Though such NRT streams can take only 100 milliseconds to reach the destination, such communication cannot be defined as real time due to the lack of determinism when using all the TCP/IP layers [6].

In contrast, RT communication is designed for cyclic exchange of IO data between controllers and devices, bypassing TCP/IP and working directly on Ethernet at OSI layer 2 (Data Link layer) to Application at OSI layer 7. This ensures low jitter and determinism, essential for automation tasks, including data exchanges ranging from hard real time in 250 microseconds, for motion control applications, to soft real time in 512 milliseconds, for regular data transmission, as well as alarm information exchange between devices and controllers [6].

C. Ethernet-APL

Ethernet-APL is a technology application in the first layer of the Ethernet protocol, the physical layer. The physical means of information transport are found within the protocol structure, according to ISO/IEC, standard ISO/IEC 7498-1 [2, 3, 7].

D. Human Machine Interface

HMI are systems normally used in automation on the factory floor, which is generally an aggressive environment. They have a robust construction, resistant to direct water, humidity, temperature and dust, according to the required Ingress Protection Index defined by the international standard IEC 60529 [8].

Applications of HMI can range from a simple dishwasher to an aircraft cockpit. In the latter case, HMI are specialized to meet the function for which they are intended. Therefore, the HMI is normally close to the production line, installed at the workstation and translates the signals coming from the PLC into graphic signals that are easier to understand [8].

E. Platinum thermoresistance

This is a type of thermoresistance that measures temperature by correlating its electrical resistance with temperature. Most sensors are made from a thin spiral wire, mounted on a ceramic or glass support. They are fragile and need to be installed in protective sheaths [9].

PT-100 is the sensor widely used in industry due to its great stability and precision. Its curve is standardized according to DIN - IEC 751 - 1965 and has a resistance of 100 Ω at 0 °C. Its works from -200 °C to 650 °C for class A and -200 °C to 850 °C for class B (most used) [9].

III. MATERIALS AND METHODS

This section presents the components used.

A. Siemens PLC – model 1214 DC/DC/DC

Siemens 1214 PLC is a programmable logic controller from the SIMATIC S7-1200 family, used in industrial automation. It has DC digital inputs and outputs that allows the connection of several sensors and actuators. It has a DC power supply,

necessary to power both the PLC and the connected devices, such as sensors and actuators. It offers 14 digital inputs, 10 digital outputs, 2 analog inputs and 1 analog output, and an Ethernet port, with PROFINET IO-Controller function [10].

B. Siemens Switch – model XF204

Siemens SCALANCE XF204 is a compact industrial Ethernet switch, designed for automation networks. It offers up to four Fast Ethernet ports to connect devices, thus ensuring communication in industrial environments. It is used to interconnect controllers, such as PLC, and other network equipment, thus offering support for the PROFINET protocol and network management functions. It uses the TIA PORTAL software to configure the network, program the PLC and manage device connections. It has a 24 Vdc power supply, which is used to power the switch and connected devices [11].

C. PROFINET HMI – model KTP400

KTP400 PROFINET HMI is a human-machine interface, with a four-inch screen, used to monitor and control industrial processes. It connects to the PLC via the PROFINET protocol that allows interaction between the operator and the system. In addition, the KTP400 model offers a touch screen, which facilitates data manipulation and allows continuous visualization of the industrial process [12].

D. Siemens PROFINET IO Device Module – model ET200 SP

SIMATIC ET 200SP is a modular and scalable IO system that includes IP20 protection for industrial automation. It supports up to 64 IO modules and 1440 bytes of memory for input and output data (IO), with speeds of up to 100 Mbps. It can be connected to higher-level PLCs using PROFINET-compatible interface modules [13, 14].

It offers several different types of interfaces, whose specific digital input and output modules can perform process automation. For this, the digital input card was used to receive the digital signal from the measurement. The digital output card was used to output the digital signal to the PLC [15, 16].

E. Ethernet-APL Converter – model Relcon A111

Ethernet-APL converter A111 is an independent protocol converter that can receive the Ethernet standard and convert it to APL. Protocols include PROFINET, Ethernet/IP and Open Platform Communications Unified Architecture (OPC UA). The purpose of the converter is only to physically convert the Ethernet physical medium into APL, without any interaction with the communication itself. The device is powered by its own 15Vdc source [17].

F. Ethernet APL temperature transmitter – model TMT86

iTEMP TMT86 is an adapter for temperature sensors. It supports the Ethernet-APL physical medium and has advanced diagnostic functions, such as corrosion monitoring. It offers network access via web server, alarm indicators and input option for PT-100, PT-1000 or thermocouple sensors [18].

IV. PRACTICAL APPLICATIONS AND RESULTS

The goal was to create a practical industrial automation application using the Ethernet-APL physical medium, integrated with the PROFINET protocol and monitored by an HMI, thus demonstrating the interoperability between the different communication protocols and equipments. The system must read the temperature from a PT-100 sensor, activate a set of light signals and pneumatic actuators, and present the information on the HMI.

PT-100 temperature sensor provides an analog signal to the TMT86 temperature transmitter, which is connected to the Ethernet-APL converter A111. The Ethernet-APL converter, the ET200 SP module and its respective input and output cards, the switch and the IO-Controller perform the PROFINET network communication. A light signal and a pneumatic piston, and the inputs represent the loads by digital switches. The study also aimed to validate the communication performed on the PROFINET network, using temporal metrics and the validation of the temperature reading from the PT-100 sensor, through the web browser. Entire system is connected to a PROFINET network, through the Siemens XF204 switch, and the KTP400 HMI, whose function is to monitor the process, as shown in Fig 1.

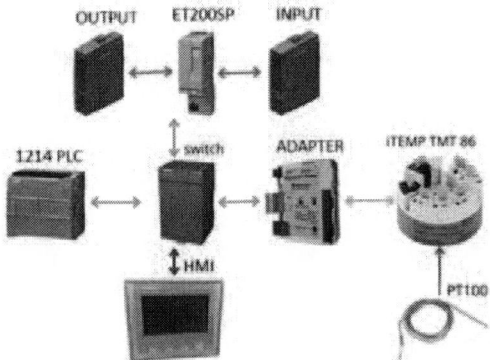

Fig. 1. General block diagram of the project.

A. PROFINET network configuration

PROFINET network was configured using a specific software. Configuration was performed in several steps. First, a new project was created and all devices were added. Then, the connection topology was defined as indicated in Fig 2.

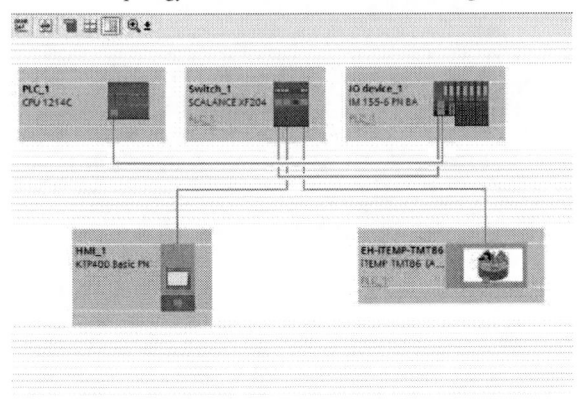

Fig. 2. Configured project topology.

Topology configuration is presented in Table I.

TABLE I.
NETWORK TOPOLOGY CONFIGURATION.

Port 1 – PLC	Port 2 – *IO-Device*
Port 1 – *IO-Device*	Port 1 – *Switch*
Port 2 – *Switch_1*	Port – HMI
Port 3 – *Switch_1*	Communication to the computer
Port 4 – *Switch_1*	EH-iTEMP-TMT86

After that, the IP network address is configured to assign a network address to each device. Then, the IP masks were associated, ensuring that all data was properly directed to the proper device.

B. Application programming

Initially it was developed a display for the HMI that shows the temperature read by the sensor. It offers two buttons, on and off. It also presents two visual signals, operation and event, as given by Fig 3.

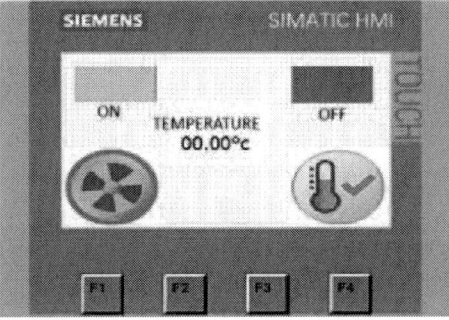

Fig. 3. HMI screen.

Project programming was performed using the Ladder language and it works as follows:

- In order to start the process, press the green button (on) on the HMI screen, to activate the auxiliary memory. If the main switch is off and the measured temperature below 35 °C, the memory responsible for the piston operation is activated. It also resets the emergency light signal output seals the activation button.

- Once activated, the piston works alternately in both directions, at a frequency of 0.5 Hz.

- When the temperature exceeds 30° C, the piston stops in its retracted position and signals a light signal.

- In order to end the process, the system can be turned off using the HMI or an emergency switch. In both cases, the piston returns to its retracted position and the light signal off.

C. Integration and results

Test consists of activating a piston when the temperature read by the PT-100 sensor is higher than 35 °C. In addition, a light signal is activated to alert the operator. After the temperature drops to below 35 ° C, the operator can restart the system.

Operator has full control of the system on the HMI. Besides the control buttons and the temperature display, the HMI screen offers two elements of operating status. One represents the piston operation and the other is a thermometer. When the

203

piston is operating, the indicator changes its color, from gray to green. When the temperature exceeds 35°C, the thermometer changes from green to red, and returns to green when the temperature drops below 35°C. Fig 4 illustrates the logic flowchart of the visual elements.

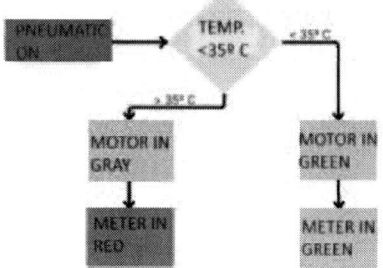

Fig. 4. Logic of the visual signaling elements of the status.

Test was conducted to validate operation of the integrated systems, through the Ethernet-APL physical layer, with temperature sensors and a web browser, as shown in Fig 5. Thus, it was possible to validate the temperature read by the PT-100 sensor, and in a web browser, which in this case was 31.17° C, without the need to configure the server.

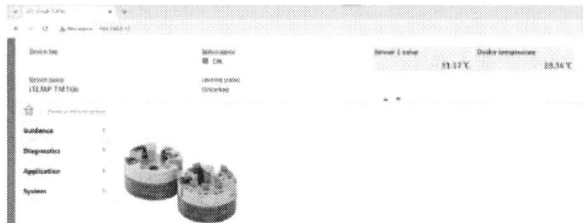

Fig. 5. Web browser.

In order to validate the communication between the components, it was used a software, connected to port three of the switch. Software showed that the maximum communication rate was 27.5 kbps, in a sample of 8 seconds, as shown in Fig 6. This measurement was only intended to validate the communication in the network, without evidencing the temperature reading or the telegrams transmitted.

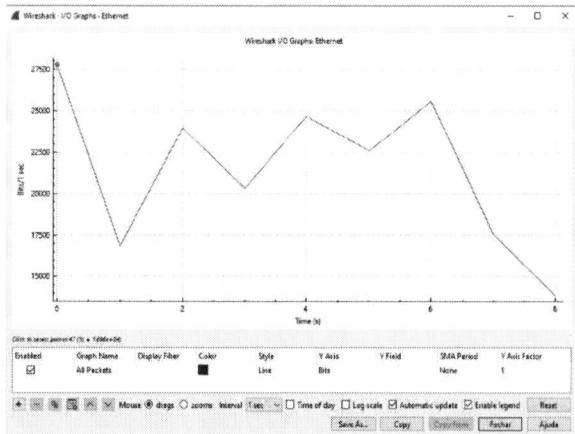

Fig. 6. Software analising.

Fig 7 shows the system components (KTP400; XF204 and ET200 SP) and their respective digital input/output cards. Fig 8 also illustrates the system components (PT-100; TMT86 and A111) and the APL cabling. Finally, Figure 9 shows the emergency light signal and the pneumatic piston system.

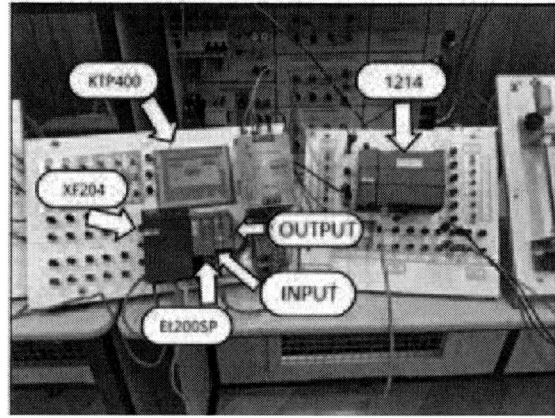

Fig. 7. Fine connections – part 1.

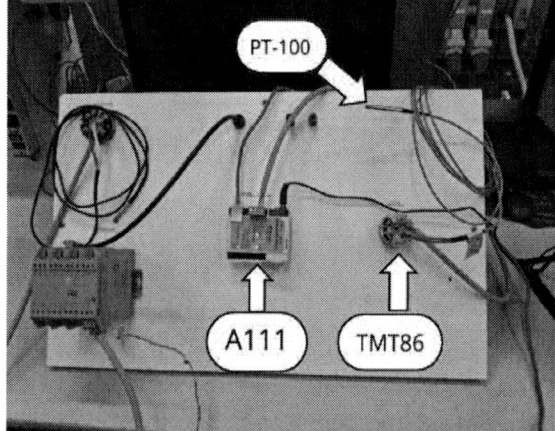

Fig. 8. Fine connections – part 2.

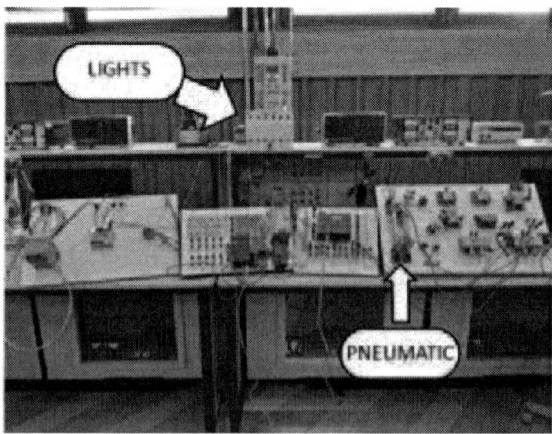

Fig. 9. Image of the entire system structure.

204

V. CONCLUSIONS

This paper addressed the importance and applicability of the PROFINET network, by using the Ethernet-APL physical layer in the industrial environment. The implementation of this network allows the integration and monitoring for hazardous applications.

In order to validate the communication and operation, it was possible to monitor network bandwidth integrating PROFINET and Ethernet-APL application. It was measured 27.5 kbps, representing 0.275% of the total bandwidth applied to Ethernet-APL (10 Mbps).

An implemented test system enables real-time temperature monitoring and integration with an HMI system. It allows fault diagnosis and adaptability to changes in operating conditions, thus opening many implementation possibilities. HMI and web browser can be used to check diagnosis and system communication parameters.

ACKNOWLEDGMENTS

CAPES, CNPq and FAPEMIG supported this paper.

REFERENCES

[1] IEEE Computer Society: IEEE Standard for Ethernet - Amendment 5: Physical Layer Specifications and Management Parameters for 10 Mb/s Operation and Associated Power Delivery over a Single Balanced Pair of Conductors, IEEE 802.3cg-2019, 2019.

[2] Karl-Heinz Niemann, "The Ethernet-APL Engineering Process. A brief look at the Ethernet-APL engineering guideline", atp Magazin, 9, 2021.

[3] Karl-Heinz Niemann, "Ethernet-APL Enginnering Guideline", Version 1.11, 06th dec. 2021.

[4] Ethernet-APL. (2021). Ethernet Advanced Physical Layer. Available in: <https://www.ethernet-apl.org/>. Access in: 25th ago. 2024.

[5] J. Jaffar, S. Michaylov, P. J. Stuckey and R. H. C. Yap, "The CLP(R) language and system: an overview," *COMPCON Spring '91 Digest of Papers*, San Francisco, CA, USA, 1991, pp. 376-381, doi: 10.1109/CMPCON.1991.128837.

[6] Manfred Popp, "Industrial Communication with Profinet", PROFIBUS Nutzerorganisation. Karlsruhe. 2014.

[7] Karl-Heinz Niemann, "Differentiation of the IT security standard series ISO 27000 and IEC 62443. A view of automation systems in the manufacturing and process industries". Whitepaper. 2021.

[8] D. Lukač, "The fourth ICT-based industrial revolution "Industry 4.0" — HMI and the case of CAE/CAD innovation with EPLAN P8," *2015 23rd Telecommunications Forum Telfor (TELFOR)*, Belgrade, Serbia, 2015, pp. 835-838, doi: 10.1109/TELFOR.2015.7377595.

[9] J. A. Prakosa, Purwowibowo and D. Larassati, "Development of Simple Method for Quality Testing of PT100 Sensors Due to Temperature Coefficient of Resistance Measurement," *2021 International Symposium on Electronics and Smart Devices (ISESD)*, Bandung, Indonesia, 2021, pp. 1-5, doi: 10.1109/ISESD53023.2021.9501552.

[10] SIEMENS. Controller SIMATIC S7-1200, CPU 1214C, 6ES7214-1AG40-0XB0. Available in: <https://mall.industry.siemens.com/mall/pt/pt/Catalog/Product/6ES7214-1AG40-0XB0>. Access in: 20th Set. 2024.

[11] SIEMENS. Industrial Switch SCALANCE XF204-2BA00 Available in: <https://mall.industry.siemens.com/mall/pt/PT/Catalog/Product/?mlfb=6GK5204-0BA00-2AF2&SiepCountryCode=PT>. Access in: 21st Set. 2024.

[12] SIEMENS. SIMATIC HMI KTP400 Basic, 6AV2124-2DC01-0AX0. Available in: <https://support.industry.siemens.com/cs/pd/379924?pdti=td&dl=en&lc=en-BR>. Access in: 14th Nov. 2024.

[13] SIEMENS. ET 200SP. Available in: <https://mall.industry.siemens.com/mall/en/WW/Catalog/Products/10170367?tree=CatalogTree>. Access in: 18th Set. 2024.

[14] SIEMENS. 6ES7155-6AR00-0AN0. Available in: <https://mall.industry.siemens.com/mall/pt/pt/Catalog/Product/6ES71556AR000AN0>. Access in: 18th Set. 2024.

[15] SIEMENS. 6ES7131-6BF01-0AA0. Available in: <https://mall.industry.siemens.com/mall/en/ww/Catalog/Product/6ES7131-6BF01-0AA0>. Access in: 18th Set. 2024.

[16] SIEMENS. 6ES7132-6BF01-0AA0. Available in: <https://mall.industry.siemens.com/mall/en/ww/Catalog/Product/6ES7132-6BF01-0AA0>. Access in 18th Set. 2024.

[17] Relcom. ETHERNET-APL Adapter. Available in: <https://www.relcominc.com/ethernet-apl-products>. Access in: 17th set. 2024.

[18] Endress+Hauser.Temperature transmitter, iTEMP TMT86, 2024. Available in: <https://www.endress.com/en/field-instruments-overview/temperature-measurement-thermometers-transmitters/itemp-tmt86?t.tabId=product-overview>. Access in: 17th Set. 2024.

Analysis of Selected Cryptographic Algorithms for Data Transmission in Airborne Networks

Szymon Baliński, Paweł Śniatała,
Maciej Sobieraj, Anna Grocholewska-Czuryło
Poznan University of Technology
Poznan, Poland
szymon.balinski@put.poznan.pl

Junfei Xie, Shangping Ren
San Diego State University
San Diego, USA
{jxie4, sren}@sdsu.edu

Abstract—**The article presents an analysis of selected cryptographic algorithms for their application in data transmission in airborne networks, as well as other devices that use microcontrollers, such as Internet of Things (IoT) devices. Different types of microcontrollers used in Unmanned Aerial Vehicle (UAV) platforms are presented. The ESP32 microcontroller is used for hardware testing. A selected set of lightweight cryptography algorithms is implemented in the microcontroller to test their computational efficiency. The tests for AEAD algorithms include: ChaChaPoly, ASCON-128, TinyJAMBU, ISAP, and PHOTON-Beetle, and for Hashing algorithms: BLAKE2s, ASCON-HASH, and PHOTON-Beetle-HASH.**

Keywords—**Lightweight Cryptography, Microcontrollers, UAV**

I. INTRODUCTION

This paper presents an analysis of a selected set of lightweight cryptographic algorithms in terms of their applications in data transmission in airborne networks. We analyze different types of microcontrollers as possible hardware platforms to apply the chosen algorithms. However, the conclusions obtained from the presented research are not only applicable to airborne networks, as illustrated in Fig. 1 but can also be a guide for the implementation of data encryption in other devices that use microcontrollers, such as Internet of Thing (IoT) devices.

The increasing adoption of IoT devices in recent years has enhanced the convenience of our daily lives, yet has also brought significant security and privacy challenges [1], [2], [3]. IoT devices are typically compact printed circuit boards embedded with a variety of sensors, which enable them

to collect and process data from their surroundings. These devices are designed to communicate wirelessly, either with larger systems, such as centralized servers or gateways, or directly with other IoT devices, using technologies such as Wi-Fi, Bluetooth, Zigbee, or other low-power communication protocols.

Functionally, IoT devices are capable of sensing environmental parameters, performing computations, storing relevant data, and transmitting information across networks. Their applications span multiple domains, including industrial automation, smart cities, agriculture, healthcare, home automation, and surveillance. In specific use cases, such as monitoring patient health, securing residential or commercial spaces, and automating home environments, the data generated by these devices can be highly sensitive. Consequently, ensuring the confidentiality, integrity, and security of this data is crucial. To address these concerns, robust data protection mechanisms, including encryption, access control, authentication protocols, and secure data transmission techniques, must be implemented to safeguard user privacy and prevent unauthorized access or cyber threats. In fact, almost the same can characterize UAVs, which often are just mobile platforms that carry sensors or IoT devices [4].

Cryptographic performance is critical in various applications, including secure communication, data protection, and authentication. By evaluating encryption and decryption speeds for both large (128 bytes) and small (16 bytes) packets, we can determine which algorithms are best suited for high-performance environments and constrained devices. This paper presents an analysis of the performance of various cryptographic algorithms, focusing on their encryption, decryption, and hashing speeds. The goal is to compare different algorithms in terms of their efficiency when handling different data sizes. The study includes Authenticated Encryption with Associated Data (AEAD) algorithms and cryptographic hash functions. We have focused on lightweight cryptography, which is suitable for implementation on microcontrollers.

The rest of the article is organized as follows. An Introduction introduces UAVs as platforms that carry different sensors and/or IoT devices. Section II presents different microcontrollers that are commonly found in UAVs. Next, a performance analysis of cryptographic algorithms implemented in

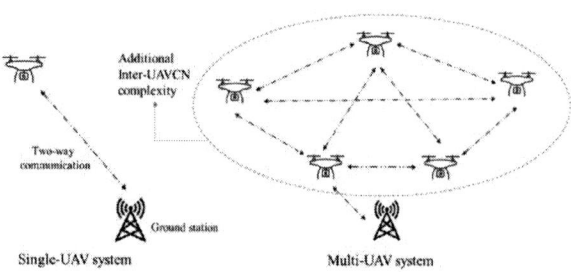

Fig. 1. Example UAV connections

the ESP32 microcontroller is presented in Section III. It includes introduction of lightweight cryptography, methodology, testing environment, and results. The last section summarizes the results achieved and briefly outlines the further planned work.

II. MICROCONTROLLERS USED IN UAVS

Unmanned Aerial Vehicles (UAVs), commonly known as drones, have revolutionized numerous industries due to their versatility and accessibility. At the heart of every UAV is a sophisticated integration of hardware components and microcontroller platforms that enable flight control, navigation, communication, and mission execution.

One of the most important elements is the communication system, which represents a critical component in multi-UAV operations, enabling coordination, data sharing, and collective intelligence among UAVs. Microcontrollers serve as the foundational building blocks for implementing these communication networks. balancing processing capabilities with power efficiency to enable effective and secure inter-UAV communication.

In the case of effective inter-UAV communication, systems require microcontrollers that balance processing power, energy efficiency, and peripheral capabilities. The following microcontroller families are commonly deployed in the implementation of various types of systems used in UAVs, each offering unique advantages for different operational requirements.

- PIC Microcontrollers (PIC18F Series). Offer a good balance between performance and power efficiency. They are often used in small UAVs which have strict power constraints. Their key advantages include: extremely low-power sleep, integrated peripherals for common communication protocols, and suitable for simple mesh networking implementations [5], [6].
- ESP32 platform (ESP32-S3 or ESP32-C3). The advantages of using that systems include: integrated RF capabilities that eliminate the need for external transceivers, support for mesh networking protocols like ESP-MESH, and up to 240 MHz clock speeds with extensive sleep modes. ESP32 is the dedicated hardware for AES encryption and SHA hashing [7], [8].
- MSP430 Family (MSP430FR Series). Their key functionalities include: ultra-low power with FRAM technology, extremely efficient sleep modes with fast and wake-up, ideal for intermittent communication scenarios, clock speeds ranging from sub-MHz to 24 MHz. These systems are popular for energy-harvesting UAV applications [9], [10].
- STM32 Family (STM32F4, STM32L4, STM32H7 series). The main features of these microcontrollers include: an extensive peripheral set supporting multiple UART, SPI, I2C, and CAN interfaces; hardware cryptographic accelerators for secure communications; DMA controllers for efficient data handling; and clock speeds ranging from 32 to 480 MHz [11], [12].

The significance of communication security has led to extensive research on scalable and modular encryption methods. For many IoT security applications, the use of lightweight and specialized cryptographic techniques is highly recommended due to their efficiency and adaptability. In cryptography, microcontrollers equipped with hardware cryptographic gas pedals or security modules are often used to encrypt data. The most popular microcontrollers used for this purpose include:

- STM32 (STMicroelectronics),
- ESP32 (Espressif Systems),
- NXP i.MX RT and LPC,
- Nordic Semiconductor nRF52, nRF53,
- Microchip PIC32 and ATSAM,
- Texas Instruments MSP430 and TM4C.

In [13], an investigation was conducted to verify the possibility of applying security measures such as strong encryption without hindering the performance of IoT devices. A comprehensive experimental performance evaluation was performed that examined the encryption of DTLS-based network traffic in STM32 Nucleo.

The analysis of the performance of AES, with and without hardware acceleration, and XTEA algorithms, comparing their memory usage, power consumption, and execution times, to determine whether XTEA is viable for resource-constrained embedded platforms, was presented in [14].

The work [15] evaluated the efficiency of ten lightweight cryptography (LWC) algorithms by comparing their power consumption, performance, and memory requirements in ARM Cortex architectures, providing practical guidance to designers implementing LWC solutions on ARM processors.

The research by [16] investigated the security of IoT data, specifically developing cryptographic algorithms optimized for the resource-limited ESP32 microcontroller. The project created a lightweight block cipher drawing inspiration from established algorithms such as AES and DES, with performance evaluations conducted to achieve an optimal balance between security strength and computational efficiency.

Taking into account the analysis of the literature on the subject, as well as the requirements for computing power and energy savings through particular microcontrollers, the authors decided to use the ESP32 microcontroller as a hardware platform, which can be used in the UAV to encrypt data transmission. This device was used in practical tests described in the next section.

III. PERFORMANCE ANALYSIS OF CRYPTOGRAPHIC ALGORITHMS ON ESP32

A. Lightweight cryptography applications

A lightweight cryptography refers to a cryptosystem with low computational cost and suitable for devices with limited resources. The concept was initiated by the National Institute of Standards and Technology (NIST) to develop a cryptographic algorithm that can work with small electronic devices in the IoT environment. Different criteria can be taken into account when choosing the appropriate algorithm for a particular application. Key criteria include:

- Security: cryptographic resistance, key and block length, analysis and verification
- Performance: capacity, delay
- Resource consumption: memory, energy
- Implementation complexity: ease of implementation, potential errors
- Resistance to side-channel attacks: physical attacks, protection measures
- Licensing and intellectual property: licensing, open source
- Compliance with standards: international standards, interoperability
- Scalability and flexibility: adaptability, support for different platforms
- Implementation experience: case studies, community and support.

B. Methodology

This evaluation focuses on benchmarking lightweight cryptography algorithms under controlled conditions. As a result of our experiments, we wanted to compare different cryptography algorithms which would be suitable to implement on a UAV platform. The study includes a range of cryptographic algorithms, divided into two categories:

- AEAD algorithms: These algorithms provide confidentiality and authenticity in encryption. The tested AEAD algorithms include ChaChaPoly, ASCON-128, TinyJAMBU, ISAP, and PHOTON-Beetle.
- Hashing algorithms: These ensure data integrity and are widely used in digital signatures and authentication mechanisms. The tested hashing algorithms include BLAKE2s, ASCON-HASH, and PHOTON-Beetle-HASH.

The evaluation metrics used to measure their performance are the encryption, decryption, and hashing times in microseconds per byte. Each algorithm is tested for two data sizes: 128 bytes (larger packets) and 16 bytes (smaller packets) to assess performance variations with different input lengths. The execution time is converted into throughput (bytes per second) to facilitate direct comparison.

C. Testing environment

The tests were carried out in a microcontroller-based environment, simulating the constraints found in embedded and IoT systems. All benchmarks were performed under the same conditions to ensure consistency in the results. Masked versions of some AEAD algorithms were also tested to compare the impact of security enhancements on performance. By following this methodology, we ensure that the results reported accurately reflect the efficiency of each cryptographic algorithm in different operational contexts. The tests were carried out on the ESP32 microcontroller, a widely used low-power system-on-chip (SoC) designed for embedded and IoT applications. The ESP32 was chosen because of its balance between performance and energy efficiency, which makes it a suitable platform for cryptographic operations in constrained

environments. The specifications of this particular microcontroller are summarized as follows:

- Processor: Dual-core Xtensa LX6 @ 240 MHz
- Memory: 520 KB SRAM
- Flash Storage: Up to 16 MB (varies by model)
- Crypto Acceleration: Hardware support for AES, SHA, and RSA
- Connectivity: Wi-Fi, Bluetooth Low Energy (BLE)

ESP32 provides built-in cryptographic acceleration, which improves the performance of algorithms such as AES and SHA. However, software-based implementations of other cryptographic schemes may exhibit different performance characteristics due to CPU limitations. The benchmarks were executed using the standard ESP-IDF framework with optimized compiler settings. The low-power nature of the microcontroller makes it suitable for real-world IoT applications where cryptographic efficiency is crucial. The results obtained from the ESP32 platform provide a realistic assessment of how cryptographic algorithms perform in embedded systems. Understanding these results allows for better selection of cryptographic methods in ESP32-based devices, optimizing security while maintaining system efficiency.

D. Results

Table I presents the performance results of the cryptographic algorithms evaluated. The metrics include encryption and decryption times in microseconds per byte and throughput in bytes per second. The results highlight significant

TABLE I
THE PERFORMANCE RESULTS OF THE EVALUATED ALGORITHMS.

Algorithm	Operation	Time/Byte (μs)	Throughput (bytes/sec)
ChaChaPoly	Encrypt 128B	0.53	1,904,450.16
ChaChaPoly	Decrypt 128B	0.63	1,583,570.46
ChaChaPoly	Encrypt 16B	1.76	569,320.73
ChaChaPoly	Decrypt 16B	1.94	516,768.05
ASCON-128	Encrypt 128B	0.83	1,212,040.87
ASCON-128	Decrypt 128B	0.89	1,120,575.70
ASCON-128	Encrypt 16B	2.33	429,799.43
ASCON-128	Decrypt 16B	2.26	442,988.33
TinyJAMBU-128	Encrypt 128B	0.89	1,125,106.58
TinyJAMBU-128	Decrypt 128B	0.99	1,010,332.23
TinyJAMBU-128	Encrypt 16B	1.55	644,641.42
TinyJAMBU-128	Decrypt 16B	1.70	588,372.29
PHOTON-Beetle-128	Encrypt 128B	9.32	107,284.45
PHOTON-Beetle-128	Decrypt 128B	9.42	106,211.73
PHOTON-Beetle-128	Encrypt 16B	16.56	60,375.84
PHOTON-Beetle-128	Decrypt 16B	16.73	59,781.80
BLAKE2s	Hash 1024B	0.21	4,775,473.47
BLAKE2s	Hash 128B	0.21	4,679,260.46
BLAKE2s	Hash 16B	0.87	1,149,817.65
ASCON-HASH	Hash 1024B	1.07	935,734.08
ASCON-HASH	Hash 128B	1.33	750,267.80
ASCON-HASH	Hash 16B	3.45	290,127.07
PHOTON-Beetle-HASH	Hash 1024B	32.34	30,918.36
PHOTON-Beetle-HASH	Hash 128B	30.57	32,717.09
PHOTON-Beetle-HASH	Hash 16B	16.33	61,230.43

differences in performance between different cryptographic schemes. Some key observations include the following:

- ChaChaPoly demonstrated the highest throughput for both encryption and decryption when processing 128-byte messages, reaching approximately 1.9 MB/s and 1.58 MB/s, respectively. However, its performance significantly deteriorated with smaller inputs (16 bytes), where throughput dropped below 600 KB/s. This indicates a sensitivity to input size that may impact real-time applications handling short messages.

- ASCON-128 offered balanced performance and relatively consistent throughput in different input sizes. For 128B input, it achieved around 1.2 MB/s in encryption and 1.1 MB/s in decryption. For 16B blocks, throughput dropped to about 430–440 KB/s. Although not as fast as ChaChaPoly, its stability and lightweight design make it suitable for embedded environments.

- TinyJAMBU-128 performed similarly to ASCON-128 for larger inputs, but was notably more efficient with smaller data. For 16-byte messages, it outperformed ASCON, reaching throughput values of approximately 645 KB/s (encryption) and 588 KB/s (decryption). These results highlight its effectiveness in constrained scenarios with frequent short data transmissions.

- PHOTON-Beetle-128 exhibited the lowest performance among all encryption schemes evaluated. Even at 128 bytes, its throughput did not exceed 110 KB/s, and at 16 bytes, it dropped to just under 60 KB/s. Despite its lightweight profile, the limited throughput may restrict its applicability to highly specialized use cases where computational performance is secondary.

- BLAKE2s clearly outperformed all other hash functions, maintaining throughput above 4.7 MB/s for 128B and 1024B inputs, and still exceeding 1.1 MB/s for 16B messages. This level of performance, combined with its cryptographic strength, makes it a strong candidate for general-purpose hashing even in lightweight applications.

- ASCON-HASH showed moderate results, achieving throughput between 935 KB/s (1024B) and 290 KB/s (16B). Its performance degrades with smaller inputs, but remains acceptable for lightweight cryptographic needs.

- PHOTON-Beetle-HASH was the slowest among all hash functions. With throughput ranging from 30 KB/s for 1024B to 61 KB/s for 16B inputs, it is clearly not designed for throughput-sensitive applications and is more appropriate in extremely resource-limited devices where size and power consumption take precedence over speed.

Performance decreases significantly for smaller data packets in all algorithms.

We have also compared masked and unmasked versions of AEAD algorithms and assessed their suitability for different security applications. Fig. 2 presents the performance of AEAD algorithms in encryption and decryption scenarios. The comparison is based on the execution time and throughput for large and small packets.

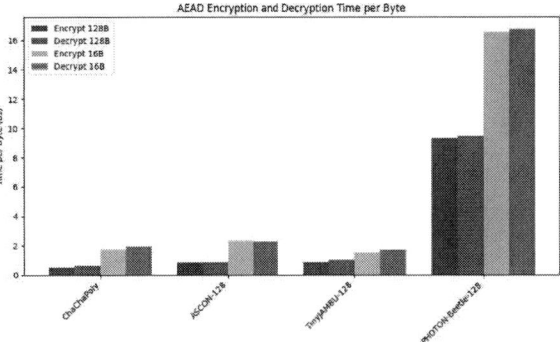

Fig. 2. AEAD algorithms in encryption and decryption scenarios

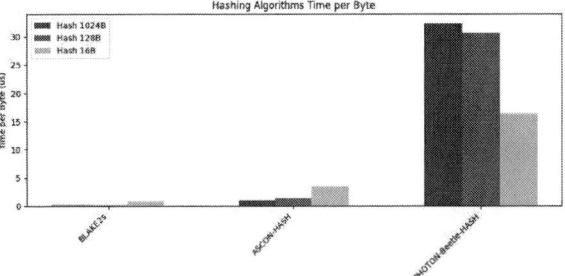

Fig. 3. Hash algorithms in encryption and decryption scenarios

Performance analysis of hashing algorithms is presented in Fig. 3 In general, the data reveal a consistent trend: The smaller the message, the higher the overhead relative to useful computation, resulting in lower throughput. This effect is visible across all categories, particularly in encryption and hashing algorithms that are not optimized for short inputs. Although some algorithms, such as ChaChaPoly and BLAKE2s, remain competitive even under these conditions, others, especially the PHOTON-Beetle family, demonstrate significant performance limitations. These insights are critical for system designers, especially in embedded or real-time systems, where both efficiency and speed must be carefully balanced against hardware constraints and security requirements.

IV. CONCLUSION

This paper has demonstrated key performance differences between lightweight cryptographic algorithms. The comparative analysis highlights ChaChaPoly and BLAKE2s as the most performant algorithms in encryption and hashing, respectively. Their high throughput and low per-byte latency make them excellent choices for applications requiring both speed and reliability. ASCON-128 and TinyJAMBU-128, while not as fast, demonstrate sufficient efficiency and are better suited for systems with severe resource constraints. In contrast, the PHOTON-Beetle family, though optimized for lightweight implementation, offers limited performance and may only be appropriate in scenarios where minimal code size or energy consumption is more critical than speed. These findings underscore the importance of context-specific algorithm selection. While high-performance primitives like ChaChaPoly and BLAKE2s offer impressive speed, lightweight alternatives like ASCON and TinyJAMBU remain essential for ultraconstrained platforms. Future research will explore optimizations to balance security and efficiency, ensuring that cryptographic solutions meet the needs of diverse applications.

ACKNOWLEDGEMENTS

This work was supported by Grant NAWA/NSF: Impress-U, ID BPN/NSF/2023/1/00005 and NSF CAREER-2048266, "Towards Networked Airborne Computing in Uncertain Airspace: A Control and Networking Facilitated Distributed Computing Framework".

REFERENCES

[1] P. S. Bangare and K. P. Patil, "Security issues and challenges in internet of things (iot) system," in *2022 2nd International Conference on Advance Computing and Innovative Technologies in Engineering (ICACITE)*, 2022, pp. 91–94.

[2] N. A. Khan, A. Awang, and S. A. A. Karim, "Security in internet of things: A review," *IEEE Access*, vol. 10, pp. 104 649–104 670, 2022.

[3] P. Śniatała, S. Iyengar, and S. K. Ramani, *Evolution of Smart Sensing Ecosystems with Tamper Evident Security*. Springer International Publishing.

[4] P. Śniatała, S. S. Iyengar, A. Bendarma, and M. Klosak, *Modern Technologies Enabling Safe and Secure UAV Operation in Urban Airspace*. IOS Press.

[5] G. Akshatha, A. Aadil, B. Baghyasri, V. V. Kubal, and K. P. Sharmila, "Microcontroller based engine control unit for uav application," in *2022 4th International Conference on Inventive Research in Computing Applications (ICIRCA)*, 2022, pp. 185–188.

[6] R. G. Sangeetha, Y. Srivastava, C. Hemanth, H. Sankar Naicker, A. P. Kumar, and S. Vidhyadharan, "Unmanned aerial surveillance and tracking system in forest areas for poachers and wildlife," *IEEE Access*, vol. 12, pp. 187 572–187 586, 2024.

[7] A. Bernier-Vega, K. Barton, I. Olson, J. Rodriguez, G. Cantu, and S. Ozcelik, "Remote data acquisition using uavs and custom sensor node technology," *Drones*, vol. 7, no. 6, 2023. [Online]. Available: https://www.mdpi.com/2504-446X/7/6/340

[8] R. Samanta, B. Saha, and S. K. Ghosh, "A low-power low-cost system for disaster locations detection using esp32 cam and tinyml," in *2025 17th International Conference on COMmunication Systems and NETworks (COMSNETS)*, 2025, pp. 907–910.

[9] S. Ghosh, K. Ghosh, S. Karamakar, S. Prasad, N. Debabhuti, P. Sharma, B. Tudu, N. Bhattacharyya, and R. Bandyopadhyay, "Development of an iot based robust architecture for environmental monitoring using uav," in *2019 IEEE 16th India Council International Conference (INDICON)*, 2019, pp. 1–4.

[10] R. Shenoy and R. Manjunatha, "Design of unmanned aerial vehicle for stability," in *2023 International Conference on Smart Systems for applications in Electrical Sciences (ICSSES)*, 2023, pp. 1–6.

[11] Z. Ren, Z. Tang, and R. Wang, "Research on key technologies of four-rotor uav flight control system based on stm32 microcontroller," in *2023 IEEE 6th International Conference on Information Systems and Computer Aided Education (ICISCAE)*, 2023, pp. 1106–1112.

[12] L. Zhang, B. Hu, S. Wang, Y. Huang, X. Zhou, and R. Liu, "Design of a quadrotor uav controller based on body sensing control," in *2024 7th International Symposium on Autonomous Systems (ISAS)*, 2024, pp. 1–6.

[13] K. Rzepka, P. Szary, K. Cabaj, and W. Mazurczyk, "Performance evaluation of raspberry pi 4 and stm32 nucleo boards for security-related operations in iot environments," *Computer Networks*, vol. 242, p. 110252, 2024. [Online]. Available: https://www.sciencedirect.com/science/article/pii/S1389128624000847

[14] S. Maitra, D. Richards, A. Abdelgawad, and K. Yelamarthi, "Performance evaluation of iot encryption algorithms: Memory, timing, and energy," in *2019 IEEE Sensors Applications Symposium (SAS)*, 2019, pp. 1–6.

[15] N. Moura, J. Lucena, E. Pereira, N. Calazans, L. Ost, F. Moraes, and R. Garibotti, "Assessment of lightweight cryptography algorithms on arm cortex-m processors," in *2023 36th SBC/SBMicro/IEEE/ACM Symposium on Integrated Circuits and Systems Design (SBCCI)*, 2023, pp. 1–6.

[16] L. C. Ni, S. Ali, A. N. A. A. Aziz, and R. A. Rashid, "Implementation of proposed cryptography algorithm on esp32-based iot system," in *2024 IEEE International Conference on Advanced Telecommunication and Networking Technologies (ATNT)*, vol. 1, 2024, pp. 1–4.

Mixed Design of Integrated Circuits and Systems – MIXDES 2025

Comparative Survey Between Industrial Communication Protocols Applied in Hazardous Areas

Alexandre B Lugli, André C Teixeira,
João Paulo C Henriques, João Pedro M P Paiva,
Júlio A Azevedo
National Institute of Telecommunications – INATEL
Santa Rita do Sapucaí, Brazil
baratella@inatel.br

Tales C Pimenta
Institute of Engineering Systems and Information Technology
Federal University of Itajubá – UNIFEI
Itajubá, Brazil
tales@unifei.edu.br

Abstract—**Industrial communication in hazardous areas requires protocols that guarantee safety, reliability and efficiency. The aim of this study is therefore to compare the characteristics of Process Fieldbus for Process Automation (PROFIBUS PA), Fieldbus Foundation and Ethernet – Advanced Physical Layer (Ethernet-APL) applied in hazardous areas, considering not only their technical differences, but also their ability to provide intrinsically safe communication inside of an industrial environment.**

Keywords—**Ethernet APL, Fieldbus Foundation, Hazardous area, PROFIBUS PA.**

I. INTRODUCTION

With the advancement of technology and the integration of industrial systems, communication between devices has been playing an increasingly fundamental role in the efficiency, monitoring and control of processes in manufacturing environments. Consequently, choosing a network protocol is a crucial decision since it requires a careful analysis of each one. The analysis must take into account the nature and features of the industry, the number and type of equipment, the distance between them and the type of industrial environment [1].

However, there are significant challenges in these applications, such as the presence of explosive atmospheres and flammable materials. These areas, known as Hazardous Areas, are categorized based on the duration that flammable substances, vapors, gases, or dust are present, as well as the potential for these substances to escape or leak, which raises concerns about the safety and operation of these environments [1, 2].

In order to mitigate these risks, industrial plants must be managed to minimize the likelihood of accidents. In this context, understanding the relationship between network protocols and the safety requirements of hazardous areas is crucial for ensuring the safe operation of industrial systems [1].

For improved integration between devices and to meet the safety measures required in Hazardous Areas, several industrial network protocols can be employed. In this paper, three protocols will be highlighted: Process Fieldbus for Process Automation (PROFIBUS PA), Fieldbus Foundation, and Ethernet – Advanced Physical Layer (Ethernet-APL).

PROFIBUS PA is an open protocol adopted by various manufacturers and is used in process automation applications [3]. Fieldbus Foundation is a local area network standard that comprises two distinct networks: H1, which is used for interconnecting instruments, and High-Speed Ethernet (HSE), which is used for integrating other networks and connecting to Programmable Logic Controllers (PLC) [2]. Ethernet-APL employs the most widely accepted wired network protocol among devices operating in the upper layers of the Open Systems Interconnection (OSI) model, thereby facilitating installation, maintenance and diagnostics [1, 3].

This paper aims to present a theoretical study of those protocols applied to Hazardous Areas and to perform a comparative analysis, highlighting their main characteristics, applications, functionalities, and limitations within those environments.

This paper is structured into five sections. Section II provides the concepts and definitions, Section III presents a comparative study between the protocols, and Section V offers the conclusions.

II. CONCEPTS AND DEFINITIONS

A. Hazardous Areas

A Hazardous Areas area is defined as a location where an atmosphere containing flammable gases, vapors, dust, or fibers is present—or where the likelihood of such an atmosphere forming is high enough to require special precautions during the construction, installation, operation, and maintenance of electrical equipment [4].

For an explosion to occur, three elements are required: fuel, oxidizer (typically oxygen), and an ignition source such as heat or a spark [4].

An ignition source may be produced by electrical equipment—for example, through the generation of sparks, the formation of an electric arc, corona effect or elevated temperatures in electrical components. The corona effect occurs in high-voltage circuits when the rupture of a space between two conductors is partial. In such cases, the voltage increases — not enough to produce an arc or spark — but sufficiently to ionize the surrounding gas, thereby establishing a small direct current known as corona current [4].

Classification of an area is divided into two attributes: group and zone [4].

The type of equipment present and the nature of the flammable mixtures determine group classification. According to the standard IEC 60079-10-01, Group I corresponds to equipment designed for use in underground mines, while Group II refers to equipment used on the surface. Group II is further subdivided based on the representative flammable gas present [4].

- IIA for substances whose representative gas is propane,
- IIB for those with ethylene,
- IIC for those with acetylene.

Group III is associated with environments containing flammable dust or fibers and is similarly subdivided into [4]:

- IIIA for combustible fibers,
- IIIB for non-conductive dust,
- IIIC for conductive dust.

Zones are assigned based on the duration of exposure to a flammable atmosphere and the probability of a flammable mixture occurring. When flammable gases are present, areas are classified as follows [4]:

- Zone 0: Locations where an explosive mixture is continuously present,
- Zone 1: Areas where such a mixture is likely to occur under normal operating conditions,
- Zone 2: Areas where the occurrence of an explosive mixture is unlikely.

Flammable mixtures have a characteristic known as the Minimum Ignition Energy (MIE). If the energy released in a given situation is below the MIE, an explosion cannot occur. Typically, the MIE ranges between 1 mJ and 1000 mJ; the lower the MIE, the higher the risk of explosion. This parameter is critical when specifying intrinsically safe circuits for operation in Zone 0 [4, 5].

Intrinsically safe equipments are categorized into three protection levels: ia, ib, and ic. Level ia comprises equipment that is incapable of generating an ignition in an explosive atmosphere under normal operating conditions, even in the presence of up to two countable faults. Equipment at this level can be used in Zones 0, 1, and 2 as well as in Zones 20, 21, and 22. Protection level ib refers to equipment that does not generate an ignition in an explosive atmosphere under normal conditions, even when one countable fault is present; it can be used in Zones 1, 2, 21, and 22. Protection level ic corresponds to equipment that is incapable of generating an explosion under normal operating conditions and is intended for use in Zones 2 and 22 [4].

B. PROFBUS PA

PROFIBUS PA is a protocol profile within the PROFIBUS family, designed to meet the specific demands of automation and process control while offering intrinsically safe transmission for use in hazardous areas [6, 7].

Its protocol architecture is based on the OSI reference model as defined by the international standard ISO 7498. OSI model comprises seven layers; however, PROFIBUS PA utilizes only three levels: the physical layer (layer 1), the data link layer (layer 2), and the application layer (layer 7) [6].

At the physical layer, PROFIBUS PA employs Manchester Encoded Bus Powered (MBP) technology, which enables intrinsically safe communication in explosive environments by using a single pair of twisted wires for both data transmission and power supply [6].

Manchester encoding is a synchronous transmission technique with a fixed rate of 31.25 kbps, as given by standard by IEC 61158-2. In this method, binary data is not transmitted directly as a series of logical "0"s and "1"s; instead, it is converted by MBP into a different format. Specifically, a logical "0" is represented by a transition from "1" to "0" at the midpoint of the bit period, while a logical "1" is indicated by a transition from "0" to "1" [6].

Manchester-coded signals contain frequent level transitions, which enable a receiver to extract the clock signal using a Digital Phase Locked Loop (DPLL) and accurately decode each bit. The high density of these transitions ensures reliable DPLL operation and proper clock extraction [6, 7].

According to the IEC 61158-2 standard, the physical medium consists of a twisted pair of wires. Although the standard does not provide detailed cable technical specifications, it is recommended to use Type A cable because of its enhanced reliability in data transmission [6].

PROFIBUS PA supports star, linear, and tree topologies. Unlike the star configuration, linear and tree topologies include spurs and a trunk; therefore, the distances between branches must be limited to ensure effective communication. Table I indicates the maximum allowable branch distances based on the number of devices connected to each branch [6].

TABLE I.
MAXIMUM PERMITTED DISTANCE BASED ON DERIVATIONS [6].

Number of derivations	1 Equip.	2 Equip.	3 Equip.	4 Equip.
25-32	1 m	1 m	1 m	1 m
19-24	30 m	1 m	1 m	1 m
15-18	60 m	30 m	1 m	1 m
13-14	90 m	60 m	30 m	1 m
1-12	120 m	90 m	60 m	30 m

Total cable length for a PROFIBUS PA segment is measured from the Decentralized Peripherals / Process Automation (DP/PA) coupler output to the furthest point in the segment, including any branches. In non-safety areas, this length can reach up to 1900 meters, whereas in intrinsically safe areas it is limited to 100 meters, with individual branch lengths restricted to 30 meters [6].

C. Fieldbus Foundation

Fieldbus Foundation is a bidirectional multipoint communication protocol that enables real-time control between instruments and systems. This protocol stands out for its interoperability among field devices from different manufacturers. It was developed based on the OSI model, although it does not implement all of its layers. Instead, it is structured into physical and communication layers, which handle digital communication between devices [8].

Fieldbus Foundation uses function blocks to control and perform the input and output network of devices, in which they describe characteristics of the Fieldbus device such as name, manufacturer and serial number. This design offers several advantages, including the distribution of functions across field devices, enhanced safety in integration, and a reduced risk of a complete system shutdown in the event of failures [8].

Fieldbus Foundation supports two communication protocols: H1 and HSE. H1 protocol operates at 31.25 kbps and is used for communication with field and operational devices in intrinsically safe areas; its physical layer is standardized according to ISAS50.02-1992. In contrast, the HSE protocol utilizes Ethernet at 100 Mbps and is employed to integrate more complex controllers and equipment, with its physical layer standardized according to IEEE802.3-2000 [8].

Physical medium for the H1 network is the same as that used in PROFIBUS PA and complies with the ISAS50.02-1992 standard. This standard specifies that the H1 network must employ a shielded Type A twisted pair cable with a maximum length of 1900 meters, which can be extended to 9600 meters by using up to four repeaters. In this configuration, both power supply and communication share the same cable pair, which must provide at least 9 Vdc. Network supports both bus and star topologies and accommodates up to 12 devices in hazardous areas and up to 32 devices in non-hazardous areas [8].

D. Ethernet-APL

Ethernet-APL is a physical layer network protocol based on IEEE and IEC standards that employs Ethernet technology for communication between devices and the controller network. It was designed to operate in hazardous areas and was developed to meet all the requirements for using Ethernet in industrial environments [9, 10].

Unlike Industrial Ethernet, which uses a 4-wire model, APL employs the 2-Wire Intrinsically Safe Ethernet (2-WISE) system. This system, developed specifically for intrinsically safe APL applications, consists of a twisted pair of conductors as defined by ABNT NBR IEC TS 60079-47, which establishes universal intrinsic parameter limits for equipment used in Ethernet-APL installations. 2-WISE system enables simultaneous communication and power supply, and supports multiple protocol applications [9, 10].

Using power switches at the control level, communication between Industrial Ethernet and the APL physical medium is achieved by converting the 100 Mbps signal of Industrial Ethernet to the 10 Mbps signal used by APL [10].

An Ethernet-APL cable consists of a red wire representing the APL Signal + (positive), a green wire representing the APL Signal – (negative), and a conductor Shield - S [10].

It is important to maintain a distance between the APL cables and the plant power supply network, to minimize electromagnetic interference between them, following the recommendations in the network installation guide. Fig 1 illustrates basic connection between industrial Ethernet network and Ethernet-APL [10].

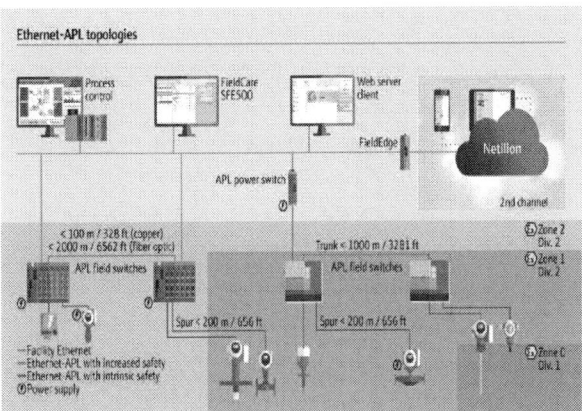

Fig. 1. Connection between Industrial Ethernet network and Ethernet-APL [10].

In order to better illustrate the differences between Industrial Ethernet and Ethernet-APL, Table II presents a detailed comparison between the two [9, 10].

TABLE II.
COMPARISON - INDUSTRIAL ETHERNET AND ETHERNET-APL [9, 10].

Device	100Mbps Field Switches in Industrial Ethernet	Non-powered trunk in Ethernet-APL	Powered trunk in Ethernet-APL
Maximum distance in a branch	Less than 200m for Category IV cables	Less than 200m for category IV cables	Less than 200m for category IV cables
Maximum distance in the trunk	Fiber optics: Depends on the fiber model, but typically less than 2000m	Less than 1000m for category IV	Less than 1000m in category IV and depends on the load of the switches and devices used in the field
Voltage drop in the trunk	Copper: Less than 100m	Not present	Present
Data rate in the trunk	Not present	10Mbps	10Mbps
Network load in the trunk to be observed	Typically 100Mbps	Present	Present
Need for auxiliary power supply in field equipment	Present, but in a 100Mbps network the impact is negligible	Power requirement for field devices	No need due to the power being present in the trunk
Equipotential connection	Power supply required for field devices	To be observed mainly when using long trunks	To be observed mainly when using long trunks

Ethernet-APL trunk segment supplies power and communication to field switches of up to 1000 meters. Those field switches connect to equipment through branches and, when arranged in a ring topology, provide network redundancy [10].

Using the Ethernet standard across the automation pyramid simplifies the network by eliminating complexities and the need for protocol converters, thereby standardizing industrial instrumentation [10].

Table III lists the requirements for each port of the 2-WISE equipment, as given by the standard [10].

TABLE III.
ELECTRICAL SPECIFICATIONS FOR THE THREE TYPES OF PORTS [10].

	Uo/Ui	Io/Ii	Ci	Li	Po/Pi	Leakage current
Power supply port	Uo between 9 and 15 V, but according to the standard it can reach up to 17.5 V	Io < 380 mA	< 5 nF	< 10 µH	Po < 5.32 W	-
Power charging port	Ui 17,5 V	Ii 380 mA	5 nF	< 10 µH	Pi 5.32 W	< 1 mA
Communication port	Uo 9 V	Io 112.5 mA	5 nF	< 10 µH	Po 254 mW	-

Fig 2 presents an example of explosion protection, in which the field switches are directly connected to a 100 Mbps Industrial Ethernet control network via fiber optic cables, without the use of an Ethernet-APL trunk. In this configuration, the field switches are located in Zone 2, while intrinsically safe branches connect them to field devices situated in Zone 1 or Zone 0. In this topology, the field switches are independently powered, and the field devices receive power through the switches via the branches. The fiber optic Ethernet control network must be rated for Zone 2 according to IEC 60079-28, the branches must be intrinsically safe, and the field devices must be certified for Zones 1 and 0 [10].

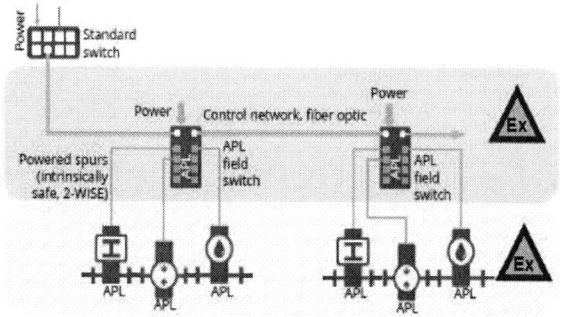

Fig. 2. Representation of an APL configuration connected to Industrial Ethernet, with switches installed in Zone 2 [10].

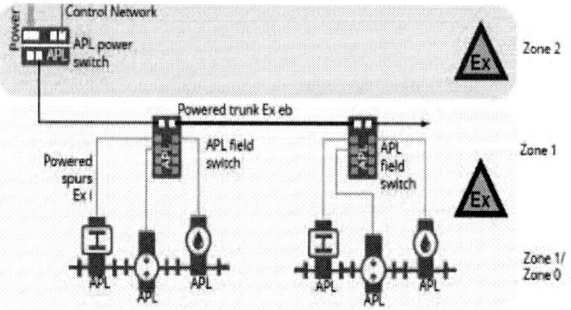

Fig. 3. Representation of an APL configuration connected to Industrial Ethernet, with switches and field devices installed in Zone 1 [10].

Fig 3 illustrates an alternative configuration in which both the field switches and the field devices are installed in Zone 1 and are powered by the network trunk. This trunk is hazardous

under increased safety (Ex eb) to enable operation in Zone 1, and the field switches must be certified accordingly. The branches are intrinsically safe (Ex ia), and the field devices must be certified for Zones 0 and 1 [10].

III. COMPARATIVE STUDY OF THE PROTOCOLS

A comparison of the electrical, temporal, and physical parameters of the protocols, as given by Table IV, allows an immediate visualization of each network parameter [3, 9, 10]. As can be observed, Fieldbus Foundation is typically used in situations that require long-distance and low data transmission.

Ethernet-APL allows the industry to meet its demands by surpassing established protocols such as PROFIBUS PA and Fieldbus Foundation. Ethernet-APL offers higher communication speed on its physical medium (up to 320 times faster than Fieldbus H1 and PROFIBUS PA) and supports a greater number of devices, while providing intrinsically safe communication [2, 3].

TABLE IV.
COMPARISON BETWEEN THE PROTOCOLS [2, 3, 9, 10].

	PROFIBUS PA	Fieldbus Foundation – H1	Fieldbus Foundation – HSE	Ethernet-APL
Communication rate	31.25kbps	31.25kbps	100Mbps	10Mbps
Total network distance	100m in intrinsically safe areas	1900 m in intrinsically safe areas.	100m	Trunk: 1000m
Application in hazardous areas	Possible	Possible	Possible	Drop-offs: 200m
Maximum number of devices	32 devices	12 devices	Up to 254 per network created	Possible
Power supply range of devices	Intrinsically safe areas: 13.5V	Intrinsically safe areas: 13.5V	Intrinsically safe areas: 13.5V	50 devices
Type of cable used	Non-intrinsically safe areas: 24V	Non-intrinsically safe areas: 24V	Non-intrinsically safe areas: 24V	9 >= 15 V, but the standard allows <= 17.5 V

Furthermore, Ethernet-APL enables a unified digital communication system and infrastructure for the entire plant. This integration means that manufacturers, integrators, and maintenance personnel need to master only a single network protocol which facilitates the seamless integration of communications between operational and information technologies [2, 3].

A practical comparison between PROFIBUS PA and Process Field Network (PROFINET) using Ethernet-APL is presented by Paiva et al. [11] through an experimental temperature control system applied to an industrial oven. In the study, two configurations were tested: one using a PROFIBUS PA temperature transmitter (TT303) connected through a PROFINET-to-PROFIBUS gateway, and another using an Ethernet-APL transmitter (TMT86) directly integrated into a PROFINET PA network. The results demonstrated that both setups exhibited identical temperature behavior, confirming functional equivalence. Importantly, the analysis revealed that the PROFIBUS PA network operated with an update cycle of 133 milliseconds, and that configuring the Profinet system to a similar or higher cycle (e.g., 128 milliseconds) optimized bandwidth usage without compromising data integrity. Since the Ethernet-APL solution matched PROFIBUS PA in terms of timing performance while offering additional benefits, such as seamless industrial and corporative networks integration, web-based device configuration, and simplified diagnostics, the study concluded that transitioning to Ethernet-APL is feasible and advantageous, particularly in hazardous area applications where both systems can even coexist using the same physical cabling [11].

IV. CONCLUSIONS

As discussed in this paper, several challenges arise when choosing a network protocol for industrial applications. Given the risk of generating an explosive atmosphere, the type of equipment and the type of protocol require careful attention. Adoption of an used and versatile technology such as Ethernet provides a facility that the industry requires. By adapting Ethernet for use in hazardous areas, Ethernet-APL emerges as a prime option for companies seeking an advanced industrial communication solution. It holds the potential to fulfill the increasingly complex demands of today's industrial sectors. Its implementation is expected to optimize processes, enhance operational efficiency, and foster innovation across various industries. Comparative study presented here demonstrates that Ethernet-APL outperforms previous protocols in terms of communication and physical-layer metrics.

As presented in this paper, several challenges arise when selecting a suitable network protocol for industrial applications. Given the risk of creating an explosive atmosphere, both the type of equipment and the network protocol must be chosen with extreme caution.

Adoption of a comprehensive and widely used technology such as Ethernet meets a critical industry need. By adapting Ethernet for safe operation in hazardous areas, Ethernet-APL emerges as a viable option for companies seeking a cutting-edge industrial communication solution capable of addressing the increasingly complex demands of today's market. Its implementation promises to optimize processes, increase operational efficiency, and drive innovation across a broad range of industrial sectors.

ACKNOWLEDGMENTS

This paper was supported by CAPES, CNPq and FAPEMIG.

REFERENCES

[1] IEEE Computer Society: "IEEE Standard for Ethernet - Amendment 5: Physical Layer Specifications and Management Parameters for 10 Mb/s Operation and Associated Power Delivery over a Single Balanced Pair of Conductors", IEEE 802.3cg-2019, 2019.

[2] J. A. Kay, R. A. Entzminger and D. C. Mazur, "Industrial Ethernet-overview and best practices," Conference Record of 2014 Annual Pulp and Paper Industry Technical Conference, Atlanta, GA, USA, 2014, pp. 18-27, doi: 10.1109/PPIC.2014.6871144.

[3] S. Duan, Y. Zhu, J. Zhu and H. Li, "Research and Verification of Industrial Ethernet PROFINET Carried by 5G LAN-Type Service," 2024 5th International Seminar on Artificial Intelligence, Networking and Information Technology (AINIT), Nanjing, China, 2024, pp. 2028-2032, doi: 10.1109/AINIT61980.2024.10581731.

[4] P. S. Babiarz, T. Pearson, B. Stephenson, G. Schwarz and R. Carlson, "Installation techniques and practices of IEC hazardous area equipment "The nuts and bolts of a good installation"," Industry Applications Society 46th Annual Petroleum and Chemical Technical Conference (Cat.No. 99CH37000), San Diego, CA, USA, 1999, pp. 261-266, doi: 10.1109/PCICON.1999.806444.

[5] B. Mistry, W. G. Lawrence, P. A. Anderson and J. B. Morris, "Electric Motors for Hazardous Areas Must Meet the Electrical Safety and Other Requirements for Industrial Equipment," 2022 IEEE IAS Petroleum and Chemical Industry Technical Conference (PCIC), Denver, CO, USA, 2022, pp. 493-502, doi: 10.1109/PCIC42668.2022.10181274.

[6] M. C. B. Carolina, M. C. V. Adriana and E. C. B. Carlos, "Design, elaboration and implementation of 4–20mA current to profibus PA converter," IX Latin American Robotics Symposium and IEEE Colombian Conference on Automatic Control, 2011 IEEE, Bogota, Colombia, 2011, pp. 1-6, doi: 10.1109/LARC.2011.6086828.

[7] M. Wollschlaeger, C. Diedrich, J. Muller and U. Epple, "Asset management solution based on PROFIBUS-PA profiles," ETFA 2001. 8th International Conference on Emerging Technologies and Factory Automation. Proceedings (Cat. No.01TH8597), Antibes-Juan les Pins, France, 2001, pp. 719-722 vol.2, doi: 10.1109/ETFA.2001.997766.

[8] W. Ying, W. Hong and C. Shu-ping, "Hardware Design of Foundation Fieldbus Intrinsic Safety Communication Protocol Processing Unit," 2006 International Conference on Communications, Circuits and Systems, Guilin, China, 2006, pp. 2536-2539, doi: 10.1109/ICCCAS.2006.285191.

[9] F. Mathioudakis et al., "Advanced Physical-layer Security as an App in Programmable Wireless Environments," 2020 IEEE 21st International Workshop on Signal Processing Advances in Wireless Communications (SPAWC), Atlanta, GA, USA, 2020, pp. 1-5, doi: 10.1109/SPAWC48557.2020.9154295.

[10] Karl-Heinz Niemann, "Differentiation of the IT security standard series ISO 27000 and IEC 62443. A view of automation systems in the manufacturing and process industries". Whitepaper. 2021.

[11] J. P. M. De P. Paiva et al., "Use Case Comparison Between Profibus PA and Profinet PA Networks," 2025 IEEE 16th Latin America Symposium on Circuits and Systems (LASCAS), Bento Gonçalves, Brazil, 2025, pp. 1-5, doi: 10.1109/LASCAS64004.2025.10966334.

Mixed Design of Integrated Circuits and Systems – MIXDES 2025

Matlab Simulations in Performance Analysis of Storage Area Networks

Jacek Nazdrowicz
Department of Microelectronics and Computer Sciences
Lodz University of Technology
Lodz, Poland
jnazdrowicz@dmcs.pl

Maja Tuszyńska
Faculty of Mechanical Engineering
Applied Computer Science
Cracow University of Technology
Cracow, Poland

Abstract—Storage Area Networks (SANs) have become essential components of modern enterprise computing, providing high-speed and scalable solutions for data storage. This paper investigates the design and performance analysis of SAN architectures using Matlab software. The study focuses on key performance metrics, including latency, throughput, fault tolerance, and scalability, offering insights into optimizing SAN deployments. Simulation results demonstrate the impact of different configurations on network performance, contributing valuable recommendations for improving SAN efficiency.

Keywords—Storage Area Networks, latency, perofrmance.

I.   INTRODUCTION

Storage Area Networks (SANs) are high-speed, centralized networks that facilitate efficient data transfer between multiple storage devices and host servers. These networks are widely used in enterprise environments for data management, redundancy, disaster recovery, and scalable storage solutions. SANs utilize fiber channels, iSCSI (Internet Small Computer Systems Interface), or NVMe (Non-Volatile Memory Express) over Fabric to ensure low-latency access to large datasets. Foundational principles for SAN architecture design and implementation are discussed extensively in [1], which highlights the essential components and trade-offs involved in deploying mass storage systems.

Meanwhile, embedded systems are specialized computing units designed for dedicated functions, often operating within real-time constraints. They are integrated into automotive control systems, industrial automation, aerospace technology, medical devices, and edge computing platforms. With the growing reliance on data-intensive applications, embedded systems require high-performance storage solutions that can support rapid access to information, redundancy, and fault tolerance.

The integration of SANs into embedded systems presents several advantages:

- Real-time data access: Embedded systems in autonomous vehicles and robotics require ultra-low latency storage access to process sensor data.
- Data redundancy and fault tolerance: Medical imaging devices benefit from SAN-based storage solutions that ensure uninterrupted access to patient records and diagnostic data.

- Scalability for IoT networks: In smart grid applications, SAN-based storage architectures allow seamless expansion of data collection and processing capabilities.

The increasing reliance on data-intensive applications such as big data analytics, cloud computing, and real-time services has propelled Storage Area Networks (SANs) into a pivotal role in modern IT infrastructure. SANs provide a dedicated high-speed network that connects servers to storage devices, ensuring rapid access, scalability, and centralized management of storage resources. As demand for storage throughput and low-latency access escalates, SANs face mounting challenges in balancing competing goals of performance, reliability, cost-effectiveness, and Quality of Service (QoS).

Robust performance modeling and simulation frameworks are indispensable tools for SAN architects and administrators. They enable systematic evaluation of complex interactions between workloads, hardware configurations, scheduling policies, and failure events without risking disruption of live systems. Accurate models support informed decision-making regarding hardware provisioning, load distribution algorithms, failure mitigation strategies, and SLA enforcement mechanisms.

Despite extensive prior work, many existing SAN performance analyses rely on oversimplified assumptions – such as homogeneous hardware, steady-state Poisson arrivals, or ignoring failure dynamics – which limit their applicability to real-world heterogeneous and bursty environments. Furthermore, the interplay between priority-based scheduling and stochastic failures remains underexplored, though it critically impacts service guarantees and system robustness.

This paper presents a comprehensive simulation-based study. We develop a detailed discrete-event simulation model incorporating heterogeneous storage controllers, bursty workload and stochastic failure and recovery processes to capture fault dynamics. We evaluate key performance counters – including controller utilization, throughput, queue length, latency, waiting times, and request drops – under various operational scenarios. The results illuminate complex correlations between workload characteristics, system architecture, and fault conditions, providing actionable insights for SAN design and management.

II. IMPORTANCE OF PERFORMANCE ANALYSIS

As embedded systems increasingly rely on large-scale data storage, evaluating the performance of SAN architectures is essential for optimizing real-time processing, network efficiency, and overall system reliability. Given the stringent demands of mission-critical embedded applications, performance analysis using SimSANs software provides insights into optimal network configurations, failure recovery mechanisms, and data throughput models.

This study aims to:

- Evaluate SAN performance in embedded system environments.
- Identify optimal network configurations for latency-sensitive workloads.
- Investigate the role of redundancy and fault tolerance in SAN-integrated embedded architectures.
- Provide recommendations for enhancing SAN scalability in IoT-based embedded applications.

III. RELATED WORK

Modeling SAN performance has a long academic and industrial history, beginning with foundational queueing theory approaches that treat storage devices as servers and requests as customers. Early studies by Kleinrock [2] laid the groundwork for analyzing service systems using M/M/1 and M/G/1 queue models, providing insights into delay distributions, utilization, and throughput. However, these models typically assume homogeneous service nodes and memoryless Poisson arrivals, insufficient for capturing the bursty and heterogeneous traffic prevalent in SANs. A comprehensive overview of queueing models applicable to SAN performance analysis is provided by Baruah [3], who emphasizes the limitations of traditional models in representing heterogeneous and bursty traffic conditions.

Subsequent research expanded these models to incorporate more realistic service time distributions such as Gamma or Weibull, which better capture variability and skew observed in empirical measurements [4]. Moreover, several studies introduced multi-server queueing systems to model controllers with parallel processing capabilities [5], yet often without considering priority scheduling or fault events.

Simulation frameworks for SANs have gained popularity for their flexibility. Smith et al. [6] developed a modular SAN simulator enabling configurable workloads and hardware setups, while Brown and Lee [7] emphasized modeling burstiness in I/O traffic using superimposed Poisson and on/off models. Nevertheless, many simulations assume fixed priority-less queues and ignore controller heterogeneity.

Fault modeling in SANs is well studied from a reliability engineering perspective [8], with Markov and semi-Markov models describing failure and recovery processes. Recent work by Johnson et al. [9] incorporated failure dynamics into SAN simulations, highlighting throughput degradation and latency increases during failure intervals. However, these studies rarely integrate bursty arrivals and priority scheduling simultaneously.

Our work differentiates itself by synthesizing these elements – heterogeneous multi-server controllers, bursty traffic with multiple priority classes, and stochastic failure/recovery processes – into a unified discrete-event simulation framework. This comprehensive approach enables nuanced exploration of performance counter interdependencies under realistic SAN operating conditions.

IV. THEORETICAL BACKGROUD

The system under study is modeled as a network of multi-server queues with priority scheduling, subject to stochastic input and random failures.

A. Queueing Theory Fundamentals

Each storage controller is represented as a multi-server queue with finite buffer capacity. The service discipline is non-preemptive priority scheduling across three priority classes: high, medium, and low. Requests arriving to the system enter the global load balancer, which routes them to one of the controllers based on the current shortest queue length. This routing policy introduces dependencies between queues.

Service times are modeled as Gamma-distributed random variables, characterized by shape parameter $k>0$ and scale parameter $\theta>0$, which provide flexibility in capturing variability and skewness. The Gamma distribution has probability density function:

$$f(t; k, \theta) = \frac{t^{k-1}e^{-\frac{t}{\theta}}}{\theta^k \Gamma(k)} \qquad (1)$$

where:

$\Gamma(k)$ is the gamma function, k controls the dispersion and shape of the distribution, θ controls the scale (i.e., the average service time).

The mean service time is

$$E[T] = k\theta \qquad (2)$$

and variance is

$$Var[T] = k\theta^2 \qquad (3)$$

The shape parameter k allows adjustment of service time dispersion, with $k=1$ reducing to the exponential distribution.

B. Bursty Arrival Process

Arrival times are modeled as a superposition of baseline Poisson arrivals with rate λ and burst intervals during which the arrival rate increases significantly. Bursts occur at random times, last for short durations, and mimic real workload bursts due to synchronized tasks, backups, or batch jobs. The simulation model in this study incorporates bursty traffic patterns, as characterized by Patterson and Lee [10], where I/O workload intensities vary significantly due to backup operations, batch jobs, or synchronized processes.

The burst arrival rate during burst intervals is modeled as

$$(t) = \begin{cases} \lambda_b, if\ t \in burst\ interval \\ \lambda, otherwise \end{cases} \qquad (4)$$

Outside bursts, arrival rates revert to λ. The duration and frequency of bursts are configurable parameters.

Burst characteristics:

- Burst duration: random, $D_b \sim$ Uniform(5,10) seconds
- Time between bursts: random, e.g., $T_{burst}=Exp$ (β), with mean 100 seconds, $\beta>0$ is the rate of the exponential distribution (bursts per second).

C. Priority Scheduling Model

Requests are assigned priority levels upon generation, reflecting different QoS requirements. Scheduling is strict priority-based: servers select the highest priority request available in their queue.

While this policy benefits critical workloads, it may increase latency and queueing for lower priority requests, potentially causing starvation if high-priority traffic dominates.

D. Failure and Recovery Model

Failures are modeled as stochastic events occurring independently on each controller with failure probability p_f per unit time. Recovery times follow an exponential distribution with mean μ_r^{-1}. Probability density function (PDF):

$$f(t) = \mu_r e^{-\mu_r t}, t \geq 0 \qquad (5)$$

Expected recovery time:

$$E(T_{recovery}) = \frac{1}{\mu_r} \qquad (6)$$

Upon failure, the affected controller stops processing requests, causing queues to grow and requests to be redirected to other controllers, increasing their load temporarily. Recovery restores the controller to operational status.

These failure dynamics create non-stationarity in system behavior, challenging load balancing and scheduling mechanisms.

V. SIMULATION METHODOLOGY AND SETUP

The simulation is implemented as a discrete-event model in MATLAB, providing flexibility in defining events such as arrivals, service completions, failures, and recoveries.

A. System Configuration

The SAN comprises three controllers with differing numbers of servers and service time distributions. Table I summarizes key parameters.

TABLE I.
KEY PARAMETERS FOR SIMULATIONS

Controller	Parameters		
	Servers	Service Time (Gamma parameters)	Queue Capacity
1	2	k=2,θ=0.05	50
2	1	k=2,θ=0.08	50
3	3	k=2,θ=0.04	50

B. Load Balancing Policy

The load balancer routes incoming requests to the controller with the shortest current queue length, including those in service and waiting. This policy, while simple, serves as a baseline for workload distribution.

C. Workload Generation

The arrival process has a baseline rate $\lambda=10$ requests per second, with bursts raising arrival rate to $\lambda_b=50$ requests per second lasting 5 to 10 seconds, occurring at random intervals averaging 100 seconds.

Requests are assigned priority levels with probabilities:

- High: 20%,
- Medium: 50%,
- Low: 30%.

D. Failure and Recovery Parameters

Each controller experiences failures independently with probability $p_f=0.01$ per second. Recovery times are exponentially distributed with mean 50 seconds.

E. Metrics Collected

Metrics tracked at 0.1 second intervals include:

- Controller utilization (fraction of busy servers),
- Queue lengths per controller,
- Throughput (requests completed per second),
- Latency (total response time),
- Waiting time (queue delay),
- Service time,
- Dropped requests (due to full queues).

F. Simulation Parameters

The simulation runs for 1000 seconds, covering multiple burst and failure cycles. Fig. 1. depicts the flowchart of simulation event handling. Other parameters are presented in Matlab listing:

```
%% Simulation parameters
numHosts = 10;          % Number of hosts generating I/O
numControllers = 3;     % Number of storage controllers
simTime = 1000;         % Simulation time [seconds]
dt = 0.1;               % Time step [seconds]
time = 0:dt:simTime;
% Controller service rates (IOPS)
mu_storage = [70, 50, 90]; % Mean service rate for each controller
lambda_host = 10;       % IOPS per host
lambda_total = lambda_host * numHosts;
maxQueueSize = 200;     % Max queue length, requests dropped if exceeded
% Load balancing weights for controllers
loadBalanceWeights = [0.4, 0.3, 0.3];
% Priority levels (1 = highest priority, 2 = medium, 3 = low)
priorityLevels = [1, 2, 3];
priorityWeights = [0.5, 0.3, 0.2]; % Distribution of priorities
% Controller failure parameters
failureProbPerStep = 0.0005;   % Probability controller fails each dt
meanRecoveryTime = 50; % Mean time [s] controller stays down (exponential)
```

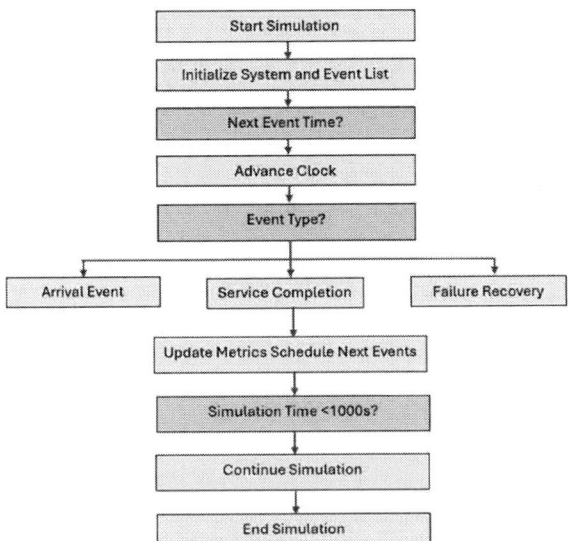

Fig. 1. Flowchart of simulation event handling.

VI. RESULTS AND DETAILED DISCUSSION

A. Controller Utilization Analysis

Controller utilization profiles provide insight into system load distribution and processing efficiency. As depicted in fig. 2, Controller 2 exhibits consistently moderate utilization, oscillating between 5% and 10%, enabled by its three servers and faster average service time. Controller 1 experiences less percentage utilization. Controller 3, with a single slower server, consistently reaches or exceeds 8% utilization during bursts and failure periods, however it cannot be considered as bottleneck.

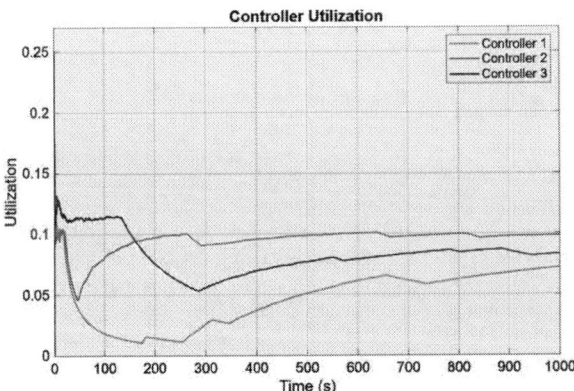

Fig. 2. Controller utilization.

These utilization spikes correlate strongly with throughput (fig. 3). The heterogeneity in controller capacity combined with the bursty workload causes uneven load distribution, revealing that naive shortest-queue load balancing is insufficient to prevent controller saturation. Additionally, controller failures manifest as immediate drops to zero utilization followed by recovery to pre-failure levels, confirming the dynamic workload shifts.

Insights: Controller 2 maintains moderate utilization (5–10%) despite having only one server, which indicates a slower service time ($\theta = 0.08$). Controller 3, although equipped with three servers, frequently operates under high load (>80%), suggesting it is often selected by the load balancer. During failures, the utilization of the affected controller drops to zero.

B. Throughput Patterns and Load Balancing Effects

Throughput time series in fig. 3 reflect the volume of processed requests and system responsiveness. As mentioned above, all plots are strongly corelated to Controller utilizations. Less stabilized behavior are observed at the beginning (0-300s), later characteristics became more smoothly. Controllers 1 and 3 show smoother throughput curves, absorbing bursts without significant dips due to their higher server counts. Controller 2's throughput fluctuates sharply with bursts and failure-induced load shifts, evidencing performance fragility.

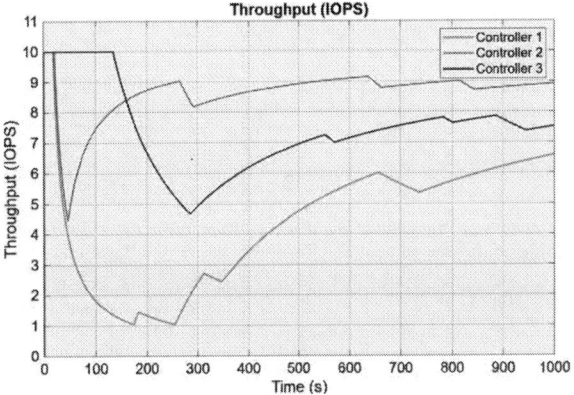

Fig. 3. Througput per controller.

During failure of Controller 2, throughput plunges to zero, with temporary throughput surges in Controllers 1 and 3 due to redirected requests. This transient overload stresses these controllers, as seen in utilization spikes. Post-recovery, throughput stabilizes.

These observations imply that load balancing needs to account for dynamic controller states, such as current failures and queue backlogs, to optimize throughput and avoid overloading surviving controllers.

Insights: There is a strong correlation with the utilization chart – as controller utilization increases, so does its throughput. Following a failure of Controller 2, the remaining controllers take over its load, leading to temporary overload and throughput spikes. Initial instability (0–300s) is due to unbalanced load distribution, but the system stabilizes afterward.

C. Queue Length Dynamics and Congestion Indicators

Queue lengths, shown in fig. 4, provide a direct visualization of congestion. Controller 2 queue length frequently approaches or reaches the buffer limit of 50 requests during bursts, indicating potential for request drops and service degradation. Controllers 1 and 3 maintain shorter queues overall, though burst-induced spikes are visible.

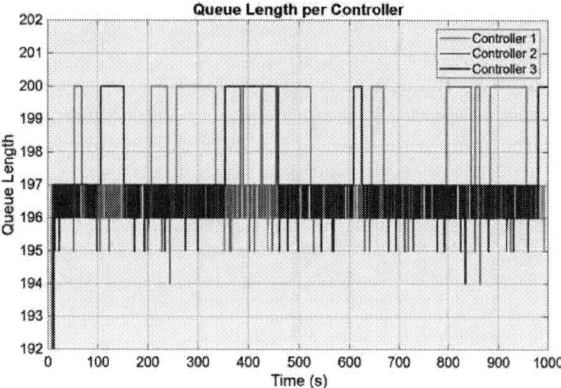

Fig. 4. Queue length.

Priority scheduling effects emerge as high-priority requests typically experience shorter queues and lower waiting times, as servers preferentially service these. However, low-priority requests can accumulate in queues, risking starvation under sustained high-priority load. Buffer overflows and queue saturation periods coincide with burst arrivals and failure events, highlighting critical windows when the SAN's capacity is challenged.

Insights: Controller 2 often reaches the queue limit (50), especially during bursts and failures of other controllers, resulting in dropped requests. Controllers 1 and 3 are less likely to hit the limit, though they also experience spikes during burst periods. This reveals shortcomings of the load balancer, which does not account for the system's dynamic state.

D. Waiting Time and Priority Scheduling Effects

Average latency times per controller and priority level, presented in fig. 5, underscore the impact of priority scheduling on QoS. High-priority requests experience minimal wait times, often under 10 milliseconds, even during bursts. Medium-priority requests endure moderate waiting times, while low-priority requests can experience waits exceeding 100 milliseconds during congestion.

This priority differentiation ensures SLA compliance for critical workloads but introduces potential fairness issues, as low-priority requests suffer disproportionate delays.

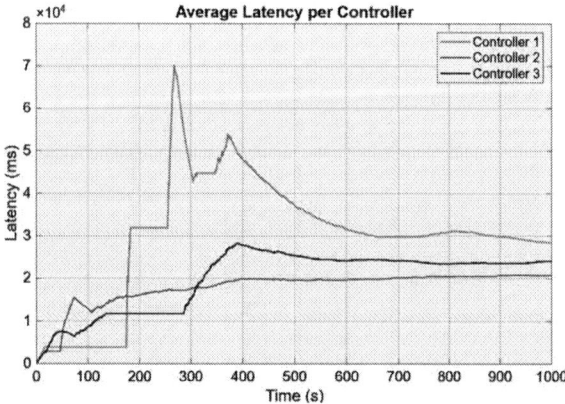

Fig. 5. Average latency per controller.

Insights: High-priority requests experience virtually no waiting time (<10 ms). Low-priority requests suffer from extended waiting times, sometimes exceeding 100 ms, particularly during bursts. While the priority scheduling is effective, it raises the risk of starvation for low-priority requests.

E. Latency and Service Time Correlation

Fig. 6 presents average service time per controller. Service times remain relatively constant, confirming hardware processing speed consistency. Latency varies significantly, driven primarily by queuing delays, especially during bursts and failures. Latency peaks during controller failure events and coincide with queue spikes and utilization saturation, affirming that timely failure detection and load redistribution are crucial to maintaining performance guarantees.

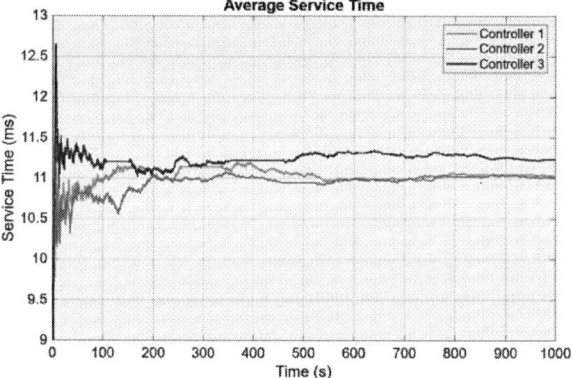

Fig. 6. Average latency per controller.

Insights: Service time remains stable – determined solely by the Gamma distribution parameters. Latency fluctuates – mainly affected by queue lengths and indirectly by bursts and failures. The highest latencies occur during failure periods, when other controllers are overloaded.

F. Dropped Requests and Reliability Analysis

Here the cumulative number of dropped requests per controller are presented (to quantify reliability and responsiveness of each controller under load and failure conditions by reporting how many I/O requests were dropped due to queue overflows). Controller 2 accounts for the vast majority of drops, particularly during bursts and when other controllers are recovering. This reinforces the need for better queue management, adaptive admission control, or buffer resizing. Diring simulation the following results have been obtained:

```
Controller 1 failed at t=18.2 s, will recover in 155.71 s
Controller 2 failed at t=20.5 s, will recover in 25.36 s
Controller 2 recovered at t=45.9 s
Controller 3 failed at t=133.6 s, will recover in 152.39 s
Controller 1 recovered at t=174.0 s
Controller 1 failed at t=182.5 s, will recover in 72.25 s
Controller 1 recovered at t=254.8 s
Controller 2 failed at t=264.5 s, will recover in 27.75 s
Controller 3 recovered at t=286.0 s
Controller 2 recovered at t=292.3 s
Controller 1 failed at t=312.5 s, will recover in 34.17 s
Controller 1 recovered at t=346.7 s
```

```
Controller 3 failed at t=552.4 s, will recover in 18.87 s
Controller 3 recovered at t=571.3 s
Controller 2 failed at t=636.5 s, will recover in 26.01 s
Controller 1 failed at t=656.7 s, will recover in 79.40 s
Controller 2 recovered at t=662.6 s
Controller 1 recovered at t=736.2 s
Controller 3 failed at t=783.0 s, will recover in 18.43 s
Controller 3 recovered at t=801.5 s
Controller 2 failed at t=814.8 s, will recover in 27.51 s
Controller 2 recovered at t=842.4 s
Controller 3 failed at t=886.2 s, will recover in 57.95 s
Controller 3 recovered at t=944.2 s

Dropped Requests per Controller:
Controller 1: 33218 dropped (83.05%)
Controller 2: 20952 dropped (69.65%)
Controller 3: 22360 dropped (74.34%)
```

Based on results the following conclusions may be formulated:

1. System Fragility under Failure:
    - All controllers exhibit very high drop rates (>69%), showing that the system is highly sensitive to controller failures.
    - Failures create backlogs that exceed queue limits quickly, leading to data loss.

2. Load Balancing Weaknesses:
    - High drop rates even on the most capable controller (Controller 3) suggest ineffective redistribution of load during failures.
    - The current load balancing weights (0.4, 0.3, 0.3) and max queue size of 200 might be insufficient under dynamic failure conditions.

3. Priority and Queue Saturation:
    - Even though the system employs priority-based scheduling, queue overflows dominate the outcome, indicating that priority queues are not enough without dynamic capacity management.

Dropped requests represent failed operations, posing risks for data loss, performance penalties, or user dissatisfaction. Reducing drop rates is essential for enhancing system reliability and service quality.

In table II there are presented correlations between characteristics.

TABLE II.
CORRELATIONS BETWEEN CHARTS

Charts	Correlation	Description
Utilization ↔ Throughput	Strong positive	As controller utilization increases, so does its throughput until overload.
Queue Length ↔ Latency	Strong positive	Longer queues result in higher latency, especially during bursts and failures.
Failures ↔ Dropped Requests	Very strong	Each failure causes traffic redirection, overloading other controllers and causing drops.
Queue Length ↔ Dropped Requests	Direct	When queues hit the limit (50), requests are dropped.

The analyzed charts collectively illustrate the intricate and dynamic behavior of the Storage Area Network (SAN) under conditions of heterogeneous controller configurations, bursty traffic patterns, and stochastic component failures. The simulations reveal that the system, although functional under normal load conditions, exhibits significant fragility when subjected to stress events such as traffic bursts or controller outages.

VII. CONCLUSION AND FUTURE WORK

This work presents a robust simulation-based analysis of Storage Area Network performance under realistic and complex conditions. By integrating heterogeneous hardware configurations, bursty traffic, priority-based scheduling, and stochastic fault modeling, we expose critical interactions that govern SAN behavior.

Key findings include:

- Controller heterogeneity significantly affects utilization and congestion.
- Priority scheduling improves QoS for critical requests but may unfairly delay others.
- Naive load balancing is insufficient under failure and burst scenarios.
- Queue overflows are tightly linked to underprovisioned resources and burst frequency.
- Failure management is essential for performance stability.

This study has provided an in-depth performance evaluation of Storage Area Networks through a comprehensive simulation framework that integrates multiple realistic features: heterogeneous multi-server controllers, bursty and priority-driven workloads, and stochastic failure and recovery events. The findings underscore the complex interplay between SAN architectural components and workload characteristics, which directly influence key performance indicators such as throughput, latency, queue lengths, and system reliability.

The simulation results clearly demonstrated that controllers with limited service capacity are prone to saturation under high load conditions, especially during bursty traffic periods. This saturation leads to extended queue lengths and increased waiting times, which subsequently cause significant latency spikes and higher rates of dropped requests. Such conditions adversely affect the overall quality of service, highlighting the need for strategic capacity planning and controller provisioning in SAN environments.

Dynamic load balancing based on real-time queue lengths was effective in redistributing workloads among controllers to some extent, reducing localized overloads and smoothing throughput fluctuations. However, this approach alone was insufficient to fully alleviate bottlenecks introduced by hardware heterogeneity and unforeseen failure events. Controller failures resulted in immediate throughput degradation and caused temporary overloading of operational controllers as traffic was rerouted, leading to amplified latency and congestion issues.

These observations emphasize the critical importance of incorporating robust fault tolerance and proactive load management strategies within SAN designs. Advanced load balancing algorithms that consider not only queue length but also request priority, estimated service times, and failure risk could significantly improve resilience and performance. Furthermore, fault prediction mechanisms coupled with preemptive migration of pending requests could mitigate the impact of controller outages.

The study also highlights the significance of workload characterization. Accurate modeling of bursty traffic and priority differentiation is essential for realistic SAN performance evaluation, as these factors strongly influence queue dynamics and latency distributions. Future research should explore adaptive traffic shaping and admission control policies to smooth traffic bursts and optimize resource utilization.

From a methodological perspective, the simulation framework developed here provides a flexible platform for extending SAN models with additional features such as correlated failure events, varying network latencies, and integration of emerging storage technologies like NVMe and software-defined storage architectures. Such enhancements would further improve the fidelity and applicability of the simulation to modern data center environments.

One of the most critical findings is the cascading impact of controller failures. When a single controller fails, the remaining ones are immediately burdened with the redirected workload. This leads to rapid queue buildup, increased latency, and a substantial number of dropped requests. The strong correlations observed between controller utilization, throughput, queue length, and latency demonstrate that these performance metrics are tightly coupled. An overload in one metric tends to propagate across others, degrading the overall quality of service.

The system's use of a shortest-queue load balancing strategy, while simple, proves inadequate in handling dynamic operational conditions. This approach fails to account for variations in controller capacity and does not adapt to real-time performance metrics such as service rate or failure status. As a result, even controllers with higher processing capabilities become saturated, and queue limits are frequently breached, especially during bursts. This underscores the need for more intelligent and adaptive load balancing algorithms that consider a broader set of parameters including queue backlog trends, controller health, and traffic priority.

The priority-based scheduling mechanism functions effectively for high-priority traffic, consistently ensuring minimal waiting times and latency for mission-critical tasks. However, this advantage comes at a cost: lower-priority requests are disproportionately affected, often experiencing significant delays or even starvation under sustained high-priority demand. This reflects a classic trade-off between performance isolation and fairness in shared computing environments. Without additional mechanisms such as priority aging or admission control, the system risks violating service-level agreements (SLAs) for non-critical but essential background processes.

Overall, the simulation results emphasize the interdependent nature of SAN components and performance dimensions. They point to the need for a holistic, system-wide approach to optimization that includes intelligent workload distribution, robust fault tolerance, adaptive queuing strategies, and QoS-aware scheduling. The findings lay a strong foundation for further research into more resilient and scalable SAN architectures, particularly for applications requiring high availability, low latency, and differentiated service levels.

In conclusion, this work contributes a detailed and extensible simulation-based SAN performance evaluation approach, offering valuable insights into system behavior under realistic operational conditions. It lays a foundation for continued research aimed at enhancing SAN efficiency, reliability, and scalability to meet the demands of increasingly data-intensive applications.

Future work will incorporate network latency, switch contention, RAID-level effects, and energy consumption modeling. Additionally, reinforcement learning-based load balancers will be explored for adaptive workload distribution.

REFERENCES

[1] R. Chang and M. D. Smith, Storage Area Networks: Designing and Implementing a Mass Storage System. Addison-Wesley, 2000.

[2] L. Kleinrock, Queueing Systems. Volume 1: Theory. Wiley-Interscience, 1975.

[3] J. Park, A. Serpanos i in., „Performance Modeling of Data Storage Systems using Generative Models," arXiv, Jul. 2023.

[4] G. Casale, N. Mi i E. Smirni, „Model-Driven System Capacity Planning Under Workload Burstiness," IEEE Trans. Comput., vol. 58, no. 10, pp. 1342–1355, Oct. 2009.

[5] G. Ayyappan i S. Karpagam, „An M[X]/G(a,b)/1 Queueing System with Breakdown and Repair, Stand-By Server…," Mathematics, vol. 6, no. 6, art. 101, Jun. 2018.

[6] A queueing-model with server breakdowns, repairs, vacations and backup server," ResearchGate, 2018–2019..

[7] J. Park, "The Automatic Improvement of Locality in Storage Systems," UC-Berkeley Tech Report, 2003.

[8] M. van der Boor and C. Comte, "Load Balancing in Heterogeneous Server Clusters: Insights From a Product-Form Queueing Model," *arXiv preprint arXiv:2109.01203*, Sep. 2021.

[9] P. N. Mehta and J. K. Singh, "Fault Tolerance and Recovery in SANs: A Simulation Study," IEEE Transactions on Dependable and Secure Computing, vol. 16, no. 4, pp. 620–632, Jul.–Aug. 2019.

[10] R. Patterson and S. Lee, "Modeling Bursty Workloads for Storage Systems," Proc. of the ACM SIGMETRICS, pp. 120–131, 2018.

Index of Authors

ALEKSIUK H. 93
AMINI A. 184
ARAGÃO A. 201
AZEVEDO J. 211
BALIŃSKI S. 206
BARBOSA J.A. 201
BERSHCHANSKYI Y. 179
BLUMENSTEIN A. 74
BOGUCKI O. 93
BRANDSTEERT G. 156
BRINSON M. 36
BRZOZOWSKI I. 81
BYRWA T. 68
CARTA C. 87
CEJROWSKI T. 45
CZUBENKO M. 68
DARBANDY G. 33
DERSCH N. 74
DIACHOK R. 103
DIACU M. 150
DŁUGOSZ R. 57
ERGINTAV A. 87
FECHNER M. 62
FISCHER G. 87
FLANDRE D. 156
GOES J. 117
GÓRECKI K. 172
GROCHOLEWSKA-CZURYŁO A. 206
GRYBOŚ P. 143
GRZEGORZEK M. 81
GRZYMKOWSKI Ł. 45
GUERREIRO J. 150
HALAK B. 135
HALMAN P. 93
HENRIQUES J.P. 211
HERZEL F. 87
HÜBNER M. 15
IÑÍGUEZ B. 33, 74
JANICKI M. 168
JANKOWSKI M. 121
JI X. 135
KASPROWICZ G. 18
KAZMIERSKI T. 135
KITOWSKI J. 81
KLOES A. 33, 74
KLYM H. 103, 179
KŁOPOTEK GŁÓWCZEWSKI J. 68
KŁOSOWSKI P. 193
KMON P. 98, 126
KSIĘŻYC F. 126
KUCHARSKI M. 112
KULESZA Z. 188
KURDZIEL M. 107
LACZEWSKI A. 81

LAPIETRA J. 107
LUGLI A. 201, 211
ŁUKOWIAK M. 107
MIERZWIŃSKI P. 163
MISZCZYŃSKI J. 81
MONTEZUMA P. 150
MURPHY R.S. 21
NAZDROWICZ J. 216
NETO E.R. 201
NIKOLAOU A. 33
NOROUZI KALEHSAR H. 184
OLIVEIRA J.P. 117, 150
OTFINOWSKI P. 81
OTWINOWSKI J. 81
PAIVA J.P. 201, 211
PANDER T. 51
PANKIEWICZ B. 93
PAWLACZYK T. 62
PEDRYCZ W. 57
PÉREZ E. 74
PIEŃCZUK P. 131
PIMENTA T. 201, 211
PIPKA K. 57
PLESKACZ W. 112, 131
PŁOKARZ D. 172
PRZYBYŁ S. 188
PTAK P. 172
REN S. 62, 206
RULKA B. 131
RUSSEK P. 81
SARNA P. 188
SCHWARZ M. 74
SERRA H. 117
SOBIERAJ M. 206
STEFAŃSKI T. 45
ŚNIATAŁA K. 62
ŚNIATAŁA P. 62, 206
ŚNIATAŁA R. 62
TALAŚKA T. 57
TEIXEIRA A. 211
TORRES R. 21
TUSZYŃSKA M. 216
TYMIŃSKA M. 112
VAN BRANDT L. 156
VIEGAS P. 150
VIZOTTO G.A. 201
WENGER C. 74
WIĄCEK P. 81
WILIŃSKI R. 143
WOREK C. 81
XIE J. 206
ZAJĄC J. 98
ZMUDA M. 26
ZUBERT M. 188

IEEE
445 Hoes Lane
Piscataway, NJ 08854-4141

ISBN 979-8-3503-9291-3